T0140475

# Advances in Intelligent Systems and Computing

Volume 669

**Series editor**

Janusz Kacprzyk, Polish Academy of Sciences, Warsaw, Poland
e-mail: kacprzyk@ibspan.waw.pl

The series "Advances in Intelligent Systems and Computing" contains publications on theory, applications, and design methods of Intelligent Systems and Intelligent Computing. Virtually all disciplines such as engineering, natural sciences, computer and information science, ICT, economics, business, e-commerce, environment, healthcare, life science are covered. The list of topics spans all the areas of modern intelligent systems and computing such as: computational intelligence, soft computing including neural networks, fuzzy systems, evolutionary computing and the fusion of these paradigms, social intelligence, ambient intelligence, computational neuroscience, artificial life, virtual worlds and society, cognitive science and systems, Perception and Vision, DNA and immune based systems, self-organizing and adaptive systems, e-Learning and teaching, human-centered and human-centric computing, recommender systems, intelligent control, robotics and mechatronics including human-machine teaming, knowledge-based paradigms, learning paradigms, machine ethics, intelligent data analysis, knowledge management, intelligent agents, intelligent decision making and support, intelligent network security, trust management, interactive entertainment, Web intelligence and multimedia.

The publications within "Advances in Intelligent Systems and Computing" are primarily proceedings of important conferences, symposia and congresses. They cover significant recent developments in the field, both of a foundational and applicable character. An important characteristic feature of the series is the short publication time and world-wide distribution. This permits a rapid and broad dissemination of research results.

More information about this series at http://www.springer.com/series/11156

Bijaya Ketan Panigrahi · Munesh C. Trivedi
Krishn K. Mishra · Shailesh Tiwari
Pradeep Kumar Singh
Editors

# Smart Innovations in Communication and Computational Sciences

Proceedings of ICSICCS 2017, Volume 1

 Springer

*Editors*
Bijaya Ketan Panigrahi
Department of Electrical Engineering
Indian Institute of Technology Delhi
New Delhi, Delhi
India

Shailesh Tiwari
CSED
ABES Engineering College
Ghaziabad, Uttar Pradesh
India

Munesh C. Trivedi
Department of Computer
   Science and Engineering
ABES Engineering College
Ghaziabad
India

Pradeep Kumar Singh
Department of Computer
   Science and Engineering
Jaypee University of Information
   Technology
Waknaghat, Solan, Himachal Pradesh
India

Krishn K. Mishra
Department of Computer
   Science and Engineering
Motilal Nehru National Institute
   of Technology
Allahabad, Uttar Pradesh
India

ISSN 2194-5357    ISSN 2194-5365  (electronic)
Advances in Intelligent Systems and Computing
ISBN 978-981-10-8967-1    ISBN 978-981-10-8968-8  (eBook)
https://doi.org/10.1007/978-981-10-8968-8

Library of Congress Control Number: 2018937328

This Springer imprint is published by the registered company Springer Nature Singapore Pte Ltd.
part of Springer Nature
The registered company address is: 152 Beach Road, #21-01/04 Gateway East, Singapore 189721,
Singapore

# Preface

The International Conference on **Smart Innovations in Communications and Computational Sciences (ICSICCS 2017)** has been held at Moga, Punjab, India, during **June 23–24, 2017**. The ICSICCS 2017 has been organized and supported by the **North West Group of Institutions, Moga, Punjab, India**.

The main purpose of ICSICCS 2017 is to provide a forum for researchers, educators, engineers, and government officials involved in the general areas of communication, computational sciences, and technology to disseminate their latest research results and exchange views on the future research directions of these fields.

The field of communications and computational sciences always deals with finding the innovative solutions to problems by proposing different techniques, methods, and tools. Generally, innovation refers to find new ways of doing usual things or doing new things in a different manner but due to increasingly growing technological advances with speedy pace *Smart Innovations* are needed. Smart refers to *"how intelligent the innovation is?"* Nowadays, there is massive need to develop new *"intelligent"* *"ideas, methods, techniques, devices, tools"*. The proceedings cover those systems, paradigms, techniques, technical reviews that employ knowledge and intelligence in a broad spectrum.

ICSICCS 2017 received around 350 submissions from around 603 authors of 9 different countries such as Taiwan, Sweden, Italy, Saudi Arabia, China, and Bangladesh. Each submission has gone through the plagiarism check. On the basis of plagiarism report, each submission was rigorously reviewed by at least two reviewers. Even some submissions have more than two reviews. On the basis of these reviews, 73 high-quality papers were selected for publication in proceedings volumes, with an acceptance rate of 20.8%.

This proceedings volume comprises 40 quality research papers in the form of chapters. These chapters are further subdivided into different tracks named as *"Smart Computing Technologies," "Intelligent Communications and Networking,"* and *"Computational Sciences."*

We are thankful to the speakers Prof. B. K. Panigrahi, IIT Delhi; Dr. Dhanajay Singh, Hankuk (Korea) University of Foreign Studies (HUFS), Seoul, South Korea; and Dr. T. V. Vijay Kumar, JNU, Delhi; delegates, and the authors for their participation and their interest in ICSICCS as a platform to share their ideas and innovation. We are also thankful to Prof. Dr. Janusz Kacprzyk, Series Editor, AISC, Springer, and Mr. Aninda Bose, Senior Editor, Hard Sciences, Springer, India, for providing continuous guidance and support. Also, we extend our heartfelt gratitude and thanks to the reviewers and Technical Program Committee Members for showing their concern and efforts in the review process. We are indeed thankful to everyone directly or indirectly associated with the conference organizing team leading it toward the success.

We hope you enjoy the conference proceedings and wish you all the best!

Organizing Committee
ICSICCS 2017

# Organizing Committee

## Chief Patron

S. Lakhbir Singh Gill (Chairman)

## Patron

S. Prabhpreet Singh Gill (Managing Director)
S. Dilpreet Singh Gill (Executive Member)

## Advisory Committee

Prof. Dr. J. S. Hundal, MRSPTU, Punjab, India
Prof. Dr. A. K. Goel, MRSPTU, Punjab, India
Prof. Gursharan Singh, MRSPTU, Punjab, India
Dr. Buta Singh, IKGPTU, Punjab, India
Dr. B. S. Bhatia, SGGSWU, Punjab, India
Dr. D. S. Bawa, Rayat & Bahra Group of Institutes, Hoshiarpur, Punjab, India
Prof. R. S. Salaria, Rayat & Bahra Group of Institutes, Hoshiarpur, Punjab, India

## Principal General Chair

Dr. N. K. Maheshwary

## Conference Co-chair

Dr. R. K. Maheshwary
Dr. Mohita

## Finance Chair

Mr. Rishideep Singh (HoD CSE)

## Publicity Chair

Prof. Surjit Arora

## Publication Chair

Prof. K. S. Panesar (Mechanical)

## Registration Chair

Ms. Navjot Jyoti (AP CSE)

## Organizing Chair

Dr. R. K. Maheshwary (Dean)

## Technical Program Committee

Prof. Ajay Gupta, Western Michigan University, USA
Prof. Babita Gupta, California State University, USA
Prof. Amit K. R. Chowdhury, University of California, USA
Prof. David M. Harvey, G.E.R.I., UK
Prof. Ajith Abraham, Director, MIR Labs
Prof. Madjid Merabti, Liverpool John Moores University, UK
Dr. Nesimi Ertugrual, University of Adelaide, Australia
Prof. Ian L. Freeston, University of Sheffield, UK
Prof. Witold Kinsner, University of Manitoba, Canada
Prof. Anup Kumar, M.I.N.D.S., University of Louisville
Prof. Sanjiv Kumar Bhatia, University of Missouri, St. Louis

Prof. Prabhat Kumar Mahanti, University of New Brunswick, Canada
Prof. Ashok De, Director, NIT Patna
Prof. Kuldip Singh, IIT Roorkee
Prof. A. K. Tiwari, IIT (BHU) Varanasi
Mr. Suryabhan, ACERC, Ajmer, India
Dr. Vivek Singh, IIT (BHU), India
Prof. Abdul Quaiyum Ansari, Jamia Millia Islamia, New Delhi, India
Prof. Aditya Trivedi, ABV-IIITM Gwalior
Prof. Ajay Kakkar, Thapar University, Patiala, India
Prof. Bharat Bhaskar, IIM Lucknow, India
Prof. Edward David Moreno, Federal University of Sergipe, Brazil
Prof. Evangelos Kranakis, Carleton University
Prof. Filipe Miguel Lopes Meneses, University of Minho, Portugal
Prof. Giovanni Manassero Junior, Universidade de São Paulo
Prof. Gregorio Martinez, University of Murcia, Spain
Prof. Pabitra Mitra, Indian Institute of Technology Kharagpur, India
Prof. Joberto Martins, Salvador University (UNIFACS)
Prof. K. Mustafa, Jamia Millia Islamia, New Delhi, India
Prof. M. M. Sufyan Beg, Jamia Millia Islamia, New Delhi, India
Prof. Jitendra Agrawal, Rajiv Gandhi Proudyogiki Vishwavidyalaya, Bhopal, MP, India
Prof. Rajesh Baliram Ingle, PICT, University of Pune, India
Prof. Romulo Alexander Ellery de Alencar, University of Fortaliza, Brazil
Prof. Youssef Fakhri, Faculté des Sciences, Université Ibn Tofail
Dr. Abanish Singh, Bioinformatics Scientist, USA
Dr. Abbas Cheddad, UCMM, Umeå Universitet, Umeå, Sweden
Dr. Abraham T. Mathew, NIT Calicut, Kerala, India
Dr. Adam Scmidit, Poznan University of Technology, Poland
Dr. Agostinho L. S. Castro, Federal University of Para, Brazil
Prof. Goo-Rak Kwon, Chosun University, Republic of Korea
Dr. Alberto Yúfera, Seville Microelectronics Institute, IMSE-CNM, NIT Calicut, Kerala, India
Dr. Adam Scmidit, Poznan University of Technology, Poland
Prof. Nishant Doshi, S V National Institute of Technology, Surat, India
Prof. Gautam Sanyal, NIT Durgapur, India
Dr. Agostinho L. S. Castro, Federal University of Para, Brazil
Dr. Alberto Yúfera, Seville Microelectronics Institute, IMSE-CNM
Dr. Alok Chakrabarty, IIIT Bhubaneswar, India
Dr. Anastasios Tefas, Aristotle University of Thessaloniki
Dr. Anirban Sarkar, NIT Durgapur, India
Dr. Anjali Sardana, IIIT Roorkee, Uttarakhand, India
Dr. Ariffin Abdul Mutalib, Universiti Utara Malaysia
Dr. Ashok Kumar Das, IIIT Hyderabad
Dr. Ashutosh Saxena, Infosys Technologies Ltd., India
Dr. Balasubramanian Raman, IIT Roorkee, India

Dr. Benahmed Khelifa, Liverpool John Moores University, UK
Dr. Björn Schuller, Technical University of Munich, Germany
Dr. Carole Bassil, Lebanese University, Lebanon
Dr. Chao Ma, Hong Kong Polytechnic University
Dr. Chi-Un Lei, University of Hong Kong
Dr. Ching-Hao Lai, Institute for Information Industry
Dr. Ching-Hao Mao, Institute for Information Industry, Taiwan
Dr. Chung-Hua Chu, National Taichung Institute of Technology, Taiwan
Dr. Chunye Gong, National University of Defense Technology
Dr. Cristina Olaverri Monreal, Instituto de Telecomunicacoes, Portugal
Dr. Chittaranjan Hota, BITS Hyderabad, India
Dr. D. Juan Carlos González Moreno, University of Vigo
Dr. Danda B. Rawat, Old Dominion University
Dr. Davide Ariu, University of Cagliari, Italy
Dr. Dimiter G. Velev, University of National and World Economy, Europe
Dr. D. S. Yadav, South Asian University, New Delhi
Dr. Darius M. Dziuda, Central Connecticut State University
Dr. Dimitrios Koukopoulos, University of Western Greece, Greece
Dr. Durga Prasad Mohapatra, NIT Rourkela, India
Dr. Eric Renault, Institut Telecom, France
Dr. Felipe Rudge Barbosa, University of Campinas, Brazil
Dr. Fermín Galán Márquez, Telefónica I+D, Spain
Dr. Fernando Zacarias Flores, Autonomous University of Puebla
Dr. Fuu-Cheng Jiang, Tunghai University, Taiwan
Prof. Aniello Castiglione, University of Salerno, Italy
Dr. Geng Yang, NUPT, Nanjing, People's Republic of China
Dr. Gadadhar Sahoo, BIT Mesra, India
Prof. Ashokk Das, International Institute of Information Technology, Hyderabad, India
Dr. Gang Wang, Hefei University of Technology
Dr. Gerard Damm, Alcatel-Lucent
Prof. Liang Gu, Yale University, New Haven, CT, USA
Prof. K. K. Pattanaik, ABV-Indian Institute of Information Technology and Management, Gwalior, India
Dr. Germano Lambert-Torres, Itajuba Federal University
Dr. Guang Jin, Intelligent Automation, Inc.
Dr. Hardi Hungar, Carl von Ossietzky University Oldenburg, Germany
Dr. Hongbo Zhou, Southern Illinois University Carbondale
Dr. Huei-Ru Tseng, Industrial Technology Research Institute, Taiwan
Dr. Hussein Attia, University of Waterloo, Canada
Prof. Hong-Jie Dai, Taipei Medical University, Taiwan
Prof. Edward David, UFS—Federal University of Sergipe, Brazil
Dr. Ivan Saraiva Silva, Federal University of Piauí, Brazil
Dr. Luigi Cerulo, University of Sannio, Italy

Dr. J. Emerson Raja, Engineering and Technology of Multimedia University, Malaysia

Dr. J. Satheesh Kumar, Bharathiar University, Coimbatore

Dr. Jacobijn Sandberg, University of Amsterdam

Dr. Jagannath V. Aghav, College of Engineering, Pune, India

Dr. JAUME Mathieu, LIP6 UPMC, France

Dr. Jen-Jee Chen, National University of Tainan

Dr. Jitender Kumar Chhabra, NIT Kurukshetra, India

Dr. John Karamitsos, Tokk Communications, Canada

Dr. Jose M. Alcaraz Calero, University of the West of Scotland, UK

Dr. K. K. Shukla, IIT (BHU), India

Dr. K. R. Pardusani, Maulana Azad NIT, Bhopal, India

Dr. Kapil Kumar Gupta, Accenture

Dr. Kuan-Wei Lee, I-Shou University, Taiwan

Dr. Lalit Awasthi, NIT Hamirpur, India

Dr. Maninder Singh, Thapar University, Patiala, India

Dr. Mehul S. Raval, DA-IICT, Gujarat, India

Dr. Michael McGuire, University of Victoria, Canada

Dr. Mohamed Naouai, Tunis El Manar University and University of Strasbourg, Tunisia

Dr. Nasimuddin, Institute for Infocomm Research

Dr. Olga C. Santos, aDeNu Research Group, UNED, Spain

Dr. Pramod Kumar Singh, ABV-IIITM Gwalior, India

Dr. Prasanta K. Jana, IIT Dhanbad, India

Dr. Preetam Ghosh, Virginia Commonwealth University, USA

Dr. Rabeb Mizouni, KUSTAR, Abu Dhabi, UAE

Dr. Rahul Khanna, Intel Corporation, USA

Dr. Rajeev Srivastava, CSE, IIT (BHU), India

Dr. Rajesh Kumar, MNIT Jaipur, India

Dr. Rajesh Bodade, MCT, Mhow, India

Dr. Rajesh Kumar, MNIT Jaipur, India

Dr. Rajesh Bodade, Military College of Telecommunication Engineering, Mhow, India

Dr. Ranjit Roy, SVNIT, Surat, Gujarat, India

Dr. Robert Koch, Bundeswehr University München, Germany

Dr. Ricardo J. Rodriguez, Nova Southeastern University, USA

Dr. Ruggero Donida Labati, Università degli Studi di Milano, Italy

Dr. Rustem Popa, "Dunarea de Jos" University of Galati, Romania

Dr. Shailesh Ramchandra Sathe, VNIT Nagpur, India

Dr. Sanjiv K. Bhatia, University of Missouri, St. Louis, USA

Dr. Sanjeev Gupta, DA-IICT, Gujarat, India

Dr. S. Selvakumar, National Institute of Technology, Tamil Nadu, India

Dr. Saurabh Chaudhury, NIT Silchar, Assam, India

Dr. Shijo M. Joseph, Kannur University, Kerala

Dr. Sim Hiew Moi, University of Technology, Malaysia

Dr. Syed Mohammed Shamsul Islam, University of Western Australia
Dr. Trapti Jain, IIT Mandi, India
Dr. Tilak Thakur, PED, Chandighar, India
Dr. Vikram Goyal, IIIT Delhi, India
Dr. Vinaya Mahesh Sawant, D. J. Sanghvi College of Engineering, India
Dr. Vanitha Rani Rentapalli, VITS Andhra Pradesh, India
Dr. Victor Govindaswamy, Texas A&M University, Texarkana, USA
Dr. Victor Hinostroza, Universidad Autónoma de Ciudad Juárez
Dr. Vidyasagar Potdar, Curtin University of Technology, Australia
Dr. Vijaykumar Chakka, DA-IICT, Gandhinagar, India
Dr. Yong Wang, School of IS & E, Central South University, China
Dr. Yu Yuan, Samsung Information Systems America, San Jose, CA
Eng. Angelos Lazaris, University of Southern California, USA
Mr. Hrvoje Belani, University of Zagreb, Croatia
Mr. Huan Song, Super Micro Computer, Inc., San Jose, USA
Mr. K. K. Patnaik, IIITM Gwalior, India
Dr. S. S. Sarangdevot, Vice Chancellor, JRN Rajasthan Vidyapeeth University, Udaipur
Dr. N. N. Jani, KSV University, Gandhinagar
Dr. Ashok K. Patel, North Gujarat University, Patan, Gujarat
Dr. Awadhesh Gupta, IMS Ghaziabad
Dr. Dilip Sharma, GLA University, Mathura, India
Dr. Li Jiyun, Donghua University, Shanghai, China
Dr. Lingfeng Wang, University of Toledo, USA
Dr. Valentina E. Balas, Aurel Vlaicu University of Arad, Romania
Dr. Vinay Rishiwal, MJP Rohilkhand University, Bareilly, India
Dr. Vishal Bhatnagar, Dr. Ambedkar Institute of Technology, New Delhi, India
Dr. Tarun Shrimali, Sunrise Group of Institutions, Udaipur
Dr. Atul Patel, C.U. Shah University, Wadhwan, Gujarat
Dr. P. V. Virparia, Sardar Patel University, VV Nagar
Dr. D. B. Choksi, Sardar Patel University, VV Nagar
Dr. Ashish N. Jani, KSV University, Gandhinagar
Dr. Sanjay M. Shah, KSV University, Gandhinagar
Dr. Vijay M. Chavda, KSV University, Gandhinagar
Dr. B. S. Agarwal, KIT, Kalol
Dr. Apurv Desai, South Gujarat University, Surat
Dr. Chitra Dhawale, Nagpur
Dr. Bikas Kumar, Pune
Dr. Nidhi Divecha, Gandhinagar
Dr. Jay Kumar Patel, Gandhinagar
Dr. Jatin Shah, Gandhinagar
Dr. Kamaljit I. Lakhtaria, AURO University, Surat
Dr. B. S. Deovra, B.N. College, Udaipur
Dr. Ashok Jain, Maharaja College of Engineering, Udaipur
Dr. Bharat Singh, JRN Rajasthan Vidyapeeth University, Udaipur

Dr. S. K. Sharma, Pacific University, Udaipur
Dr. Naresh Trivedi, Ideal Institute of Technology, Ghaziabad
Dr. Akheela Khanum, Integral University, Lucknow
Dr. R. S. Bajpai, Shri Ramswaroop Memorial University, Lucknow
Dr. Manish Shrimali, JRN Rajasthan Vidyapeeth University, Udaipur
Dr. Ravi Gulati, South Gujarat University, Surat
Dr. Atul Gosai, Saurashtra University, Rajkot
Dr. Digvijai sinh Rathore, BBA Open University, Ahmedabad
Dr. Vishal Goar, Government Engineering College Bikaner
Dr. Neeraj Bhargava, MDS University, Ajmer
Dr. Ritu Bhargava, Government Women Engineering College, Ajmer
Dr. Rajender Singh Chhillar, MDU, Rohtak
Dr. Dhaval R. Kathiriya, Saurashtra University, Rajkot
Dr. Vineet Sharma, KIET, Ghaziabad
Dr. A. P. Shukla, KIET, Ghaziabad
Dr. R. K. Manocha, Ghaziabad
Dr. Nandita Mishra, IMS Ghaziabad
Dr. Manisha Agarwal, IMS Ghaziabad
Dr. Deepika Garg, IGNOU, New Delhi
Dr. Goutam Chakraborty, Iwate Prefectural University, Iwate Ken, Takizawa, Japan
Dr. Amit Manocha, Maharaja Agrasen University, HP, India
Prof. Enrique Chirivella-Perez, University of the West of Scotland, UK
Prof. Pablo Salva Garcia, University of the West of Scotland, UK
Prof. Ricardo Marco Alaez, University of the West of Scotland, UK
Prof. Nitin Rakesh, Amity University, Noida, India
Prof. Mamta Mittal, G. B. Pant Government Engineering College, Delhi, India
Dr. Shashank Srivastava, MNNIT Allahabad, India
Prof. Lalit Goyal, JMI, Delhi, India
Dr. Sanjay Maurya, GLA University, Mathura, India
Prof. Alexandros Iosifidis, Tampere University of Technology, Finland
Prof. Shanthi Makka, JRE Engineering College, Greater Noida, India
Dr. Deepak Gupta, Amity University, Noida, India
Dr. Manu Vardhan, NIT Raipur, India
Dr. Sarsij Tripathi, NIT Raipur, India
Prof. Wg Edison, HeFei University of Technology, China
Dr. Atul Bansal, GLA University, Mathura, India
Dr. Alimul Haque, V.K.S. University, Bihar, India
Prof. Simhiew Moi, Universiti Teknologi Malaysia
Prof. Rustem Popa, "Dunarea de Jos" University of Galati, Romania
Prof. Vinod Kumar, IIT Roorkee, India
Prof. Christos Bouras, University of Patras and RACTI, Greece
Prof. Devesh Jinwala, SVNIT Surat, India
Prof. Germano Lambert-Torres, PS Solutions, Brazil
Prof. Byoungho Kim, Broadcom Corp., USA

# Contents

# About the Editors

**Dr. Bijaya Ketan Panigrahi** is working as a Professor in the Electrical Engineering Department, IIT Delhi, India. Prior to joining IIT Delhi in 2005, he has served as a Faculty in Electrical Engineering Department, UCE Burla, Odisha, India, from 1992 to 2005. He is a Senior Member of IEEE and Fellow of INAE, India. His research interest includes application of soft computing and evolutionary computing techniques to power system planning, operation, and control. He has also worked in the field of biomedical signal processing and image processing. He has served as the editorial board member, associate editor, and special issue guest editor of different international journals. He is also associated with various international conferences in various capacities. He has published more than 100 research papers in various international and national journals.

**Dr. Munesh C. Trivedi** is currently working as a Professor in the Computer Science and Engineering Department, ABES Engineering College, Ghaziabad, India. He has rich experience in teaching the undergraduate and postgraduate classes. He has published 20 textbooks and 80 research publications in different international journals and proceedings of international conferences of repute. He has received Young Scientist Visiting Fellowship and numerous awards from different national as well as international forums. He has organized several international conferences technically sponsored by IEEE, ACM, and Springer. He has delivered numerous invited and plenary conference talks throughout the country and chaired technical sessions in international and national conferences in India. He is on the review panel of IEEE Computer Society, International Journal of Network Security, Pattern Recognition Letter and Computer & Education (Elsevier's Journal). He is an Executive Committee Member of IEEE UP Section, IEEE India Council, and also IEEE Asia Pacific Region 10. He is an Active Member of IEEE Computer Society, International Association of Computer Science and Information Technology, Computer Society of India, International Association of Engineers, and a Life Member of ISTE.

**Dr. Krishn K. Mishra** is currently working as a Visiting Faculty in the Department of Mathematics & Computer Science, University of Missouri, St. Louis, USA. He is an alumnus of Motilal Nehru National Institute of Technology Allahabad, India, which is also his base working institute. His primary areas of research include evolutionary algorithms, optimization techniques, and design and analysis of algorithms. He has also published more than 50 publications in international journals and in proceedings of international conferences of repute. He is serving as a program committee member of several conferences and also editing few Scopus and SCI-indexed journals. He has 15 years of teaching and research experience during which he made all his efforts to bridge the gaps between teaching and research.

**Dr. Shailesh Tiwari** is currently working as a Professor in the Computer Science and Engineering Department, ABES Engineering College, Ghaziabad, India. He is also administratively heading the department. He is an alumnus of Motilal Nehru National Institute of Technology Allahabad, India. He has more than 16 years of experience in teaching, research, and academic administration. His primary areas of research are software testing, implementation of optimization algorithms, and machine learning techniques in software engineering. He has also published more than 50 publications in international journals and in proceedings of international conferences of repute. He is also serving as a program committee member of several conferences and also editing few Scopus and E-SCI-indexed journals. He has organized several international conferences under the sponsorship of IEEE and Springer. He is a Fellow of Institution of Engineers (FIE), Senior Member of IEEE, Member of IEEE Computer Society, and Former Executive Committee Member of IEEE Uttar Pradesh Section. He is also a member of reviewer and editorial board of several international journals and conferences. He has also edited several books published under various book series of Springer.

**Dr. Pradeep Kumar Singh** is currently working as an Assistant Professor (Senior Grade) in the Department of Computer Science & Engineering, Jaypee University of Information Technology (JUIT), Waknaghat, India. He has 10 years of vast experience in academics at reputed colleges and universities of India. He has completed his Ph.D. in Computer Science & Engineering from Gautam Buddha University (State Government University), Greater Noida, UP, India. He received his M.Tech. (CSE) with distinction from Guru Gobind Singh Indraprastha University, New Delhi, India. He has obtained his B.Tech. (CSE) from Uttar Pradesh Technical University (UPTU), Lucknow, India. He is having Life Membership of Computer Society of India (CSI) and promoted to Senior Member Grade from CSI. He is a Member of ACM, IACSIT, Singapore, and IAENG, Hong Kong. He is associated with many IEEE International Conferences as TPC member, reviewer, and session chair. He is an Associate Editor of International Journal of Information Security and Cybercrime (IJISC) a scientific peer-reviewed journal from Romania. He has published nearly 50 research papers in various international journals and conferences of repute. He has organized various theme-based special sessions during the international conferences also.

# Part I
# Smart Computing Technologies

# A Neural Network-Based Novel Detection Technique for Analysis of Pollutant Levels in Air

Jagjot Singh Khokhar, Monika Arora, Gurvinder Kaur,
Garima Mahendru, Monica Kaushik, Nirdosh and Shikha Singh

**Abstract** The dispersions or suspensions of the particles in solid and the liquid forms in atmosphere are coined as aerosols. These suspensions are the matter of concern in recent times as the ecosystem and the human health are at risk. These atmospheric aerosols are defined in broader terms; to be more precise, the term particulate matter (PM) is used to define the suspended solid-phase matter in the atmosphere. It is the mixture of the diverse elements. Further pollutants like $SO_2$ and $NO_2$ are largely found in the industrial waste. The evidences reveal that sulfate and organic matter are the two main contributing factors for annual $PM_{10}$ concentration, and its consequences are like health problems and ecological imbalance which are correlating and pointing especially toward the particulate matter. In this paper, the average concentration of various pollutants like $SO_2$, $NO_2$, $PM_{10}$, and SPM in air have been predicted efficiently. The detailed analysis of different models and its effects on the environment have been examined with the help of neural network tool.

**Keywords** Particulate matter · Neural network · Air pollution
Human health

J. S. Khokhar (✉) · M. Arora · G. Kaur · G. Mahendru · M. Kaushik · Nirdosh · S. Singh
Department of Electronics and Communication, Amity University, Noida, India
e-mail: khokharjagjotsingh92@gmail.com

M. Arora
e-mail: monika4dec@gmail.com

G. Kaur
e-mail: gurvinder.kaur248@gmail.com

G. Mahendru
e-mail: gmahendru@amity.edu

M. Kaushik
e-mail: mkaushik@amity.edu

© Springer Nature Singapore Pte Ltd. 2019
B. K. Panigrahi et al. (eds.), *Smart Innovations in Communication and Computational Sciences*, Advances in Intelligent Systems and Computing 669, https://doi.org/10.1007/978-981-10-8968-8_1

# 1    Introduction

In the environmental engineering literature, the terms aerosols and PM are often used interchangeably although the former one has broader definition and scope. The aerosols come from the wider range of anthropogenic and natural sources from the earth, whereas airborne PM is not a single pollutant; rather, it is the mixture of many subclasses of pollutants [1]. The fate and transport of gas-phase components in atmosphere is closely linked to the aerosols in contrast to the PM which are either directly emitted by emission sources or are formed due to reaction between the gases; for example, reaction between ammonia and oxides of sulfur or nitrogen results in PM [2].

As per the numerous epidemiologic studies, PM having aerodynamic diameter <10 μm ($PM_{10}$) or <2.5 μm ($PM_{2.5}$) is of major concern in public health issues. The fact that respirable suspended particulate matter is more dangerous to health than larger particulate up to 100 micron is well established. It is important to remember though that the ratio of RSPM to SPM will be specific to an area and the measurement of the one should be able to infer the other if the ratio has been experimentally determined. The various kinds of the study show that there are mainly two objectives of atmospheric aerosols: firstly, the direct impact on the health as a result of near exposure on the surface of the earth and secondly, the role of aerosols in the physical processes and atmospheric chemicals and the way it is affecting the local climate and the global climate [3]. Further in a narrow way, the recent studies show that concentration of $PM_{10}$ and $PM_{2.5}$ airborne aerosols in urban areas shows the good ratio in traffic-related pollutants and other combustion processes [4]. The long-term exposure to $PM_{10}$ leads to inflammation in lungs. The lungs and heart get affected due to the inhalation of air particles.

Rambagh, located five kilometer northeast of Taj Mahal, Agra, experiences semiarid climate that borders on humid subtropical climate. It has mild winters with hot and dry summers and monsoons. The recent surveys show that Agra is ahead of the capital, Delhi or Kanpur in terms of the black carbon levels in atmosphere. There is also an increase in ratio of particulate matter beyond the permissible values. In this paper, the recent data for 2015–2016 (August) has been gathered across the Rambagh which includes the concentration of $PM_{10}$ μg/m$^3$, levels of $SO_2$ and $NO_2$, and the weather conditions of a place. In the following sections, the statistical analysis, methodologies, and results have been further analyzed.

# 2    Statistical Analysis

In this research work, many statistical indexes such as Bayesian regularization (BR), Levenberg–Marquardt algorithm (LM), and scaled conjugate gradient (SCG) have been used for the evaluation and accuracy of the performance and results [5].

Bayesian classification is the technique to construct the classifiers. Classifiers are nothing but the models that assign the class labels to the problem instance. Levenberg–Marquardt algorithm is also known as DLS that is damped least squares used for solving generic curve fitting problems by finding the local minimum which may not be the global minimum [6]. Scaled conjugate gradient is feed-forward and supervised algorithm for neural networks; the feed forward here means that in connections there is no loop between the units. The general equations corresponding to each are mentioned below:

Bayesian regularization:

$$x = arg \max_{b \in \{1,....B\}} p(C_b) \prod_{i=1} n\, p(y_i | C_b) \tag{1}$$

Levenberg–Marquardt algorithm:

$$H(\beta) = \sum_{j=1} m\left[x_j - f\left(y_j, \beta\right)\right]^2 \tag{2}$$

Scaled conjugate:

$$S_k = \frac{\dot{E}(W_k + \sigma_k P_k) - \dot{E}(W_k)}{\sigma_k} \tag{3}$$

Bayesian regularization, Levenberg–Marquardt algorithm, and scaled conjugate are the various algorithmic parameters and functions used in the neural networks. BR can eliminate or reduce the need for lengthy cross-validations, and it is more robust than the standard back-propagation methods, whereas to solve nonlinear least squares problems, the LM technique is considered to be the standard one as it shows lower performance in terms of predictive ability. On the other hand, SCG needs O (n) of memory where n represents the number of weights in the network although it uses second order of information from neural networks [7]. Among these three, the BR is considered to be the optimal one as it develops the nonlinear relationships and it has more predictive abilities. To get refined results, the data was tested through five hidden layers, and on observing the results, it can be seen that the BR shows least mean square error (Table 1). In further sections, a brief introduction of neural networks is cited and the results in regard to BR have been shown and explained in the sections after it.

**Table 1** No. of hidden layers and mean square values

| S. No. | No. of hidden layers | Mean square error | No. of iterations |
|--------|---------------------|-------------------|-------------------|
| 1      | 5                   | 0.014719          | 43                |
| 2      | 10                  | 0.013359          | 148               |
| 3      | 15                  | 0.013966          | 50                |
| 4      | 20                  | 0.013095          | 96                |
| 5      | 25                  | 0.013677          | 69                |
| 6      | 30                  | 0.012769          | 57                |

## 2.1  *Neural Networks*

It is one of the concepts, which has been inspired by the functionality of the human brain and its performance in identification of phenomena. Neurons (a single neuron shown in Fig. 1) are placed in different layers in multilayer neuron network [5]. Input layer being the first layer receiving information and till its capability with other neurons, it transfers the information in the form of input signals to the other next layers. Neuron weight is the communication ability of each neuron with other neurons. The number of neurons in each layer depends on the weight of neuron and the previous layers' neurons. In addition to the input layer, the neural network also consists of the hidden layers and the output layers. Some of the advantages of using artificial neural networks are its record-breaking accuracy on a wide variety range of problems, less requirement of formal statistical training, offering various multiple training algorithms, and having the implicit ability of detecting nonlinear complex relationships between independent and dependent variables. In this, neuron is the main processor and adding neurons to hidden layers will reduce calculation error but will be more time consuming for calculations [6]. Hence, deciding on the logical proportion for choosing the number of neurons for hidden layers and processing is obligatory. In the next section, the methodology used in the paper and results have been discussed.

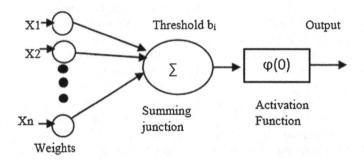

**Fig. 1** Single neuron

## 3 Data and Methodology

The data set used in this paper is the recent one of the place Rambagh located in Agra. The values of $SO_2$, $NO_2$, $PM_{10}$, and suspended particulate matter (SPM) for the one and a half years that is from January to December of 2015 and from January to August of 2016. Central Pollution Control Board (CPCB), Ministry of Environment and Forest, Government of India, has provided this data on the alternate-day basis also including the local weather conditions, for example, cloudy day or sunny day [8]. The values are in $\mu g/m^3$ unit. There are three portions of the data that also act as training set, validation set, and testing set in neural networks having 70%, 15%, and 15% weightage, respectively. These values can be changed, but these particular values give better results. For this paper, 186 sample spaces of data have been used. Figure 2 shows the basic view of the neural network consisting of inputs, hidden layers, and the output.

The nftool (neural fitting tool) of MATLAB has been used in our proposed work to determine the performance and the results. In this, the number of input data to layer and the no. of hidden layers have to be defined. In this paper, 5, 10, 15, 20, 25, and 30 hidden layers have been used one by one to get the output and their mean square errors have been compared to get the performance measures. Since layer 30 gives most refined results with least value of error, the data has been trained till 30 layers.

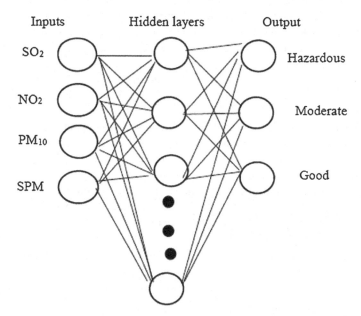

**Fig. 2** Neural network

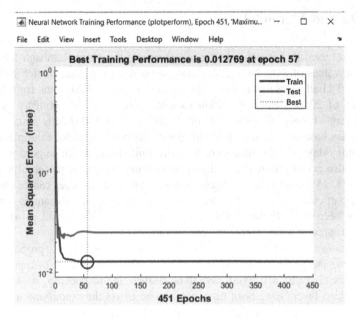

**Fig. 3** Performance graph showing mean square error

Figure 2 shows the neural network marked with the inputs such as the values of $SO_2$, $NO_2$, $PM_{10}$, and the SPM, the hidden layers, and the output having three desired values such as hazardous (1), moderate (0.5), and good (0). These are levels of the $PM_{10}$ in which the range of its values 200 and above is considered to be hazardous level, range from 100 to 200 is moderate, and 0–100 is good [8]. These levels further suggest how much harmful is the $PM_{10}$ for the human life and the ecosystem explained in the next section. Table 1 shows the number of hidden layers with their mean square.

Table 1 and Figure 3 show that the validation, test data sets, and the performance of training with respect to epochs of $PM_{10}$ level prediction in the atmosphere. As it is clearly seen the best performance measure is at S. No. 6 that is with the 30 hidden layers and with minimum mean square value that is 0.012769. Further Figs. 4 and 5 show the regression curve between the target data and the output of levels of the pollutant level prediction in the atmosphere and the error histogram of pollutant levels in the atmosphere.

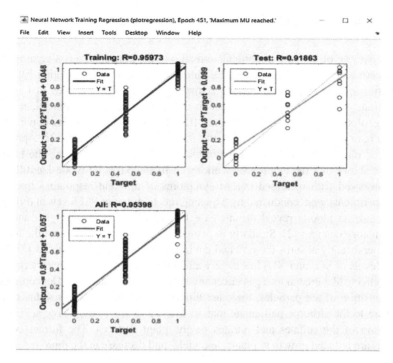

**Fig. 4** Regression curve for PM$_{10}$ levels in atmosphere

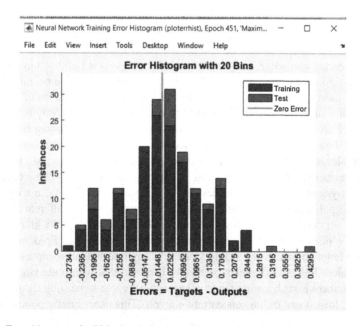

**Fig. 5** Error histogram for PM$_{10}$ levels in atmosphere

# 4   Impact on Human Health and Ecosystem

The recent rate of increase in particulate matter in the atmosphere is a great matter of concern as it is largely affecting the human health and the vegetation [9]. The deposition on the vegetation surface is mainly having three sources that is dry deposition, wet deposition, and the occult deposition, and on the other hand, the particles of size between 2.5 micrometer–10 Âµm are removed by the upper track of respiratory system of humans but the particles of size less than 2.5 Âµm get deposited on the bronchi walls in the bronchi system or bronchi tree. The further analysis shows that the employees working in municipal solid-waste landfill area were observed with increased rate of symptoms of different respiratory troubles. The experiments were conducted by exposing rats to different PM levels at different times, and the reports reveal the increase in size of the lungs' weights and the inflammatory changes [2]. Similarly in plants, the most exposed part are the leaves which persistently absorbs the polluted environment and the dust particles [6]. The rise in levels of $SO_2$ and $NO_2$ has largely affected the regional weather patterns, the global effects like greenhouse gas effect and the heritage monuments. Depending on the deposition of the particles, there are differing phytotoxic responses due to the exposure to the airborne particulate matter. It has also led to the heavy acidifying deposition of the sulfates and nitrates on the plant surfaces. The further consequences are reduced growth in plants, less yield, and decrease in the reproduction of plants.

# 5   Conclusion

In this paper, the predictive analysis of the pollutant levels at Rambagh in Agra has been done through neural networks. The different parameters of the neural network have been used to get the accurate results. The highest concentration of $SO_2$ and concentration of $PM_{10}$ are found in industrial areas and that of $NO_2$ is found in commercial areas. So, the topic of concern, it is slowly deteriorating the human health and the ecosystem. In this paper, analysis has been presented using nftool to predict the levels of the various pollutants of a place and the various risks involved. Since the airborne PM is characterized by the diverse effects on climate, human health, ecosystem, etc. Further studies can be done in a manner like analyzing the levels of air pollutants causing pollution and the consequences in different areas, for example, commercial areas, industrial areas, residential areas, and the greenbelts of a particular place in different seasons and weather conditions. Epidemiological studies on large pollutants are unable to identify any threshold concentration below which ambient PM has no effects on health. It is said that within the large human population, there is such a wide range in susceptibility that some subjects are at risk even at the lowest end of the concentration levels. This paper briefly points out the affects, and therefore, it is a high time that steps should be taken to curb the problem before it destroys life system completely. In further researches, the different parts will be studied and the solutions will be proposed to reduce the effects.

# References

1. Chate, D. M. and Murugavel, P., Atmospheric aerosol formation and its growth during the cold season in India. *J. Earth Syst. Sci.*, 2010, 119(4), 471–477 for forest trees. Environmental Pollution, 82: 167–180

2. Sunder Raman, R., Ramachandran, S. and Rastogi, N., Source identification of ambient aerosols over an urban region in western India. *J. Environ. Monit,* 2010, 12(6), 1330–1340.S

3. P. Raktim, K. Ki-Hyun, H. Yoon-Jung, and J. Eui-Chan. "The pollution status of atmospheric carbonyls in a highly industrialized area". Journal of Hazardous Materials, 153(3):1122–1135, 2008.

4. Gianola, D.; Okut, H.; Weigel, K.A.; Rosa, G.J.M. Predicting complex quantitative traits with Bayesian neural networks: A case study with Jersey cows and wheat. BMC Genet. 2011, 12, 1–37.

5. M.I.A. Lourakis. A brief description of the Levenberg-Marquardt algorithm implemented by levmar, Technical Report, Institute of Computer Science, Foundation for Research and Technology - Hellas, 2005.

6. Bui, D.T.; Pradhan, B.; Lofman, O.; Revhaug, I.; Dick, O.B. Landslide susceptibility assessment in the HoaBinh province of Vieatnam: A comparison of the Levenberg–Marqardt and Bayesian regularized neural networks.Geomorphology 2012, 171, 12–29.

7. Tiwari, S., Srivastava, A.K., Bisht, D.S., Bano, T., Singh, S., Behura, S., Srivastava, M.K., Chate, D.M. and Padmanabhamurty, B;.Black carbon and chemical characteristics of PM10 and PM2.5 at an urban site of North India. Journal of Atmospheric Chemistry., 2009, 62(3), 193–209.

8. http://cpcb.nic.in/National_Ambient_Air_Quality_Standards.php Makridakis, S., Wheelwright, S. C. & Hyndman, R. J. (1998), Forecasting: methods and applications, 3rd edn, John Wiley & Sons, New York.

9. M. Younger, H.R. Morrow-Almeida, S. M. Vindigni, and A. L. Dannenberg. "The Built Environment, Climate Change, and Health: Opportunities for Co-Benets". American Journal of Preventive Medicine, 35(5):517–526 2008.

# A Text Preprocessing Approach for Efficacious Information Retrieval

**Deepali Virmani and Shweta Taneja**

**Abstract** The information retrieval is the task of obtaining relevant information from a large collection of databases. Preprocessing plays an important role in information retrieval to extract the relevant information. In this paper, a text preprocessing approach text preprocessing for information retrieval (TPIR) is proposed. The proposed approach works in two steps. Firstly, spell check utility is used for enhancing stemming and secondly, synonyms of similar tokens are combined. In this paper, proposed technique is applied to a case study on International Monetary Fund. The experimental results prove the efficiency of the proposed approach in terms of complexity, time and performance.

**Keywords** Information retrieval · Information extraction · Preprocessing
Stemming

## 1 Introduction

Text mining [1, 2] is the method of mining or extracting useful information from text documents. This method involves information extraction and information retrieval [3]. Information extraction [4] is the process of automatically extracting useful information. It makes use of methods and techniques of natural language processing. Information retrieval is the activity of searching and manipulating large collection of databases. Information retrieval systems are used worldwide with millions of people relying on them to promote business, education and entertainment [5, 6]. The following Fig. 1 shows the process of text mining.

D. Virmani (✉) · S. Taneja
Department of Computer Science, Bhagwan Parshuram Institute of Technology,
New Delhi, India
e-mail: deepalivirmani@gmail.com

S. Taneja
e-mail: shweta_taneja08@yahoo.co.in

© Springer Nature Singapore Pte Ltd. 2019
B. K. Panigrahi et al. (eds.), *Smart Innovations in Communication
and Computational Sciences*, Advances in Intelligent Systems
and Computing 669, https://doi.org/10.1007/978-981-10-8968-8_2

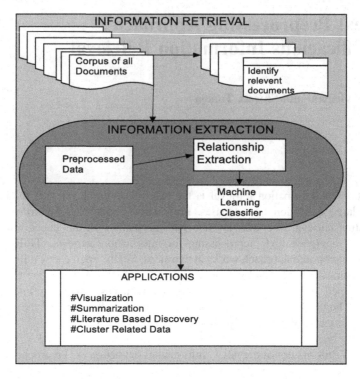

**Fig. 1** Framework of text mining

Text preprocessing is an initial phase in text mining. There are various pre-processing techniques to categorize text documents. These are filtering, splitting of sentences, stemming, stop words removal and token frequency count. Filtering has a set of rules for removing duplicate strings and irrelevant text [2]. Stemming is done to convert a word to its root. Stop words are those that have little or no importance in data mining. Tokens are the special words, characters or symbols, etc. Tokenization is the process of fragmenting strings to small elements called tokens which include words, symbols, phrases, symbols, keywords [1].

The organization of paper is as follows: The literature review in the field of preprocessing techniques is given in Sect. 2. Then in Sect. 3, our proposed pre-processing technique is explained. This is followed by a Case Study in Sect. 4. Lastly, the conclusion is addressed in Sect. 5.

## 2 Literature Review

Researchers have performed different studies to improve the accuracy of methods proposed for preprocessing technique in text mining. This section discusses the related work done.

In [6], authors have introduced a new method which eliminates stop words. This can lead to loss of information that may affect the performance of the text mining algorithm. So, they have proposed a method in which a set of stop words are chosen to improve the accuracy of text mining.

Some authors in [4] have presented a simple approach to make preprocessing technique more efficient by using improved stemming algorithm. Porters stemming algorithm is the most popular algorithm. The authors have defined some constraints to improve it. By doing this, the problem of named entity can be overcome.

In another work [7], an efficient tokenization approach has been described. It is based on training vectors, and results show the proficiency of proposed algorithm. Tokenization satisfies user's information needs more precisely and is an effective part of information retrieval.

In another work, authors [5] have proposed different steps for document pre-processing. Firstly, stop words are removed. Then porter stemmer algorithm is used for stemming. Afterwards, Word net thesaurus is used to study relationships between important terms. This is followed by forming a data matrix. Thirdly, term selection approaches are used to select terms from the documents depending on their minimum threshold value.

In [8], the authors present a survey on different preprocessing techniques for text mining. It focuses on different stemming algorithms giving their drawbacks and applications.

In this work, a new text preprocessing approach text preprocessing for information retrieval (TPIR) is proposed. The proposed approach works in two steps. Firstly, spellcheck utility is used for enhancing stemming, and secondly, synonyms of similar tokens are combined.

## 3 Proposed Text Preprocessing Technique (TPIR)

The goal of our work is to improve the efficiency of text preprocessing to enhance the accuracy of text mining. The proposed technique consists of the following phases which can be explained as shown in Fig. 1. This proposed technique works in four phases. In the first phase, text documents are split into paragraphs or sentences depending upon user's application and then syntactical information is extracted by tokenization. In the second phase, feature space is reduced by combining synonyms of similar tokens. In the third phase, the dimensions of feature space are reduced by separating the stop words from tokens. In the fourth phase, enhanced stemming approach using spell check utility is applied [4].

### 3.1 Information Extraction by Tokenization

The text documents are split into paragraphs and sentences, and syntactical information is extracted from them using tokenization. The tokens are extracted from

**Fig. 2** Proposed framework
of TPIR

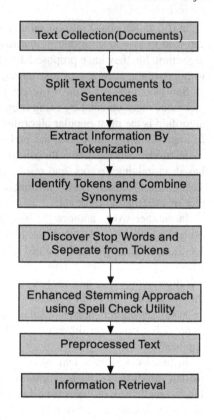

these documents. These tokens are further sent for processes like parsing or text mining. During this phase, unrelated symbols are also removed [1].

Generally, tokens are generated at the level of words. But, in certain situations it is not possible to define a "word". For example, punctuation and whitespace shall not be counted as tokens. Special characters like spaces, line breaks or punctuation characters are used to separate the tokens [2] (Figs. 2 and 3).

**Fig. 3** Combination of synonyms

Input: A token T with synonyms

Output: Resulting set of tokens
1. Let $T_i = \{ x_{i1}, x_{i2}, x_{i3}, \ldots x_{ik} \}$ be a set of t tokens represented in one dimension
2. Let $T_j = \{ y_{j1}, y_{j2}, y_{j3} \ldots y_{jl} \}$ be another set of tokens represented in other dimension
3. If any $x_{ik} = y_{jl}$ and $y_{jl} = x_{ik}$ for any k,l
   Then $T_i = T_j$
4. If $T_i = T_j$, then $T(syn)_{i,j} = T_i + T_j$

---

**Algorithm 1: Tokenization**
Input: Set of Documents
Output: Tokens
1. Input a set of documents ($D_i$) where i=1,2,3......
2. For each document ($D_i$):
  2.1 Identify word ($E_i$)= $D_i$ where i=1,2,3,.... // apply extract word process for all the documents i=1,2,3..
  2.2 Store $D_i[n]$ = $E_i$ where i=1,2,3.. // maintain document vector to store extracted words for every input document
3. For each extracted word ($E_i$):
  3.1 Stop Word ($S_i$)= $E_i$ where i=1,2,3...// remove all stop words
  3.2 Stemming ( $ST_i$)=$S_i$ where i=1,2,3...// identify all stem words
4. For each stem word ($ST_i$):
  4.1 FreqCount ($FC_i$)= $ST_i$ where i=1,2,3...// find total occurrences
    4.2 Return ($ST_i$)

---

**Fig. 4** Tokenization process

## 3.2 Combine Synonyms of Similar Tokens

It is a very important step, and it leads to reduction in the dimensionality of the feature space. The tokens which represent the same meaning are combined so that no two words occur that have similar meaning. This is done to reduce the dimensionality of the features. This step also keeps track of the frequency of the highest occurring word and returns the keyword with the highest frequency [9] (Figs. 4 and 5).

## 3.3 Separate Stop Words from Tokens

Stop words [10] are the words that carry little or no information. All the stop words present in the text document are identified and are separated from tokens obtained (Figs. 6 and 7).

**Fig. 5** "CONSUME" term: process of stemming

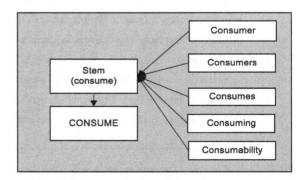

**Fig. 6** Definition of two test
cases

| |
|---|
| **Test Case A-** The traditional text preprocessing technique was used with stemming and stop words removal. |
| **Test Case B-** The proposed TPIR method was used with enhanced stemming approach using Spell Check Utility. |

## 3.4  Stemming with Spell Check Utility

Stemming finds the stem of a word. This method removes different suffixes, reduces the number of words, to match exact stems of words. This helps in the saving of memory space and time [11].

The traditional stemming process has drawbacks like that in many circumstances, it is not required for a suffix to be removed, and it is not possible to match something with nothing. To overcome certain drawbacks in stemming, our technique uses the utility of spell check [4]. This module solves various problems like named-entity recognition. This utility is implemented using Boyer–Moore's algorithm which is used to implement this utility. It is an effective algorithm for matching string patterns. If this method detects any named entities, then that word is removed from the stemming process, and therefore accuracy is increased.

## 4  Case Study

Our proposed TPIR is executed on the original abstract of a study done on The International Monetary Fund. We have developed two test cases to check the efficiency and accuracy of our proposed TPIR and the traditional preprocessing method. Thus, our experiments conducted have the following cases as shown below.

| |
|---|
| The International Monetary Fund is an international organization which has its headquarters in Washington D.C. It works to foster monetary cooperation. This organization facilitates international trade. It promotes sustainable economic growth. The foundation of this organization was laid in Bretton Woods Conference. It came into formal existence in the year 1945. This fund has improved the economy of member countries and has made them stable and secure. |

**Fig. 7** Original abstract of a study on International Monetary Fund

**Fig. 8** Tokens and their synonyms

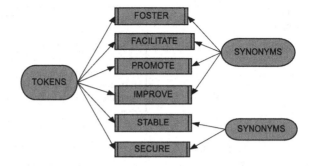

## 4.1 Information Extraction by Tokenization

In this step, the tokens are obtained from the abstract of a study. Figure 8 shows the representation of tokens after lexical analysis. The enhanced preprocessing approach combines tokens that have similar meanings and generates the count of tokens thereby reducing the feature space.

## 4.2 Preprocessed Text

See Figs. 9, 10, 11, 12 and 13.

**Fig. 9** Preprocessed abstract of the study with traditional method

**Test Case A-** Internation Monetary Fund internation organizate headquarters Washing D.C. works foster monetary cooperate organizate facilitates internation trade promotes sustainable economic growth foundation organizate was laid in Brett Woods Conference came formal existence year 1945 fund improved economy member countries made stable secure.

**Fig. 10** Preprocessed
abstract of the study TPIR

**Test Case** B- International Monetary Fund international
organization headquarters Washington D.C. works foster
monetary cooperation organization facilitates international
trade promotes  sustainable economic growth  foundation
organization was laid in Bretton Woods Conference came
formal existence year 1945  fund improved economy
 member countries made   stable secure.

**Fig. 11** Named entities
present in the abstract

| **Named Entities in the Abstract** |
| --- |
| International Monetary Fund- Organization |
|  |
| Washington D.C- Location |
| Bretton Woods Conference-Event |
| 1945- Time |

**Fig. 12** Preprocessed result
of test case A

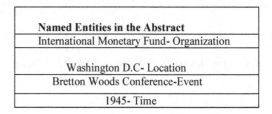

**Fig. 13** Preprocessed result
of test case B

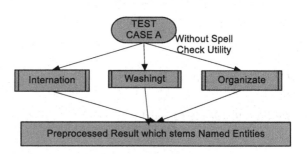

## 4.3  Analysis of Result

From comparisons, we can make out that proposed TPIR, Test case B outperforms
Test Case A in terms of parameters, namely complexity and overall time. The
accuracy level of output content is increased, and information retrieval process is
improved. In Test Case B, a number of steps are reduced, thereby feature space is
reduced, and therefore complexity for the whole preprocessing task reduces. The
results are shown in Fig. 14.

**Fig. 14** Comparison
between TPIR and traditional
preprocessing technique on
the basis of time

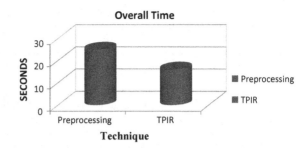

## 5 Conclusion

Preprocessing is the most important task in information retrieval for extracting important information from unstructured text data. But traditional preprocessing technique does not always generate good quality results due to named-entity recognition. In this paper, we have proposed an efficient preprocessing technique TPIR in which spell check utility is used to overcome the problem of named-entity recognition and combined tokens of similar meanings. This technique gives better results as the number of tokens is reduced, thereby reducing the storage space. It leads to improvement in accuracy as well as the overall time. Our proposed TPIR has proved to be better than traditional preprocessing technique in terms of complexity and overall time.

## References

1. Subbaiah S.: Extracting Knowledge using Probabilistic Classifier for Text Mining. In: International Conference on Pattern Recognition Informatics and Mobile Engineering, pp. 440–442 (2013).
2. Bhujade V., Jhanwe N.J.: Knowledge Discovery in Text Mining Technique Using Association Rules Extraction. In: International Conference on Computational Intelligence and Communication Systems, pp. 498–502 (2011).
3. Tari, Hakenberg J., Chen Y., Son T., Gonzalez G., Baral C.: Incremental Information Extraction Using Relational Databases. IEEE Transactions on Knowledge and Data Engineering, 24, 1, 86–99 (2012).
4. Ramasubramanian C., Ramya R.: Effective Pre-Processing Activities in Text Mining using Improved Porter's Stemming Algorithm. International Journal of Advanced Research in Computer and Communication Engineering. 2, 12, 4536–4538 (2013).
5. Patil L.H., Atique M.: A Novel Approach for Feature Selection Method TF-IDF in Document Clustering. In: IEEE International Advance Computing Conference (IACC), pp. 858–862 (2013).
6. Amarasinghe K., Hruska R.: Optimal Stop Word Selection for Text Mining in Critical Infrastructure Domain. In: IEEE Conference, pp. 179–184 (2015).
7. Singh V., Saini B.: An Effective Pre-Processing Algorithm for Information Retrieval Systems. International Journal of Database Management Systems (IJDMS), 6, 6, 13–24 (2014).

8. Nayak A.S., Kanive A.P., Chandavekar N., Balasubramani R: Survey on Pre-Processing Techniques for Text Mining. International Journal Of Engineering And Computer Science, ISSN: 2319-7242, 5, 6, pp. 16875–16879, (2016).
9. Xubu M., Guo J.: Information Extraction of Strategic Activities based on Semi-structured Text. In: International Joint Conference on Computational Sciences and Optimization, pp. 579–583 (2014).
10. Hadni M., Lachkar A., Ouatik S.A.: A New and Efficient Stemming Technique for Arabic Text Categorization. In: International Conference on Multimedia Computing and Systems (ICMCS), pp. 791–796 (2012).
11. Feilmayr C.: Text Mining-Supported Information Extraction an Extended Methodology for Developing Information Extraction Systems. In: International Workshop on Database and Expert Systems Applications, pp. 217–221 (2011).

# Reliability Growth Modeling for OSS: A Method Combining the Bass Model and Imperfect Debugging

Neha Gandhi, Himanshu Sharma, Anu G. Aggarwal
and Abhishek Tandon

**Abstract** In this highly competitive era of technology, for software to sustain in the market, it has to maintain its quality. Software reliability is one of the metrics to determine software quality. As very few efforts are spent on testing open source software, the reliability of open source software hugely depends on the number of users working on it after release. This study proposes new non-homogeneous Poisson process-based software reliability growth models incorporating factor of user growth in reliability growth of open source software. To represent user growth phenomenon in the proposed SRGMs, the Bass diffusion and Kenney's models are used. The models are proposed for scenarios of both imperfect debugging and perfect debugging. Reliability analysis is carried out on real-world failure dataset (GNOME 2.0), and a parallel comparison among all SRGMs on four goodness-of-fit criteria (mean square error, coefficient of determination, predictive ratio risk, and predictive power) is performed. It is observed that SRGMs which are considered imperfect debugging outperforms its perfect counterpart which is consistent with realistic situations.

**Keywords** Open source software · Imperfect debugging · Non-homogeneous Poisson process · Software reliability growth model

N. Gandhi (✉)
Shaheed Rajguru College of Applied Sciences for Women, University of Delhi,
New Delhi 110096, Delhi, India
e-mail: nehagandhi1990@gmail.com

H. Sharma · A. G. Aggarwal
Department of Operational Research, University of Delhi, New Delhi 110007, Delhi, India
e-mail: himanshusharma.du.or@gmail.com

A. G. Aggarwal
e-mail: anuagg17@gmail.com

A. Tandon
Shaheed Sukhdev College of Business Studies, University of Delhi, New Delhi 110095,
Delhi, India
e-mail: abhishektandon86@gmail.com

© Springer Nature Singapore Pte Ltd. 2019                                    23
B. K. Panigrahi et al. (eds.), *Smart Innovations in Communication
and Computational Sciences*, Advances in Intelligent Systems
and Computing 669, https://doi.org/10.1007/978-981-10-8968-8_3

# 1  Introduction

Open source software (OSS) is software for which source code is also released for public examination, modifications, distributions, and enhancements. The aim of OSS is to invite developers with vast knowledge base across the globe to voluntarily contribute toward a reliable and innovative product. Development of OSS follows a considerably different trend. The basic prototype of the software is developed by a person or a small group and is released in public for its evolution. Due to the early release of prototype, it does not get through a thorough testing phase and the software comes into its operational phase at a very premature stage. Therefore, it becomes crucial to analyze the reliability of such software. Software reliability growth modeling is done to analyze reliability growth trend for software. Experiencing any difference between expected and actual behavior in any software use case is considered as a defect and is reported for a fix. Software reliability growth models (SRGM) interpolate such defect data collected as a function of time by some mathematically derived models, and estimation of parameters is performed. NHPP-based models assume defect detection process to be a non-homogenous process and are the most effective type of models used for parameter estimation. Under the mathematical modeling framework of NHPP, various SRGMs have been proposed in the literature. Goel and Okumoto model [6] proposed an exponential model. The model gives a concave-shaped mean value function under the assumption of constant defect detection rate. Later, Yamada and Ohba [6] proposed a model incorporating the expertise team gets with time, leading to an S-shaped defect rate curve. Also, researches have been carried out in past to study and model reliability growth phenomenon for OSS. Tamura and Yamada [9] presented reliability model for OSS based on neural network approach. Li et al. [3] presented a reliability growth model for OSS and also suggested a solution to optimal version update problem of OSS. Tamura and Yamada [10] proposed a model based on flexible hazard rate for embedded OSS. Younis et al. [12] assessed vulnerability exploitability risk on Open Source Vulnerability Database (OSVDB) by proposing a metric called structural severity. Most studies assume that the process of debugging starts as soon as a defect is detected and every defect is fixed with certainty such that neither the defect reoccurs nor the fix introduces any more defects. But, due to the inefficiency of testing team, situations like imperfect debugging and generation of faults arise in real software development [6]. Pachauri et al. [4] proposed an SRGM in imperfect debugging conditions and cost analysis based on fuzzy theory and also suggested optimal release time. Kapur et al. [7] formulated an approach to model reliability when there is no distinction between failure observation and fault removal testing process and later extending it for the case when differentiation is there between these two due to the presence of imperfect debugging and error generation. The process of debugging is considered as a deciding factor in software reliability and is explained as follows.

## 1.1  Debugging

The process of locating defects and their rectification is known as debugging. Debugging process can be classified into two broad categories:

- **Perfect Debugging**: This type of debugging assumes that fix for each and every defect eliminates the defect permanently without causing any other defects. Most NHPP models such as GO model [6] and Yamada and Ohba S-Shaped model [6] model the reliability growth phenomenon under the assumption of perfect debugging.
- **Imperfect Debugging**: "Imperfect removal of faults is called as imperfect fault debugging" [6]. This eliminates the unrealistic assumption of perfect debugging and assumes defect removal process is subject to two possibilities:

  - The removal of defect or the fix provided does not eliminate the defect permanently.
  - The fix can lead to the introduction of new defects.

Some of the SRGMs in the literature, such as Pham and Zhang [5] model and Yamada et al. model [11], incorporated imperfect debugging in reliability modeling. In this paper, we propose NHPP-based SRGMs for open source software incorporating the number of users under both scenarios of perfect debugging and imperfect debugging. The rest of the paper is organized as follows: Sect. 2 discusses the model foundations, Sect. 3 covers the proposed model development, Sect. 4 presents the proposed model, Sect. 5 presents the numerical example and discussion on results, and Sect. 6 describes the conclusion and future scope.

## 2  Model Foundations

In this study, we analyze the reliability of an OSS by presenting an SRGM that incorporates the effect of user growth in reliability growth once the software is released in the market. In our model, we considered defect detection rate depends on the number of users which in turn is a function of time. Also, the impact of imperfect debugging is taken care in our model to represent realistic debugging process. It has been assumed that the number of detected defects depends on the number of users for an OSS. User growth in the proposed approach is modeled with innovation diffusion model of marketing. Rogers's theory of diffusion of innovation [8] explains the process of proliferation of new ideas, concepts, technology, product, i.e., innovation into society or market. This theory deals with the reasons, factors, and the rate at which these innovations spread. The basic characteristics of an innovation namely relative advantage, compatibility, complexity, trialability, and observability, can be easily observed in an OSS, and due to this reason, OSS can be viewed as an innovation. Based on the above reason, the Theory of Diffusion of

Innovation perfectly applies to OSS. To model user growth in our SRGM, we have used the following two well-known models:

- Bass diffusion model [1] and
- Kenney's model [2].

## 2.1  The Bass Diffusion Model

The Bass diffusion model [1] was given by Bass as a mathematical expression for the Theory of Diffusion of Innovation given by Rogers [8] and can be effectively used to describe the growth in the number of user with respect to time. The model depicts the process of diffusion of innovation by classifying customers or adopters in two broad categories. The first category denotes the individuals that are decisive enough to adopt an innovation based on their own judgments of merits of innovation which are known as innovators. The other category involves those individuals who adopt innovation because of word of mouth from innovators which are known as imitators. Innovators get the first knowledge of the product, and thus, the relative number of innovators decreases monotonically over time due to the product gaining awareness among the potential adopters. Imitators are influenced by the number of previous buyers, and therefore, their numbers monotonically increase till it reaches a peak, where the product is at its maximum potential, and then starts decreasing monotonically. Bass assumed that the probability of adoption at time t given that no purchase has yet been made is linearly dependent on the remaining number of potential adopters. The combined rate of first purchase by both innovators and imitators is defined by the term $\left[p + q\frac{N(t)}{N}\right]$ and increases through time as $N(t)$ increases. But, there is a decrease in the remaining number of non-adopters defined by $(\bar{N} - N(t))$ over time. The coefficient of imitation must exceed the coefficient of innovation for a software product to be successful. The following differential equation describes the above situation:

$$\frac{dN(t)}{dt} = \left(p + q\frac{N(t)}{\bar{N}}\right)(\bar{N} - N(t)) \tag{1}$$

The two components in Bass diffusion model express the two main processes in diffusion theory, i.e., the processes of innovation and imitation.

**Innovators**: $p(\bar{N} - N(t))$ represents the rate of user growth with respect to innovators at time $t$, where $p$ denotes the coefficient of innovation or external influence as innovators are basically attracted toward innovation via external factors such as media. In case of OSS, innovators are the people who adopt it at very initial versions and work on it.

**Imitators**: The other component $\left(q\frac{N(t)}{N}\right)(\bar{N}-N(t))$ represents the rate of user growth with respect to imitators at time $t$. Here, $q$ denotes the coefficient of imitation or internal influence as imitators are attracted toward innovation via electronic word of mouth (reviews) of previous users represented by $N(t)$. In case of OSS, imitators are the people who are attracted to OSS by the following reviews given by innovators; i.e., they avoid adopting initial versions.

## 2.2  Kenney's Model

In his work [2], Kenney assumed that the number of defects to be a function of the number of instruction executed and the number of instructions executed is modeled in terms of user growth phenomenon. The expression that defines the growth in the user population of software is based on the Weibull distribution given in the form of a power function:

$$N(t) = \frac{t^{k+1}}{k+1} \tag{2}$$

The model showed the high-to-low trajectory of the usage growth. This is due to the fact that the customers who are upgrading to new releases lead to an initial increase in use and then as customers become aware of the product the growth rate slows down and increases at a constant rate.

## 3  Model Development

### 3.1  Model Nomenclature

| | |
|---|---|
| $t$ | Calendar time |
| $m$, $m(t)$ | Expected number of defects removed in interval $(0, t)$ |
| $N$, $N(t)$ | Cumulative number of users in the time interval $(0, t)$ |
| $a$ | Constant denoting number of defects initially in a software |
| $b$ | Constant denoting defect removal rate |
| $p$ | Constant denoting coefficient of innovation |
| $q$ | Constant denoting coefficient of imitation |
| $\bar{N}$ | Constant representing total number of potential users of the software |
| $\gamma$ | Probability of perfect debugging |
| $\alpha$ | Proportion of error generation |

## 3.2 Assumptions

- Software defect process can be described by a NHPP.
- The number of failures during operational phase is dependent upon the number of defects remaining in the software.
- As soon as any deviation from expected behavior is encountered, it is treated as a failure and the defect causing that failure is identified. There is a possibility that the found defects have not been removed perfectly and few additional defects may get introduced.
- The number of defects removed is a function of the number of users using that software.
- The number of users is a function of time, and they are modeled by innovation diffusion Model due to Bass [1] and Kenny [2].

Using the above assumptions, the failure phenomenon can be described with respect to time as follows:

$$\frac{dm}{dt} = \frac{dm}{dN}\frac{dN}{dt} \tag{3}$$

We will discuss each component on the right-hand side of the above equation individually. The models are discussed for both the cases of perfect debugging and imperfect debugging.

**Component-1**. Here, we consider four cases to model rate of detection of defects with respect to the number of users, i.e., $\frac{dm}{dN}$.

**Case 1** The rate at which failures occur depends upon the number of defects remaining in the software. Based on this assumption, the differential equation for defect removal can be written as:

$$\frac{dm}{dN} = b(a - m) \tag{4}$$

Here, $b$ is the defect removal rate which is treated as a constant as each one of these defects has an equal probability of causing a failure.

**Case 2** Here, the defect removal is represented using an Erlang 2-stage model [6]. In the first step, a failure is isolated by the failure identification team. In the second step, the defect causing the failure is removed by another team. Also, the failure observation and the defect removal rates are taken to be a constant $b$. The differential equation for defect removal is given by the differential equation:

$$\frac{dm}{dN} = b(N)[a - m] \tag{5}$$

where

$$b(N) = \frac{b^2 N}{1 + bN} \tag{6}$$

**Case 3** The rate at which failures occur depends upon the number of defects remaining in the software. In this case, we consider the concept of imperfect debugging. Thus, based on this assumption, the differential equation for defect removal can be written as:

$$\frac{dm}{dN} = b(a(N) - m) \tag{7}$$

where,

$$a(N) = a + \frac{\alpha}{\gamma} m \tag{8}$$

**Case 4** The concept of imperfect debugging is also taken up in this case. Along with this, the defect removal is done in two steps. In the first step, a failure is isolated by the failure identification team. In the second step, the defect causing the failure is removed by another team. Also, the failure observation and the defect removal rates are taken to be constant $b$. The differential equation for defect removal is given by the differential equation:

$$\frac{dm}{dN} = b(N)[a(N) - m] \tag{9}$$

where $b(N)$ and $a(N)$ are given by Eqs. (6) and (8), respectively.

**Component-2.**

**Case 1** Here, the user growth function is derived from the Bass innovation diffusion model as given in Eq. (1). Under the initial condition, $N(0) = 0$, the solution to Eq. (1) provides the user growth function as:

$$N(t) = \bar{N} \frac{1 - e^{-(p+q)t}}{1 + \frac{q}{p} e^{-(p+q)t}} \tag{10}$$

**Case 2** In this case, we consider the user growth function proposed by Kenny as given in Eq. (2).

## 4 Proposed Model

| SRGM 1 | SRGM 2 |
|---|---|
| $m_1(t) = a\left(1 - e^{-b\bar{N}\frac{1-e^{-(p+q)t}}{1+\frac{q}{p}e^{-(p+q)t}}}\right)$ | $m_2(t) = a\left(1 - e^{-b\frac{t^{k+1}}{k+1}}\right)$ |
| **SRGM 3** | **SRGM 4** |
| $m_3(t) = a\left(1 - \left(1 + b\bar{N}\frac{1-e^{-(p+q)t}}{1+\frac{q}{p}e^{-(p+q)t}}\right)e^{-b\bar{N}\frac{1-e^{-(p+q)t}}{1+\frac{q}{p}e^{-(p+q)t}}}\right)$ | $m_4(t) = a\left(1 - \left(1 + b\frac{t^{k+1}}{k+1}\right)e^{-b\frac{t^{k+1}}{k+1}}\right)$ |
| **SRGM 5** | **SRGM 6** |
| $m_5(t) = \frac{a}{1-a}\left(1 - e^{-b\gamma(1-a)\bar{N}\frac{1-e^{-(p+q)t}}{1+\frac{q}{p}e^{-(p+q)t}}}\right)$ | $m_6(t) = \frac{a}{1-a}\left(1 - e^{-b\gamma(1-a)\frac{t^{k+1}}{k+1}}\right)$ |
| **SRGM 7** | **SRGM 8** |
| $m_7(t) = \frac{a}{1-a}\left(1 - \left(\left(1 + b\bar{N}\frac{1-e^{-(p+q)t}}{1+\frac{q}{p}e^{-(p+q)t}}\right)e^{-b\bar{N}\frac{1-e^{-(p+q)t}}{1+\frac{q}{p}e^{-(p+q)t}}}\right)^{\gamma(1-a)}\right)$ | $m_8(t) = \frac{a}{1-a}\left(1 - \left(\left(1 + b\frac{t^{k+1}}{k+1}\right)e^{-b\frac{t^{k+1}}{k+1}}\right)^{\gamma(1-a)}\right)$ |

## 5 Numerical Study

This section discusses the estimation results of proposed SRGMs on the failure dataset of open source software (GNOME 2.0) provided in the literature by Li et al. [3]. Table 1 represents the summary of results obtained after estimation process. For the estimation of parameters in this study, we have performed least squares estimation using the software Statistical Package for Social Sciences (SPSS). The parameters to be estimated are: $a$, $b$, $p$, $q$, $k$, $\alpha$, $\gamma$, and $\bar{N}$.

Also, to obtain the goodness of fit we have used the following criteria: coefficient of determination $(R^2)$, mean square error (MSE), predictive ratio risk (PRR), and predictive power (PP). The interpretation and expression of the goodness-of-fit criteria used in our study are described in Table 2.

**Table 1** Estimated parameters

| Proposed model | Growth function | Parameter estimation | | | | | | | |
|---|---|---|---|---|---|---|---|---|---|
| | | $a$ | $b$ | $\bar{N}$ | $k$ | $p$ | $q$ | $\gamma$ | $\alpha$ |
| SRGM 1 | Bass | 87.118 | 0.025 | 251.341 | – | 0.007 | 0.122 | – | – |
| SRGM 2 | Kenny | 93.31 | 0.041 | – | 0.405 | – | – | – | – |
| SRGM 3 | Bass | 85.974 | 0.184 | 100.023 | – | 0.010 | 0.027 | – | – |
| SRGM 4 | Kenny | 90.971 | 0.181 | – | 0.0001 | – | – | – | – |
| SRGM 5 | Bass | 88.353 | 0.782 | 104.707 | – | 0.012 | 0.136 | 0.043 | 0.042 |
| SRGM 6 | Kenny | 79.636 | 0.774 | – | 0.405 | – | – | 0.061 | 0.147 |
| SRGM 7 | Bass | 66.712 | 0.681 | 84.366 | – | 0.033 | 0.110 | 0.041 | 0.547 |
| SRGM 8 | Kenny | 71.174 | 0.948 | – | 0.107 | – | – | 0.111 | 0.286 |

**Table 2** Comparison criteria and their interpretation

| Comparison criteria | Interpretation | Expression |
|---|---|---|
| $R^2$ | Higher the value of $R^2$, higher the data fits the model | $R^2 = 1 - \frac{Residual\ Sum\ of\ Squares}{Total\ Sum\ of\ Squares}$ |
| MSE | The lower the value, the better the data fits the model | $MSE = \sum_{i=1}^{k} \frac{\left(m'(t_i) - y_i\right)^2}{k}$ |
| PRR | The lower the PRR value, the better the model fits the data | $PRR = \sum_{i=1}^{n} \frac{\left(m'(t_i) - y_i\right)^2}{m(t_i)}$ |
| PP | The lower the PRR value, the better the model fits the data | $PP = \sum_{i=1}^{n} \frac{\left(m'(t_i) - y_i\right)^2}{y_i}$ |

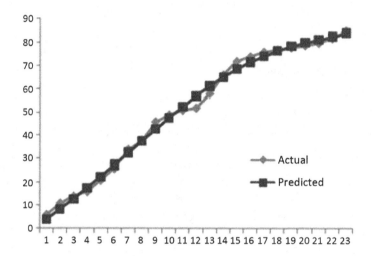

**Fig. 1** Goodness of fit for SRGM 1

The goodness-of-fit curves obtained for SRGM1 and SRGM2 are given by Figs. 1 and 2. Similarly, curves can be drawn for other SRGMs as well. The results of the goodness of fit for all eight proposed SRGMs are depicted in Table 3.

## 5.1 Discussion on Results

- From Table 1, we observe that the value of $q$ is always greater than $p$ which is consistent with the Bass diffusion model and illustrates the role of word of mouth in decision making of potential users.
- The values of the comparison criteria presented in Table 2 infer that the models inculcating both the Bass and Kenny growth functions fit the data very well.

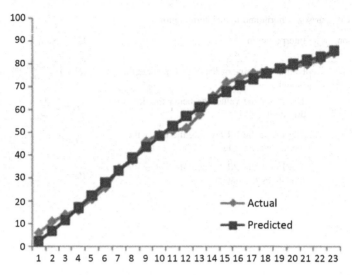

**Fig. 2** Goodness of fit for SRGM 2

**Table 3** Goodness-of-fit comparison table

| Proposed model | Growth function | Comparison criteria | | | |
|---|---|---|---|---|---|
| | | $R^2$ | MSE | PRR | PP |
| SRGM 1 | Bass | 0.994 | 5.428 | 3.65 | 3.14 |
| SRGM 2 | Kenny | 0.991 | 6.522 | 8.87 | 5.59 |
| SRGM 3 | Bass | 0.987 | 11.213 | 32.76 | 11.48 |
| SRGM 4 | Kenny | 0.988 | 8.855 | 29.04 | 10.52 |
| SRGM 5 | Bass | 0.993 | 6.276 | 3.87 | 3.28 |
| SRGM 6 | Kenny | 0.991 | 7.246 | 8.87 | 5.59 |
| SRGM 7 | Bass | 0.992 | 8.066 | 10.05 | 5.88 |
| SRGM 8 | Kenny | 0.990 | 8.705 | 15.73 | 7.55 |

- From Table 2, it can be observed that the goodness-of-fit values obtained on various comparison criteria are better for those imperfect cases (SRGM 5 and SRGM 6) than its perfect counterparts (SRGM 1 and SRGM 2). This may be due to the fact that the assumption of removing all the defects at once without causing others is quite unrealistic.
- Table 2 depicts that the models considering the Bass growth function perform better than those using Kenny's growth function, except in the comparison for SRGM 3 and SRGM 4. Since Kenny growth function follows Weibull distribution and Bass growth function follows logistic function, it has been proved in the literature that both these distributions are flexible and can take any shape according to the data. This validates the results obtained by the study presented here.

# 6 Conclusion and Future Scope

The advent of OSS brings a new software development trend in the market. Instability in volunteer's participation and very little testing before release are some of the factors that make reliability analysis crucial for OSS. Therefore, for OSS the number of users has a very critical role in debugging process and thus in reliability of software. For these reasons, it becomes very important to model user growth in reliability analysis of such software. In this study, we presented SRGMs relating the reliability growth phenomenon with user growth. The usage factor is modeled with two very popular models, i.e., Kenney's and Bass diffusion models. The failure data of a very popular OSS (Gnome 2.0) is used for reliability modeling. Eight models have been proposed, four each for perfect and imperfect debugging conditions. Four goodness-of-fit criteria are used to find out the estimating capabilities of the models. Results reveal the fact that models with the consideration of imperfect debugging situations give better results over their perfect debugging coequal. In future, studies can explore more factors that are significant for OSS to be incorporated in its reliability modeling and such research will refine the proposed model.

# References

1. Bass, F. M. (1969). A New Product Growth for Model Consumer Durables. *Management Science, 15*(5), 215–227.
2. Kenny, G. (1993). Estimating defects in commercial software during operational use. *IEEE Transactions on Reliability, 42*(1), 107–115.
3. Li, X., Li, Y. F., Xie, M., & Ng, S. H. (2011). Reliability analysis and optimal version-updating for open source software. *Information and Software Technology, 53*(9), 929–936.
4. Pachauri, B., Kumar, A., & Dhar, J. (2013). Modeling optimal release policy under fuzzy paradigm in imperfect debugging environment. *Information and Software Technology, 55* (11), 1974–1980.
5. Pham, H., & Zhang, X. (2003). NHPP software reliability and cost models with testing coverage. *European Journal of Operational Research, 145*(2).
6. Pham, H., Kapur, P., Jha, P., & Gupta, A. (2011). Software Reliability Assessment with OR Applications. Springer Verlag London Ltd.
7. P. K. Kapur, H. Pham, S. Anand and K. Yadav, "A Unified Approach for Developing Software Reliability Growth Models in the Presence of Imperfect Debugging and Error Generation," in *IEEE Transactions on Reliability*, vol. 60, no. 1, pp. 331–340, March 2011.
8. Rogers, E. M., "Diffusion of innovations", The Free Press, 1962.
9. Tamura, Y., & Yamada, S. (2008). A Component-Oriented Reliability Assessment Method For Open Source Software. *International Journal of Reliability, Quality and Safety Engineering, 15*(01), 33–53.
10. Tamura, Y., & Yamada, S. (2011). Reliability assessment based on hazard rate model for an embedded OSS porting-phase. Software Testing, Verification and Reliability, 23(1), 77–88.

11. Yamada, S., Tokuno, K., & Osaki, S. (1992). Imperfect debugging models with fault introduction rate for software reliability assessment. *International Journal of Systems Science,* 23(12), 2241–2252.
12. Younis, A., Malaiya, Y. K., & Ray, I. (2015). Assessing vulnerability exploitability risk using software properties. *Software Quality Journal, 24*(1), 159–202. https://doi.org/10.1007/s11219-015-9274-6.

# Multi-objective League Championship Algorithms and its Applications to Optimal Control Problems

Debasis Maharana, Siddhant Maheshka and Prakash Kotecha

**Abstract** Evolutionary algorithms are effective in solving complex nonlinear optimization problems with multiple conflicting objectives. League Championship Algorithm (LCA) is a recently proposed single-objective evolutionary algorithm which has shown impressive results on benchmark problems used in Conference on Evolutionary Computation (CEC). In this work, two multi-objective versions of LCA, viz. NS-LCA and ε-dominance LCA are proposed which utilizes non-dominated sorting of solutions and ε-dominance concept respectively to solve multi-objective problems. Performance of both algorithms has been tested on three multi-objective optimal control problems.

**Keywords** Multi-objective optimization · League Championship Algorithm Non-dominated sorting · ε-dominance · Optimal control problem

## 1 Introduction

Optimization is the process of determining the best decisions among available alternatives in the presence of several constraints. Most real-life optimization problems require solutions to cater multiple conflicting objectives. Unlike single-objective optimization problems, multi-objective optimization problems (MOOPs) possess a set of non-dominated solutions known as Pareto optimal solutions, which reflect the trade-off between the various objectives. The goal in MOOPs is to obtain the global Pareto front. Evolutionary algorithms (EA) are commonly used to solve optimization problems as they do not require any gradient information and are efficient in handling black-box optimization problems. EAs provide an easy and better alternative compared to other traditional MOOP solving

D. Maharana · S. Maheshka · P. Kotecha (✉)
Department of Chemical Engineering, Indian Institute of Technology Guwahati,
Guwahati 781039, Assam, India
e-mail: pkotecha@iitg.ernet.in

© Springer Nature Singapore Pte Ltd. 2019
B. K. Panigrahi et al. (eds.), *Smart Innovations in Communication and Computational Sciences*, Advances in Intelligent Systems and Computing 669, https://doi.org/10.1007/978-981-10-8968-8_4

techniques [1, 2] as they do not require any user interference in handling multiple objectives or limit on the nature of decision variables.

Single-objective evolutionary algorithms can be modified using available MOOP handling techniques so as to solve problems with more than one objective. The most popular among are (i) non-dominated sorting method [3] and (ii) ε-dominance method [4]. Some of the examples in which single-objective algorithms have been converted to multi-objective using these two procedures include Teaching Learning-based Optimization [5], Artificial Bee Colony Optimization [6], Firefly Algorithm [7], Cuckoo Search Algorithm [8], Improved Teaching–Learning-Based Optimization Algorithm [9] and Front-based Yin-Yang-Pair Optimization [10].

In this work, we extend a recently proposed single-objective algorithm LCA [11] to solve multi-objective problems using both the previously mentioned strategies. In the succeeding section, we briefly describe LCA. In Sect. 3, we propose the NS-LCA and ε-dominance LCA that can be used to solve multi-objective problems. In Sect. 4, we describe three multi-objective optimal control problems on which both the proposed variants are tested. In Sect. 5, we discuss the performances of the variants and finally conclude the article.

## 2    League Championship Algorithm

LCA [11] is a recently proposed evolutionary algorithm for solving single-objective optimization problem. It is inspired from the sports championship leagues in which teams play with each other for several weeks over the course of a season. A league schedule is generated to determine the matches that will occur in a given week. In a match between two teams, the outcome (win or loss) is determined by the playing strength (fitness) of the team. After the match, each team analyses its performance and updates its formation (solution) for next week based on the results of the previous week. The championship progresses for several iterations (based on stopping criteria such as number of function evaluations). The generation of a new solution utilizes its own results in the previous week as well as the results of its opponent. For the sake of brevity, we have provided a brief description of LCA and further details can be obtained from the literature [11].

Consider teams $i$ and $j$ playing at week $t$, with their formations $X_i^t$ and $X_j^t$ along with their playing strengths $f(X_i^t)$ and $f(X_j^t)$, respectively. Let $f_{min}$ be an ideal value (e.g. a lower bound on the optimal value). The chance of team $i$ to win over team $j$ at week $t$ is denoted by $p_i^t$ and is determined using Eq. (1):

$$p_i^t = \frac{f(X_j^t) - f_{min}}{f(X_i^t) + f(X_j^t) - 2f_{min}} \tag{1}$$

To determine the winner or loser, a random number between [0, 1] is generated; if it is less than or equal to $p_i^t$, team $i$ wins and team $j$ loses; if it is greater than $p_i^t$, then team $i$ loses and team $j$ wins. Note that the higher the value of $p_i^t$, higher is the chance of team $i$ winning. If the value of $f_{min}$ is unavailable in advance, then the value of the best function obtained so far is used. If the denominator is zero, then $p_i^t$ is assumed to be 0.5. In the subsequent discussion, let $l$ be the index of the team that will play with team $i$ at week $t + 1$ based on the league schedule, $j$ be the index of the team that has played with team $i$ at week $t$ based on the league schedule and $k$ be the index of the team that has played with team $l$ at week $t$ based on the league schedule. Let $B_i^t$ denote the best previously experienced formation by team $i$ until week $t$, yielding the best fitness. The new solutions for team $i$ are generated using the following equations:

**If** $i$ had won and $l$ had won too:

$$x_{id}^{t+1} = b_{id}^t + y_{id}^t(c_1 r_{1id}(x_{id}^t - x_{kd}^t) + c_1 r_{2id}(x_{id}^t - x_{jd}^t)) \quad \forall d = 1, \ldots, n \qquad (2)$$

**Else if** $i$ had won and $l$ had lost:

$$x_{id}^{t+1} = b_{id}^t + y_{id}^t(c_2 r_{1id}(x_{kd}^t - x_{id}^t) + c_1 r_{2id}(x_{id}^t - x_{jd}^t)) \quad \forall d = 1, \ldots, n \qquad (3)$$

**Else if** $i$ had lost and $l$ had won

$$x_{id}^{t+1} = b_{id}^t + y_{id}^t\left\{c_1 r_{1id}(x_{id}^t - x_{kd}^t) + c_2 r_{2id}(x_{jd}^t - x_{id}^t)\right\} \quad \forall d = 1, \ldots, n \qquad (4)$$

**Else if** $i$ had lost and $l$ had lost too

$$x_{id}^{t+1} = b_{id}^t + y_{id}^t(c_2 r_{1id}(x_{kd}^t - x_{id}^t) + c_2 r_{2id}(x_{jd}^t - x_{id}^t)) \quad \forall d = 1, \ldots, n \qquad (5)$$

**End if**

In the above equations, $n$ represents the number of decision variables, $r_{1id}$ and $r_{2id}$ are uniform random numbers between [0, 1], and $c_1$ and $c_2$ are the retreat coefficient and the approach coefficient, respectively. It has been reported that these equations result in solutions accelerating towards winners and away from losers. In the above equations, $y_{id}^t$ is a binary variable which indicates whether or not the $d$th element in the new formation should differ from its counterpart in the current best formation. $Y_i^t (= (y_{i1}^t, y_{i2}^t, \ldots y_{in}^t))$ is a binary array in which the number of ones is equal to $q_i^t$, which is obtained by the following formula:

$$q_i^t = ceil\left(\frac{\ln(1 - (1 - (1 - p_c)^{n - q_o + 1})r)}{\ln(1 - p_c)}\right) + q_o - 1, \quad q_i^t \in \{q_o, q_o + 1, \ldots, n\} \qquad (6)$$

A truncated geometric probability distribution has been used in the literature to give greater emphasis to smaller number of changes. Here, $r$ is a random number in

[0, 1] and $p_c < 1$ is a control parameter $(p_c \neq 0)$. A lower value of $p_c$ results in a higher value of $q_i^t$, and accordingly, a higher number of changes will occur. The term $q_0$ corresponds to the least number of changes realized during the artificial match analysis.

# 3 Multi-objective LCA

In this section, we extend LCA to solve multi-objective optimization problems based on non-dominated sorting and $\varepsilon$-dominance. The probability for selecting winners and losers is calculated using Eq. (7) to accommodate the multiple objectives.

$$p_i^t = \frac{1}{nObj} \sum_{n=1}^{nObj} E_{oi}^t \tag{7}$$

$$E_{oi}^t = \frac{f_o\left(X_j^t\right) - f_{o,\,min}}{f_o(X_i^t) + f_o\left(X_j^t\right) - 2f_{o,\,min}}, \quad \forall o = 1, \ldots, nObj \tag{8}$$

The best experienced solution ($B$) till now is determined based on the dominance between the current solution and the best experienced solution. If these two solutions are non-dominating, one of them is randomly selected, otherwise the dominating solution is selected.

## 3.1 Non-dominated Sorting Based LCA (NS-LCA)

In NS-LCA, the non-dominated strategy used in NSGA-II is adapted wherein the solutions are ranked into various fronts. The pseudo-code of the variant is given below. It has to be noted that maximum function evaluation is considered as the termination criteria for this work.

**Pseudo-code for NS-LCA**

Step 1: Initialize league size (*LS*) with an even number, retreat coefficient $c_1$, approach coefficient $c_2$ and control parameter $p_c$.

Step 2: Create a null matrix as timetable (*TT*) of size ($LS \times LS - 1$).

Step 3: Create a bisectioned list (*BL*) of size ($2 \times LS/2$). The elements of the first row are integers from 1 to $LS/2$, whereas the elements of the second row are integers from $LS$ to $(LS/2) + 1$.

Step 4: Repeat Step 4 to Step 10 for each week $w, w \in [1, LS - 1]$.

Step 5:  Repeat Step 6 for every $j$ in $BL, j \in [1, LS/2]$.

Step 6:  Assign $BL_{2,j}$ to $TT_{(BL_{1,j}),w}$ and $BL_{1,j}$ to $TT_{(BL_{2,j}),w}$.

Step 7:  Create temporary list ($TL$) of size ($2 \times LS/2$). Set $TL_{1,1}$ to 1 and $TL_{1,2}$ to $BL_{2,1}$.

Step 8:  For every $k$ in $BL$, $k \in [3, LS/2]$, assign $BL_{1,k-1}$ to $TL_{1,k}$.

Step 9:  For every $k$ in $BL$, $k \in [1, LS/2 - 1]$ assign $BL_{2,k+1}$ to $TL_{2,k}$.

Step 10:  Assign $BL_{1,LS/2}$ to $TL_{2,LS/2}$ and set $BL$ equal to $TL$.

Step 11:  Append the first column of $TT$ as its last column thereby making it of size ($LS \times LS$).

Step 12:  Initialize population members $X$ randomly within the bounds of the decision variables and calculate their fitness $X_{obj}$.

Step 13:  Initialize best population $B$, best objectives $B_{obj}$ equal to $X$ and $X_{obj}$, respectively.

Step 14:  Initialize week $w$ to 1 and assign zero to scoreboard ($S$ of size $LS \times 1$) which is used for recording winners and losers.

Step 15:  Repeat Step 16 to Step 22 till a termination criterion is not satisfied.

Step 16:  For all the matches in week $w$, decide winners and losers based on Eq. (7) and record the results in $S$.

Step 17:  Repeat Step 18 to Step 20 for each team $t$ in 1 to $LS$. If function evaluation exceeds maximum allowed evaluations, go to Step 21.

Step 18:  Generate new solution ($N^t$) using match analysis Eqs. (2)–(6).

Step 19:  Replace any bound violating members of ($N^t$) with new members created randomly within the bounds and calculate the fitness $\left( N^t_{obj} \right)$.

Step 20:  If the solution ($N^t$) dominates the best solution ($B^t$), then it replaces the best solution, otherwise if $B^t$ dominates $N^t$ then no change is made. If both are non-dominated, $B^t$ is replaced by $N^t$ with a probability of 0.5.

Step 21:  Create a temporary population which has both the old solutions ($X$) and new solutions ($N^t$). Use the sorting technique used in NSGA-II to order these solutions. Choose the best $LS$ solutions (using the front ranking and crowding distance) and store these as ($X$). Reset the scoreboard ($S$).

Step 22:  If remainder of $w$ with ($LS - 1$) is zero, i.e. season has ended, then make $w = 1$ else $w = w + 1$.

The algorithm to determine the non-dominated sorting and calculation of crowding distance (in Step 21) is available in the literature [3] and not provided here for the sake of brevity.

## 3.2 ε-Dominance-Based LCA (ε-LCA)

The procedure for ε-dominance-based Pareto selection can be found from the literature [4]. Unlike non-dominated sorting, the Pareto solutions obtained by ε-dominance concept are placed in an archive and the archive is updated periodically.

**Pseudo-code for ε-LCA**

Step 1:   Perform Step 1 to Step 14 from NS-LCA. Initialize archive by storing the non-dominated solutions from $X$.

Step 2:   Repeat Step 3 to Step 5 till termination criteria is satisfied.

Step 3:   Perform Step 16 to Step 20 from NS-LCA.

Step 4:   Update archive from the newly created solutions $(N^t)$ using archive updating procedure. Assign new solutions $(N^t)$ to $X$ and reset the score-board $(S)$.

Step 5:   If remainder of $w$ with $(LS - 1)$ is zero i.e. if season has ended, assign $w = 1$ else $w = w + 1$.

# 4   Application to Multi-objective Optimal Control

In this section, we describe three multi-objective optimal control problems which have been widely used in the literature [12–14]. The values and equations are assumed to be dimensionally consistent, and further details can be obtained from the respective literature. The decision variables for each problem are the trajectories of control variables with respect to time or a spatial variable.

## 4.1   Temperature Control of a Batch Reactor

This problem considers first-order reactions $A \xrightarrow{k_1} B \xrightarrow{k_2} C$ in a continuous stirred tank batch reactor. The concentration of the components A and B is dimensionless and is governed by the following ODEs [12, 14]:

$$\frac{dC_A}{dt} = -k_1 C_A^2; \qquad C_A(0) = 1$$

$$\frac{dC_B}{dt} = k_1 C_A^2 - k_2 C_B; \quad C_B(0) = 0 \tag{9}$$

$$k_1 = 4 \times 10^3 \times e^{-2500/T}; k_2 = 6.2 \times 10^5 \times e^{-5000/T}$$

$$298K \leq T \leq 398K$$

Here, $C_A(0)$ and $C_B(0)$ represent the initial concentrations of components $A$ and $B$ and $T$ denotes the temperature of the reactor. The motive is to determine the temperature at pre-defined intervals of time such that the concentration of component $B$ ($C_B$) is maximum and the concentration of $A$ ($C_A$) is minimized at the end of the reaction time ($t_f = 1$). The objective functions can be represented as

$$Min\ F_1 = -C_B(t_f); \quad Min\ F_2 = C_A(t_f) \tag{10}$$

## 4.2 Catalytic Mixing Problem in a Plug Flow Reactor

A case of mixing of catalysts in a plug flow reactor packed with catalyst $A$ and catalyst $B$ at steady state is considered where the reactions $S_1 \leftrightarrow S_2 \rightarrow S_3$ occur. The amount of components $S_1$ (denoted by $x_1$) and $S_2$ (denoted by $x_2$) varies with respect to the following ODEs [12, 13]:

$$\frac{dx_1}{dz} = u(z)[10x_2(z) - x_1(z)] \qquad\qquad x_1(z=0) = 1$$

$$\frac{dx_2}{dz} = -u(z)[10x_2(z) - x_1(z)] - [1 - u(z)]x_2(z) \quad x_2(z=0) = 0 \tag{11}$$

The variable $z$ represents the spatial coordinate, whereas $x_1$ ($z = 0$) and $x_2$ ($z = 0$) indicate the amount of $S_1$ and $S_2$ at $z = 0$. The term $u(z)$ represents the fraction of catalyst $A$ at a specific spatial coordinate ($z$) in the plug flow reactor. The bounds considered for decision variable $u(z)$ are between 0 and 1. The outlet of the plug flow reactor is at $z_f = 1$. In view of the higher cost for catalyst A, the objectives in the problem are to maximize the quantity of product $S_3$ in the outlet of the plug flow reactor as well as to minimize the quantity of catalyst $A$ used. The objective functions are given as

$$Min\ F_1 = -\left(1 - x_1(z_f) - x_2(z_f)\right); \quad Min\ F_2 = \int_0^{z_f} u(z)dz; \tag{12}$$

The decision variables are the fraction of the catalyst $A$ at specified locations.

### 4.3  Optimal Operations of a Fed-Batch Reactor

A continuously fed-batch reactor is considered wherein the following reactions occur [12, 14]:

$$A + B \xrightarrow{k_1} C; \quad B + B \xrightarrow{k_2} D \tag{13}$$

The system dynamics are modelled using the following ODEs:

$$\left.\begin{aligned}
\frac{d[A]}{dt} &= -k_1[A][B] - ([A]/V)u \\
\frac{d[B]}{dt} &= -k_1[A][B] - 2k_2[B]^2 + \left(\frac{b_{feed} - [B]}{V}\right)u \\
\frac{d[C]}{dt} &= -k_1[A][B] - ([C]/V)u \\
\frac{d[D]}{dt} &= -2k_2[B]^2 - ([D]/V)u \\
\frac{d[V]}{dt} &= u
\end{aligned}\right\} [A](t=0) = 0.2, V(t=0) = 0.5 \tag{14}$$

The concentration of components A, B, C and D in the reactor is represented as [A], [B], [C] and [D], respectively. The current reaction volume is represented as [V]. Reactant feed rate (u) is the decision variable and is bounded between 0 and 0.01. The initial concentration of all the components except A is zero. The rate constants are $k_1 = 0.5$ and $k_2 = 0.5$. Maximizing the concentration of product C while minimizing the concentration of by-product D at the final time ($t_f = 120$) is considered as the two objectives:

$$Min \; F_1 = -[C](t_f)V(t_f); \quad Min \; F_2 = [D](t_f)V(t_f) \tag{15}$$

## 5  Results and Discussions

Both the proposed multi-objective variants of LCA are evaluated on the three multi-objective optimal control problems using a PC with an i7 3.4 GHz processor and 16 GB RAM on MATLAB 2016a simulation environment. Ten independent runs were realized by varying the seed of the *twister* function from 1 to 10 in MATLAB to generate reproducible results. The termination criterion for all the three optimal control problems is 10,100 evaluations of the objective function. The retreat coefficient, the approach coefficient and control parameter were set to 0.7, 1 and 0.5, respectively. The league size was set to 60, whereas a value of $10^{-5}$ is used as epsilon in the ε-dom LCA. The results of the ten runs from both the variants are combined to obtain the global Pareto front.

## 5.1   Temperature Control of a Batch Reactor

The time interval is divided into 10 steps for maximizing the concentration of B and minimizing concentration of A. The optimal reactor temperature profile obtained at one of the corner points ($F_1$) using ε-LCA and NS-LCA is shown in Fig. 1.

From the Pareto solutions, it is difficult to conclude the dominance between the two proposed variants. Both ε-LCA and NS-LCA are able to determine at least one of the corner solutions. However, the solutions obtained from ε-LCA are closer to the global Pareto solutions for most of the points as compared to NS-LCA.

## 5.2   Catalyst Mixing Problem in a Plug Flow Reactor

The spatial coordinate for this problem has been divided into 10 equal parts. The trade-off between the amount of catalyst A that is utilized and the conversion can be visualized from the Pareto solutions as shown in Fig. 2. The optimal catalyst mixing profile at one of the corner points ($F_1$) can also be analysed from the given figure.

The dominance of ε-LCA method over NS-LCA can be clearly observed from the diversified Pareto set. It should be noted that all the solutions obtained by utilizing NS-LCA are dominated by ε-LCA.

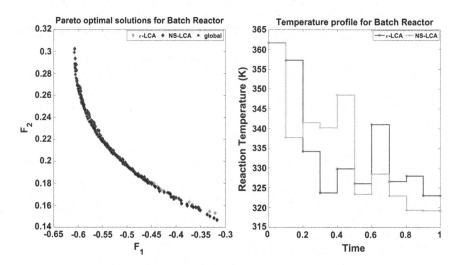

**Fig. 1** Pareto-optimal solutions and optimal temperature profile for batch reactor

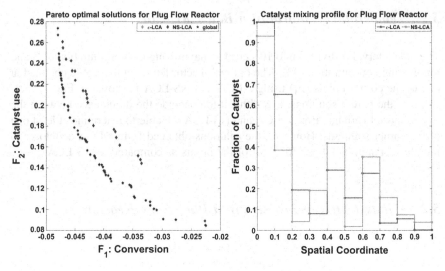

**Fig. 2** Pareto-optimal solutions and catalyst mixing profile for plug flow reactor

## 5.3    Optimal Operations of a Fed-Batch Reactor

The reactant feed rate is varied over 10 time intervals between 0 and final time of 120. Figure 3 shows the Pareto optimal set obtained using both methods and it can be observed that both of them provide approximately similar diversified solutions. Though NS-LCA is able to determine better corner points, it is difficult to provide any conclusion on the dominance among the two variants. The feed rate profile for a solution corresponding to the maximum amount of product C is shown in Fig. 3.

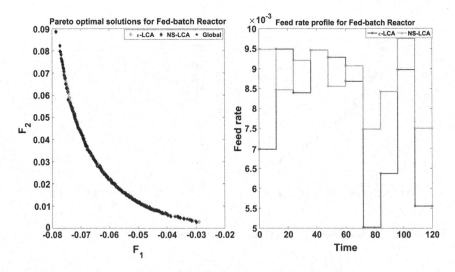

**Fig. 3** Pareto-optimal solutions and feed rate profile for fed-batch reactor

# 6 Conclusion

In this work, we have proposed two multi-objective variants for LCA, namely NS-LCA and ε-LCA. NS-LCA utilizes non-dominated sorting of solutions and crowding distance, whereas ε-LCA utilizes the ε-dominance-based approach for obtaining the Pareto optimal set of solutions. Both the approaches are tested on three multi-objective optimal control problems. The results show the dominance of ε-LCA over NS-LCA in obtaining optimal catalyst mixing profile for a plug flow reactor problem. Both the variants are able to perform equally for the optimal temperature control of a batch reactor and the optimal feed flow rate in a fed-batch reactor. Hybridization of the algorithm and utilization of other multi-objective techniques can be a future extension to this work.

# References

1. Deb, k., Multi-Objective Optimization using Evolutionary Algorithms. 2003: WILEY.
2. Kotecha, P.R., M. Bhushan, and R.D. Gudi, *Efficient optimization strategies with constraint programming*. AIChE Journal, **56**(2): p. 387–404, 2010.
3. Deb, K., et al., *A fast and elitist multiobjective genetic algorithm: NSGA-II*. IEEE Transactions on Evolutionary Computation, **6**(2): p. 182–197, 2002.
4. Deb, K., M. Mohan, and S. Mishra, Evaluating the ε-Domination Based Multi-Objective Evolutionary Algorithm for a Quick Computation of Pareto-Optimal Solutions. Evol. Comput., **13**(4): p. 501–525, 2005.
5. Zou, F., et al., *Multi-objective optimization using teaching-learning-based optimization algorithm*. Engineering Applications of Artificial Intelligence, **26**(4): p. 1291–1300, 2013.
6. Kishor, A., P.K. Singh, and J. Prakash, NSABC: Non-dominated sorting based multi-objective artificial bee colony algorithm and its application in data clustering. Neurocomputing, **216**: p. 514–533, 2016.
7. Tsai, C.W., Y.T. Huang, and M.C. Chiang. A non-dominated sorting firefly algorithm for multi-objective optimization. in 2014 14th International Conference on Intelligent Systems Design and Applications. 2014.
8. He, X.S., N. Li, and X.S. Yang. Non-dominated sorting cuckoo search for multiobjective optimization. in 2014 IEEE Symposium on Swarm Intelligence. 2014.
9. Chinta, S., R. Kommadath, and P. Kotecha, A note on multi-objective improved teaching–learning based optimization algorithm (MO-ITLBO). Information Sciences, **373**: p. 337–350, 2016.
10. Punnathanam, V. and P. Kotecha, Multi-objective optimization of Stirling engine systems using Front-based Yin-Yang-Pair Optimization. Energy Conversion and Management, **133**: p. 332–348, 2017.
11. Husseinzadeh Kashan, A., League Championship Algorithm (LCA): An algorithm for global optimization inspired by sport championships. Applied Soft Computing, **16**: p. 171–200, 2014.
12. Chen, X., W. Du, and F. Qian, Multi-objective differential evolution with ranking-based mutation operator and its application in chemical process optimization. Chemometrics and Intelligent Laboratory Systems, **136**: p. 85–96, 2014.

13. Logist, F., et al., *Multi-objective optimal control of chemical processes using ACADO toolkit.* Computers & Chemical Engineering, **37**: p. 191–199, 2012.
14. Jia, L., D. Cheng, and M.-S. Chiu, Pareto-optimal solutions based multi-objective particle swarm optimization control for batch processes. Neural Computing and Applications, **21**(6): p. 1107–1116, 2012.

# Optimization of Job Shop Scheduling Problem with Grey Wolf Optimizer and JAYA Algorithm

**Debasis Maharana and Prakash Kotecha**

**Abstract** The design of optimal job schedules on parallel machines with finite time is a combinatorial optimization problem and plays a crucial role in manufacturing and production facilities. In this work, we evaluate the performance of two recently proposed computational intelligence techniques, Grey Wolf Optimizer (GWO) and JAYA on ten datasets arising from five job shop scheduling problems with parallel machines. The computational results have shown GWO to be efficient than the JAYA algorithm for problems with higher number of orders and machines.

**Keywords** Job shop scheduling · Combinatorial optimization
Grey wolf optimizer · JAYA

## 1 Introduction

Job shop scheduling requires the allocation of tasks to available resources while satisfying various time, supply and demand constraints to optimize certain objective(s). Most scheduling problems are formulated in a framework that is amenable for solving using mathematical programming techniques. However, this requires the use of a significant number of binary variables which can scale exponentially even with a moderate increase in the number of orders and machines and can lead to issues with computational time and memory. The job shop scheduling problem (JSSP) considered in this work requires the completion of orders before their due date using the available machines. This problem has been solved in the literature using stand-alone techniques such as Mixed Integer Linear Programming (MILP) and Constraint Programming (CP) [1] but requires considerably large computational resources for problems with higher number of orders and machines and required advanced strategies to effectively solve this problem.

D. Maharana · P. Kotecha (✉)
Department of Chemical Engineering, Indian Institute of Technology Guwahati,
Guwahati 781039, Assam, India
e-mail: pkotecha@iitg.ernet.in

© Springer Nature Singapore Pte Ltd. 2019                                    47
B. K. Panigrahi et al. (eds.), *Smart Innovations in Communication
and Computational Sciences*, Advances in Intelligent Systems
and Computing 669, https://doi.org/10.1007/978-981-10-8968-8_5

Computational intelligence (CI) techniques for optimization are commonly inspired from various naturally occurring phenomena and are quite popular in solving black-box optimization problems as most of them do not require any gradient information. The use of CI techniques to solve combinatorial problem is not trivial and a careful choice of the decision variables is required to efficiently solve the problem. Though this JSSP has been solved with Genetic Algorithm (GA) [2], it has not been solved with recently developed CI techniques which have shown impressive performance on benchmark optimization problems. Some of the recently developed CI techniques are League Championship Algorithm [3], Cuckoo Search [4] and Yin-Yang Pair Optimization [5]. A characteristic of many of the evolutionary and swarm intelligence-based algorithms is that they employ several user-defined parameters. Few of the user-defined parameters include mutation probability, crossover probability and selection parameter in GA, inertia weight and velocity coefficients in Particle Swarm Optimization. The success of many of the algorithms is strongly dependent on the proper tuning of these parameters. However, there are no universal guidelines for the tuning of these parameters and a successful tuning procedure might even require deep insights into the working of the algorithm. This creates an impediment in the performance evaluation of such algorithms and also restricts the use of these techniques among the application community.

In this work, we evaluate the performance of two recently proposed algorithms, viz. Grey Wolf Optimizer (GWO) and JAYA which have been reported to be efficient and do not require any tuning parameters except for the population size and the termination criteria. In the next section, we provide the problem statement and briefly discuss the solution strategy. In Sect. 3, we have provided brief description of both the optimization algorithms. Experimental settings and results are discussed in Sects. 4 and 5, respectively. We conclude the work by summarizing the developments and discuss possible future work.

## 2   Problem Statement

In this problem, a set of $I$ independent orders need to be processed on a set of $M$ dissimilar parallel machines. The processing cost and the processing time associated with each combination of order and machine are known along with the release and due date for each order. The processing of an order can start on or after the release date and has to be completed on or before the due date. Though multiple orders can be processed on a single machine, a single order cannot be processed on multiple machines. Moreover, a machine cannot simultaneously process more than one order. The objective is to determine an optimal schedule that will enable the completion of all the orders, within their due dates, with the minimum cost.

In the formulation in the literature, the solution vector (set of decision variables) comprised of two components, viz. one to represent the machine which processes the $i$th order and the other to represent the starting time of each order on their allotted machines $(ts_i)$. Thus for each order, there would be two decision variables associated with it and hence the number of decision variables of a problem is equal to twice the number of orders. The variable corresponding to the selected machine $m$ for processing order $i$ takes an integer value between 1 and total number of machines, while $ts_i$ takes an integer value between the release date and due date of the $i$th order. An example of the solution vector, for a problem with five orders and two machines, comprising of machine number and starting time corresponding to each order is shown as

$$X = \overbrace{\underset{\substack{order\,1 \\ 2}}{} \underset{\substack{order\,2 \\ 1}}{} \underset{\substack{order\,3 \\ 1}}{} \underset{\substack{order\,4 \\ 1}}{} \underset{\substack{order\,5 \\ 2}}{}}^{Machines} \overbrace{\underset{\substack{order\,1 \\ 12}}{} \underset{\substack{order\,2 \\ 3}}{} \underset{\substack{order\,3 \\ 9}}{} \underset{\substack{order\,4 \\ 5}}{} \underset{\substack{order\,5 \\ 7}}{}}^{Starting\ time} \tag{1}$$

In this example, order 1 and 5 are processed by machine 2 at starting time 12 and 7, respectively, whereas the other three orders are processed by machine 1. The constraints are as discussed below

(a) The processing of an order should end before due date.

$$ts_i + p_{im} \le d_i \quad \forall i \in I \tag{2}$$

In this constraint, $m$ represents the machine allotted for order $i$ and $p_{im}$ denotes the processing time of the order $i$ on the assigned machine $m$.

(b) If more than one order is processed by a single machine, then the subsequent order cannot begin before the preceding order is completed.

$$ts_i + p_{im} \le ts_{i+1} \quad \forall m \in M, \forall i \in I_m, ts_i < ts_{i+1} \tag{3}$$

Here, $I_m$ represents the orders assigned to machine $m$. It has to be noted that this constraint has to be evaluated such that $ts_i < ts_{i+1}$ for all the orders $i$ assigned to machine $m$.

For the example in Eq. (1), the constraint 2 would translate into the following three constraints $ts_2 + p_{21} \le ts_4$;   $ts_4 + p_{41} \le ts_3$;   $ts_5 + p_{52} \le ts_1$.

(c) The total processing time for all the orders assigned to a machine should be less
than or equal to the difference of the last due date and the earliest release date of
all the orders assigned to that machine.

$$\sum_{i=1}^{I_m} p_{im} \leq \max_{\forall i \in I_m} (d_i) - \min_{\forall i \in I_m} (r_i) \quad \forall m \in M \qquad (4)$$

All the constraints are transformed to the form, $g(x) \geq 0$ for providing appropriate penalty against constraint violation. The penalty ($Penalty_n$) associated with each of the $N$ constraints can be determined as follows

$$Penalty_n = \begin{cases} |g_n(x)|, & \text{if } g_n(x) < 0 \\ 0 & \text{otherwise} \end{cases} \quad \forall n \in N \qquad (5)$$

*Fitness function*: The objective is to determine a schedule with a minimum total cost for processing all the orders. The total cost associated with a potential schedule is given by $C = \sum_{i=1}^{I} c_{im}$, where $c_{im}$ indicates the cost of processing order $i$ on its assigned machine $m$. The fitness function incorporating the penalties for constraint violation is given by Eq. (6)

$$f = \begin{cases} C & \text{if } \sum_{n=1}^{N} Penalty_n = 0 \\ C + \sum_{n=1}^{N} Penalty_n + \sum_{i \in I} \max_{\forall m \in M} \{c_{im}\} & \text{otherwise} \end{cases} \qquad (6)$$

For a feasible solution, the sum of all penalties will be equal to zero. An infeasible solution is penalized by the amount of violation incurred by each constraint as well as the total maximum cost. This ensures that an infeasible solution will not have a better fitness value in comparison with any feasible solution.

# 3    Computational Intelligence Techniques

A brief description of the two CI techniques is provided in this section. Readers are encouraged to refer the literature [6, 7] for additional details about these algorithms.

## 3.1  Grey Wolf Optimizer

GWO [6] is a stochastic optimization algorithm which is inspired from the hunting behaviour of a pack of grey wolves. In this algorithm, the population is termed as a pack of grey wolves with the fittest solution being termed as the alpha $(T_1)$, the second fittest as beta $(T_2)$, the third fittest as delta $(T_3)$ and the remaining solutions as omega $(\omega)$. The $\omega$ wolves follow $T_1$, $T_2$ and $T_3$ wolves, and the hunting mechanism of the pack is modelled using mathematical equations in order to generate new solutions which in turn will update the position of the wolves. The optimal solution is referred as prey. In every iteration $(t)$, the position of each solution undergoes exploration and exploitation stages and is modified with respect to the position of $T_1$, $T_2$ and $T_3$. The new solution for the subsequent iteration, $t + 1$, is generated using the following Eq. 7.

$$X_j^{t+1} = \frac{1}{3}\left(\sum_{i=1}^{3} T_{i,j}^t - a^t\left(2r_{i,j}^t - 1\right)\left(\left|2r_{i,j}^{'t}T_{i,j}^t - X_j^t\right|\right)\right), \quad \forall j \in 1, 2, \ldots, D \quad (7)$$

The terms $r_{i,j}^t$ and $r_{i,j}^{'t}$ represent two random numbers between [0, 1], and $a$ denotes an adaptive parameter which decreases linearly from 2 to 0 as the iteration proceeds. Each newly created solution is checked for bound violation, and if any of the decision variables violates the bound, then it is replaced with a new value within its bounds. The iterative process is continued till the termination criteria are satisfied.

## 3.2  JAYA Algorithm

The algorithm has been reportedly designed with a motivation to move towards the best solution and away from the worst solution. In this algorithm, each solution is updated using the following equation.

$$X_j^{N,t+1} = X_j^t + r_j^t\left(X_{j,best}^t - \left|X_j^t\right|\right) - r_j^{'t}\left(X_{j,worst}^t - \left|X_j^t\right|\right) \quad \forall j \in 1, 2, \ldots, D \quad (8)$$

In the above equation, $X_j^{N,t+1}$ represents the $j$th variable of the new solution. The notation $X_{j,best}^t$ and $X_{j,worst}^t$ represent the solutions corresponding to the best and the worst fitness values, respectively, for the current iteration $t$, whereas $r_j^t$ and $r_j^{'t}$ are two random numbers between [0, 1]. The variables which violate the bounds are replaced with values of the nearest bound. The acceptance of a new solution is based on greedy selection where the new solution will be accepted if it gives the better function value than the current solution. The iterative process is continued till the termination criteria are satisfied.

# 4 Experimental Settings

Five problems with different number of orders and machines are used in this work which have been studied in the literature with conventional algorithms. For each problem, two different datasets are used that differ in the processing time of the orders, with the processing time being longer for the first dataset, and hence lead to fewer feasible solutions. The data used in this work can be obtained from the literature [1]. The codes available online for GWO (https://goo.gl/2Mx1eX) and JAYA (https://goo.gl/0biAcH) are used in this work. For all the ten instances, population size is set to 40 times the number of jobs, while the number of iterations is set to 500. The number of function evaluations per iteration for both the techniques is equal to the size of the population, and hence, the termination criteria based on the number of iterations also ensure that the number of functional evaluations in both the cases are identical. Both the techniques have been designed to handle continuous variables. Hence in order to incorporate integer variables, we have utilized the round off strategy for handling the integer variables though other integer handling strategies have been reported in the literature [8]. All the simulations are carried with MATLAB 2016a on a desktop computer with an i7 3.4 GHz processor and 16 GB RAM. Due to the stochastic nature of CI algorithms, 51 independent runs are performed for each of the algorithms. The independency among the runs is realized by varying the seeds of the *twister* algorithm (in MATLAB) from 1 to 51.

# 5 Results and Discussion

## 5.1 Statistical Analysis

In this section, a comparative study of both the CI techniques on the ten instances is provided. The best values obtained in each independent run of both the algorithms on the ten instances are shown in Fig. 1.

It can be observed that both the algorithms are able to determine identical solutions consistently for problems P1S1 and P1S2 in all the runs. For problem P2S1, both GWO and JAYA are able to obtain the best cost for identical number of runs. In case of the problems P3S1 and P5S1, GWO is able to determine better solutions than the JAYA algorithm. However, there is deviation among the solutions determined in the multiple runs of GWO. For problem P4S1, GWO is able to determine solutions with lower cost than JAYA which was able to determine solutions closer to GWO in only one of the 51 runs. In rest of the problems, GWO is observed to be dominating JAYA for maximum number of runs. The best, mean, median and standard deviation values obtained for GWO and JAYA by accounting all the runs for each instance are shown in Table 1. GWO and JAYA are able to determine an identical best solution for four instances (P1S1, P1S2, P2S1 and

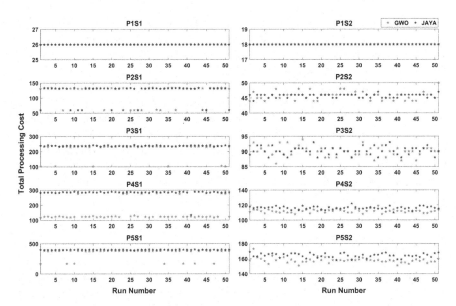

**Fig. 1** Performance of GWO and JAYA in the 51 runs

**Table 1** Statistical analysis of the 51 runs

|    |      | GWO  |       |        |      | JAYA |       |        |      |
|----|------|------|-------|--------|------|------|-------|--------|------|
|    |      | Best | Mean  | Median | SD   | Best | Mean  | Median | SD   |
| P1 | P1S1 | 26   | 26    | 26     | 0    | 26   | 26    | 26     | 0    |
|    | P1S2 | 18   | 18    | 18     | 0    | 18   | 18    | 18     | 0    |
| P2 | P2S1 | 60   | 117.9 | 132    | 28.9 | 60   | 117.9 | 132    | 28.9 |
|    | P2S2 | 44   | 45.7  | 46     | 1.3  | 44   | 45.7  | 46     | 0.6  |
| P3 | P3S1 | 102  | 217.9 | 233    | 42.3 | 233  | 237.4 | 237    | 2.5  |
|    | P3S2 | 86   | 90.1  | 90     | 1.8  | 88   | 90.2  | 90     | 1.3  |
| P4 | P4S1 | 118  | 147.0 | 123    | 57.1 | 134  | 282.7 | 286    | 21.4 |
|    | P4S2 | 108  | 111.9 | 112    | 2.2  | 110  | 116.2 | 116    | 2.0  |
| P5 | P5S1 | 167  | 352.2 | 380    | 73.8 | 392  | 399.4 | 400    | 3.6  |
|    | P5S2 | 151  | 157.8 | 158    | 3.5  | 158  | 164.2 | 164    | 2.8  |

P2S2). The mean and the best values by the two algorithms on both the instances of
P1 are identical, which implies that both the algorithms have determined the best
solutions in all the runs. In addition, the mean, median and standard deviation
values in P1S1, P1S2 and P2S1 are identical for both the algorithms. GWO is able
to determine solutions whose mean is better for problems P3S1, P4S1 and P5S1
than JAYA which is unable to provide equivalent results in most of the independent
runs. In the problems P3S2, P4S2 and P5S2, JAYA was able to obtain solutions
whose mean is inferior to the solutions obtained by GWO. However, the difference

in mean for the second dataset is lower than the first dataset for problem P3, P4 and P5. It can be observed that the standard deviation value in solutions obtained by JAYA is either lower or identical to GWO. This reveals the consistency of JAYA in determining similar results in all the runs. In view of the overall performance by both the algorithms, GWO is able to provide better or identical results in all the cases compared to JAYA. In the literature, three problems (P1, P4 and P5) have been solved with GA. The results obtained by GWO are better than GA in all three problems, whereas the results obtained by JAYA are inferior to GA. It should be noted that the solutions determined by CP for both the instances of P4 and P5S2 are better than GWO. However, the results obtained by GWO are the best solutions obtained by any CI technique. Moreover, GWO outperforms CP in P5S1.

## 5.2  Convergence Analysis

The convergence details for best objective value determined by both the algorithms on all the problems over all independent runs are shown in Figs. 2, 3 and 4. The convergence profile of GWO and JAYA is analysed from the best objective values obtained at 1, 5, 10, 20, 30, 40, 50, 60, 70, 80, 90 and 100 per cent of the total functional evaluations. In problem P1S1, both the algorithms are able to provide best objective value in all the specified intervals. In the case of problem P1S2, JAYA has converged to the final solution quickly compared to GWO. In both the instances of P1 and P2, JAYA as well as GWO have converged to identical

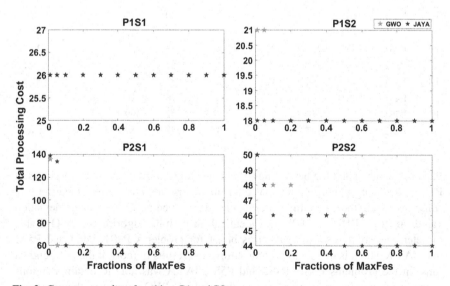

**Fig. 2** Convergence plot of problem P1 and P2

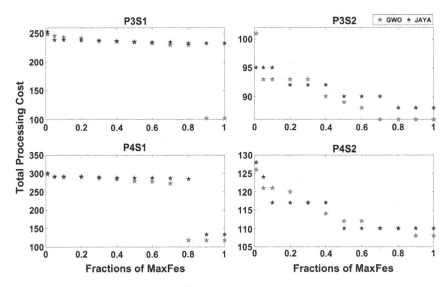

**Fig. 3** Convergence plot of problem P3 and P4

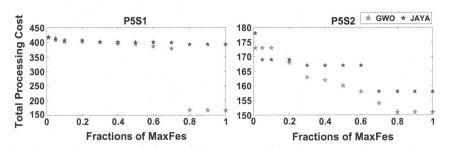

**Fig. 4** Convergence plot of problem P5

solutions. For the second instance of P2, JAYA has shown a faster convergence than GWO.

On analysing the cases of P3 and P4, it is observed that GWO is able to determine better solutions compared to JAYA. For P3S2 and both the cases of P4, JAYA was able to converge to a solution near to GWO, whereas GWO was able to obtain a significantly better solution in P3S1.

For both the instances P5, GWO has converged to a better solution compared to JAYA. For the instance P5S1, the solution determined by GWO has much lower processing cost.

Thus, it can be concluded that JAYA and GWO are able to determine equivalent solutions for lower dimensional job shop scheduling problems, whereas the performance of GWO is superior in higher dimensional problems.

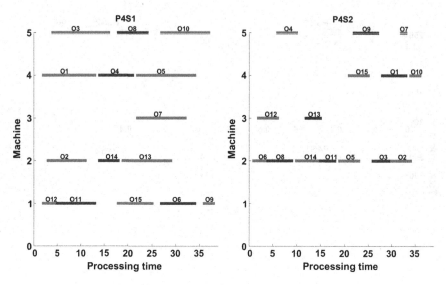

**Fig. 5** Optimal schedule determined by JAYA for P4

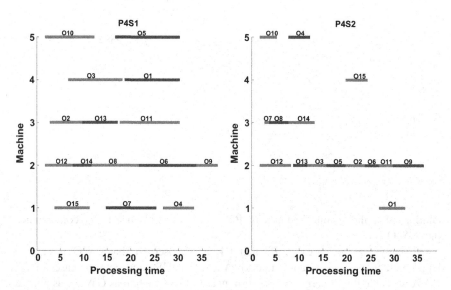

**Fig. 6** Optimal schedule determined by GWO for P4

Solution schedules using JAYA and GWO for two instances of problem P4 are shown in Fig. 5 and Fig. 6, respectively. The problem constitutes 15 jobs that can be processed with 5 machines. From the given schedule for P4S1, it can be observed that the schedule determined by JAYA has assigned maximum number of jobs to machine 1, whereas GWO has assigned machine 2 for processing maximum jobs. It can be observed from the schedule of JAYA that the order 4 (represented by o4) is processed on machine 4 with a cost of 9 despite the fact that it could have been processed on machine 3 with a lower cost of 8.

For the instance P4S2, it can be observed that both GWO and JAYA have assigned highest number of jobs (8 by GWO and 7 by JAYA) to machine 2 which resulted in a processing cost of 51 and 44, respectively. The total processing cost obtained by GWO is 108, but it can be observed that a better solution could have been discovered by assigning order 1 and order 15 to machine 3 and decrease the total processing cost to 105. This will also eliminate the need for machine 1 and machine 4. Thus, it can be observed that this problem can be solved as a multi-objective optimization problem [9] which can also accommodate the cost of the machines.

## 6  Conclusion

In this work, we have evaluated the performance of two CI techniques on a set of five problems with two datasets. The optimization results conclude GWO to be superior in obtaining better results compared to JAYA algorithm. Both the algorithms perform identically for lower dimensional problems. As the dimension of problem set increases, GWO outperforms JAYA in all cases. This reveals the potential of GWO in solving JSSPs. This work can be extended by incorporating multiple objectives and more sophisticated constraint handling techniques.

## References

1. V. Jain and I. E. Grossmann, "Algorithms for Hybrid MILP/CP Models for a Class of Optimization Problems," *INFORMS J. on Computing,* vol. 13, pp. 258–276, 2001.
2. P. R. Kotecha, M. Bhushan, and R. D. Gudi, "Constraint Programming and Genetic Algorithm," in *Stochastic Global Optimization: Techniques and Applications in Chemical Engineering,* G. P. Rangaiah, Ed., ed: World Scientific, 2010.
3. A. Husseinzadeh Kashan, "League Championship Algorithm (LCA): An algorithm for global optimization inspired by sport championships," *Applied Soft Computing,* vol. 16, pp. 171–200, 2014.
4. A. H. Gandomi, X.-S. Yang, and A. H. Alavi, "Cuckoo search algorithm: a metaheuristic approach to solve structural optimization problems," *Engineering with Computers,* vol. 29, pp. 17–35, 2013.
5. V. Punnathanam and P. Kotecha, "Yin-Yang-pair Optimization: A novel lightweight optimization algorithm," *Engineering Applications of Artificial Intelligence,* vol. 54, pp. 62–79, 2016.
6. S. Mirjalili, S. M. Mirjalili, and A. Lewis, "Grey Wolf Optimizer," *Advances in Engineering Software,* vol. 69, pp. 46–61, 2014.
7. R. Rao, "Jaya: A simple and new optimization algorithm for solving constrained and unconstrained optimization problems," *International Journal of Industrial Engineering Computations,* vol. 7, pp. 19–34, 2016.

8. V. Punnathanam, R. Kommadath, and P. Kotecha, "Extension and performance evaluation of recent optimization techniques on mixed integer optimization problems," in *2016 IEEE Congress on Evolutionary Computation (CEC)*, pp. 4390–4397, 2016.
9. P. R. Kotecha, M. D. Kapadi, M. Bhushan, and R. D. Gudi, "Multi-objective optimization issues in short-term batch scheduling," in *17th IFAC World Congress*, Seoul, Korea, 2008.

# Discovery of High Utility Rare Itemsets Using PCR Tree

**Bhavya Shahi, Suchira Basu and M. Geetha**

**Abstract** Data mining is used to extract interesting relationships between data in a large database. High utility rare itemsets in a transaction database can be used by retail stores to adapt their marketing strategies in order to increase their profits. Even though the itemsets mined are infrequent, since they generate a high profit for the store, marketing strategies can be used to increase the sales of these items. In this paper, a new method called the PCR tree method is proposed to generate all high utility rare itemsets while keeping the algorithm time-efficient. The proposed method generates the itemsets in one scan of the database. Results show that the time taken by the proposed method is nearly half that of the existing method, i.e. the UPR tree.

**Keywords** Data mining · High utility · Infrequent itemsets · Support
Rare itemsets

## 1 Introduction

Association rule mining is used to discover relationships that exist within large databases. In data mining, high utility itemsets are those that generate a profit value greater than a minimum value provided by the user. A rare itemset will have support value lower than a minimum support provided by the user. Mining rare,

B. Shahi · S. Basu (✉) · M. Geetha
Department of Computer Science and Engineering, Manipal Institute of Technology,
Manipal Academy of Higher Education, Manipal, Karnataka, India
e-mail: suchira.basu95@gmail.com

B. Shahi
e-mail: bhavya754@gmail.com

M. Geetha
e-mail: geetha.maiya@manipal.edu

© Springer Nature Singapore Pte Ltd. 2019
B. K. Panigrahi et al. (eds.), *Smart Innovations in Communication and Computational Sciences*, Advances in Intelligent Systems and Computing 669, https://doi.org/10.1007/978-981-10-8968-8_6

high utility itemsets are an important area of research today, since it has several applications in market analysis. Usually, there is more focus on mining high-frequency datasets which can be used to increase profits for stores by placing certain products together, or by recommending certain products based on the customer's previous choices. However, even though certain itemsets are rare, they can generate high profits for a store. For example, a refrigerator may be bought with a dishwasher. Using this data, the store can then change their marketing strategy in order to increase the sale of these rare itemsets. They can also be used to distinguish customers based on their shopping habits and change their marketing according to these habits.

## 2 Motivation

The goal of the UPR tree [1] is to extract high utility rare itemsets from transactional databases, by considering user preferences like quantity, cost, profit. The method is divided into two phases. In the first phase, the transaction utility, transaction-weighted utility and support of each item are calculated. The utility pattern rare tree (UPR tree) is formed with the itemsets having utility value greater than the threshold value. In the second phase, these itemsets are mined based on the minimum support threshold given by the user.

The UPR tree method traverses through the database twice, once for each phase. Moreover, the UPR tree does not contain the transactions which have low utility or/ and are rare itemsets. Therefore, for different support and utility values, rare itemsets have to be extracted, and then different UPR tree will have to be created.

## 3 Related Work

Data mining is the process of discovering patterns, anomalies, associations and significant structures in large amounts of data such as databases, data warehouses. Most surveys or observations done in data mining are based on frequent itemsets, i.e. only itemsets that are found frequently are mined. But recently due to advancements in technology, medicine and other fields, high utility and rare itemsets have caught the interest of the data mining community. Fournier-Viger et al. [2] proposed an algorithm called FHM + for mining high utility itemsets which introduced the concept of length constraints using length upper-bound (LUR) constraints. This method mines interesting itemsets up to a particular length and ignores those that are very long, since these itemsets may be uninteresting to the user. Pillai, J. et al. [3] proposed the HURI algorithm which has two phases. First it generates all rare itemsets and then removes those that do not have high utility. J. Pillai et al. [4] further worked on the fuzzy HURI algorithm which extracted rare high utility itemsets by fuzzification of the utility values. Gan W. et al. [5] proposed a high utility itemset mining algorithm that uses multiple minimum utility thresholds and proposed a novel tree structure named MIU-tree. Goyal V. et al. [6] proposed the UP-Rare growth algorithm that uses utility and frequency of itemsets together along with the proposed UP-Tree data structure.

Ryang, H. et al. [7] developed a new algorithm called MHU-Growth for mining high utility rare itemsets by considering different minimum support values for items of differing nature. They used a novel tree structure named MHU-Tree. Lin et al. [8] proposed the high utility pattern tree (HUP tree) which integrated the two-phase procedure used for utility mining along with the downward closure property to generate a tree structure. Tsang et al. [9] proposed the RP tree structure for mining rare association rules. This tree also contained a component storing the information gain for each rule, which helped in identifying the most useful rules.

# 4 Proposed Method

The method uses a data structure called Pattern Count Rare (PCR) Itemset to discover all high utility rare itemsets in one scan of the database. This data structure is similar to reduced pattern count tree discussed in the paper [10].

Every node of the PCR tree consists of the following parts:

- item_number.
- frequency of occurrence on itemset in database pointer to child node.
- pointer to sibling.
- a left-child right-sibling tree is created where each transaction is stored.

## 4.1 Definitions

- Item Utility

Utility of an item in a transaction database is the product of its profit and quantity in that particular transaction.

- High utility itemset

It is an itemset whose transaction utility is higher than a user-specified minimum utility threshold.

- Transaction Utility value of a transaction (TU)

It is the sum of utility values of all items in a transaction.
TU = Item profit * quantity of all items in the transaction

- Transaction-Weighted Utility of an itemset X (TWU(X))

It is the sum of the transaction utilities of all the transactions containing X.
TWU(X) = Sum of TU of all transactions in which the item is present.

- High transaction-weighted utility itemset

It is an itemset in which TWU(X) is no less than minimum utility threshold.

- Rare Itemset

It is an itemset whose support is below the minimum user-specified threshold.

- Support

The number of times an item occurs in the database/Total number of transactions in the database.

## 4.2   Data flow Diagram for PCR Tree

In the PCR tree method, every transaction is read from the database and inserted into the tree. When the transaction is read, its transaction utility [11] and transaction-weighted utility are calculated and support for each item present in the transaction is updated. For each transaction, the items present in it are sorted alphabetically, and then the transaction is inserted into the PCR tree structure. If any transaction matches an existing branch in the PCR tree, the frequency for each node in that branch is incremented, and no new branch is created. If the transaction does not match any existing branch, a new branch of the PCR tree is created. The PCR tree created is shown in Fig. 2. After this, the conditional path for each leaf node is calculated and stored. The power set of each conditional path along with its corresponding transaction utility is calculated, and the items in the power set with transaction utility greater than minimum utility and support less than minimum support are saved as high utility rare itemsets. Figure 1 shows the data flow of the PCR tree method and depicts the flow of the pseudocode.

**Fig. 1** Data flow diagram

## 4.3 Pseudocode for PCR Tree

```
WHILE !EOF
   FOR each transaction
      Calculate transaction utility
      Update transaction weighted utility and support for each item
      Sort all items in transaction in ascending order of item number
      Insert each and every transaction into PCR tree
            IF an existing branch in the tree matches the transaction
                Increment frequency value
            ELSE
                  Create new branch in tree, with each item in the transaction as one
      node
   END FOR LOOP
END WHILE LOOP
FOR each node in tree
   Find conditional path
   Find power set of conditional path
   FOR each itemset in power set
            IF itemset is rare AND itemset is of high utility
                    store itemset in linked list
            END IF
   END FOR LOOP
END FOR LOOP
```

This method requires only one scan of the database to discover rare itemsets. The proposed method is illustrated with an example.

## 4.4 Example

In this example, a sample dataset of 5 transactions is taken, with 5 items in total. The sample database is shown in Table 1. Profit values for each item in the sample database are shown in Table 2. The procedure to mine high utility rare itemsets is then shown. The support and transaction-weighted utility [11] values calculated for each item is shown in Table 3. The minimum support is considered to be 0.4 and the minimum utility is 250. The tree that is created is shown in Fig. 2.

**Table 1** Sample dataset

| Transaction ID | Transaction | Transaction utility |
|---|---|---|
| 1 | (A, 6), (B, 6), (C, 8), (D, 12), (E, 2) | 116 |
| 2 | (E, 6), (C, 6) | 42 |
| 3 | (A, 6), (D, 3) | 24 |
| 4 | (C, 7), (B, 3), (D, 1) | 56 |
| 5 | (B, 8), (C, 5) | 62 |

**Table 2** Profit values for each item

| A | B | C | D | E |
|---|---|---|---|---|
| 3 | 2 | 6 | 2 | 1 |

**Table 3** Calculated support and transaction-weighted utilities

| | A | B | C | D | E |
|---|---|---|---|---|---|
| Support | 0.4 | 0.6 | 0.8 | 0.6 | 0.4 |
| TWU | 140 | 234 | 276 | 196 | 158 |

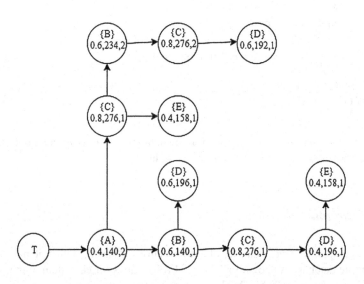

**Fig. 2** PCR tree structure for sample database

The sample of rare high utility itemsets generated for the given example will be:

$$\{A, B, D\}, \{B, D\}. \{A, D\}, \{D, E\}, \{B, E\}, \{A, B, E\}, \{A, E\}, \{B, C, E\},$$

## 5 Experimental Analysis

The algorithm was implemented in C and was executed on a Intel Core i5 processor. The dataset used consists of online retail data from a UK-based online retail store [12, 13]. It consists of the following attributes:

- Invoice Number
- Item
- Stock Code
- Quantity
- Unit price
- Description of the product
- Invoice Date
- Customer ID
- Country

The Invoice Number is used to differentiate between transactions, while Stock Codes are used to differentiate between items. Invoice Date, Item Description, Customer ID and country are considered as irrelevant attributes.

The UPR tree and PCR tree methods have been compared based on time and memory consumed. The UPR tree method is chosen here for comparison since the UPR tree method also discovers high utility rare itemsets and is also a tree-based method. The number of transactions that have been considered has been varied from 100 to 1000.

*Case* 1:

The minimum support threshold is taken as 1%, and the minimum utility is considered to be 50. Table 4 shows the comparison between the time taken to generate high utility rare itemsets by the two methods. Figure 3 shows the graph comparing the two methods. Approximately, there is a 20% decrease in time consumed by the PCR tree as compared to the UPR tree. Also, it can be noticed in the graphs that there is an increase in the difference of execution time of both methods when the number of transactions is increased. Thus for large number of transactions, PCR tree method will be much more profitable.

Table 5 shows the memory consumed in bytes by the two methods. The number of transactions has been varied from 100 to 1000.

**Table 4** Execution time of UPR tree and PCR tree for case 1

| No. of transactions | PCR tree time (in s) | UPR tree time (in s) |
|---|---|---|
| 100 | 0.585 | 0.65 |
| 250 | 6.34 | 5.70 |
| 500 | 20.06 | 23.64 |
| 1000 | 65.95 | 94.53 |

**Fig. 3** Graph for time taken by varied number of transactions with support threshold of 1% and minimum utility of 50

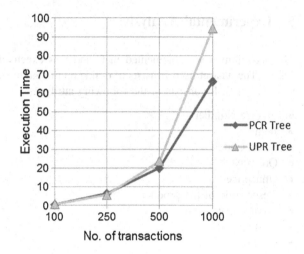

**Table 5** Memory consumed by each method for case 1

| No of transactions | PCR tree memory (in kB) | UPR tree memory (in kB) |
| --- | --- | --- |
| 100 | 28.64 | 19.453 |
| 250 | 83.34 | 59.218 |
| 500 | 157.68 | 121.406 |
| 1000 | 318.093 | 254.687 |

The memory consumed by the PCR tree method is higher than the UPR method since the PCR tree stores all the items in the transaction without removing the low utility or frequent items. Moreover, more data about the item is being stored in the tree such as transaction-weighted utility and support.

Approximately, there is a 35% increase in memory consumed by the PCR tree as compared to the UPR tree. Since, the tree created by the PCR tree contains all the transactions and also more data about each node, the memory consumed is more than the UPR tree. This does not become a huge setback because when and if we change the support values of any item, we would not have to go through the entire database again (Fig. 4).

*Case* 2:

Minimum utility values were varied from 50 to 150 while keeping minimum support equal to 1%. The number of transactions considered in each test is 1000. Table 6 shows the time taken (in seconds) by each method to generate the itemsets. Figure 5 shows the graph comparing the results.

Approximately, there is a 7% decrease in time consumed by the PCR tree as compared to the UPR tree. Also, it can be noticed in the graphs that there is a decrease in the difference of execution time of both methods when the number of

**Fig. 4** Graph showing the memory (in bytes) consumed for varied number of transactions with minimum support as 1% and minimum utility as 50

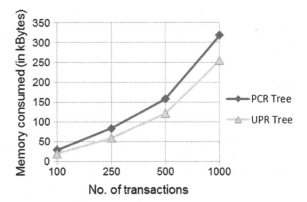

**Table 6** Execution time of UPR tree and PCR tree for case 2

| Minimum utility | PCR tree time (in s) | UPR tree time (in s) |
|---|---|---|
| 50 | 65.94 | 93.38 |
| 75 | 42.69 | 47.90 |
| 100 | 37.73 | 39.53 |

**Fig. 5** Graph for minimum utility versus execution time with minimum support of 1%

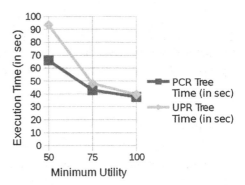

transactions is increased. Thus for small number of transactions, PCR tree method will be much more profitable.

*Case* 3:

Minimum support values were varied from 1 to 0.30% while keeping minimum utility equal to 50. The number of transactions considered in each test is 1000. Table 7 shows the time taken (in seconds) by each method to generate the itemsets. Figure 6 shows the graph comparing the results.

Approximately, there is a 24% decrease in time consumed by the PCR tree as compared to the UPR tree. Also, it can be noticed in the graphs that there is a

**Table 7** Execution time of UPR tree and PCR tree for case 3

| Minimum support (%) | PCR tree time (in s) | UPR tree time (in s) |
| --- | --- | --- |
| 1.00 | 65.94 | 93.38 |
| 0.70 | 64.82 | 85.12 |
| 0.50 | 64.47 | 82.50 |
| 0.30 | 57.34 | 68.06 |

**Fig. 6** Graph for minimum support versus execution time with minimum utility of 50

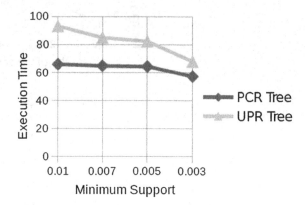

decrease in the difference of execution time of both methods when the number of transactions is increased. But still the difference in execution time varies a lot thus making PCR tree method more profitable than UPR tree for all cases.

## 6 Conclusion

This paper proposes the PCR tree algorithm which stores the entire database in a compact manner and discovers high utility rare itemsets in a single scan of the database. From the experimental analysis, it is observed that this algorithm is time-efficient when compared to UPR tree method. However, since the UPR Tree method stores only that information which is necessary for the mining of rare itemsets, while the PCR tree method stores all information, the PCR tree method is less space-efficient than the UPR tree method. Hence, in the future a more space-efficient algorithm will be developed.

# References

1. S.A.R. Niha, Uma N Dulhare,: Extraction of high utility rare itemsets from transactional database, International Conference on Computing and Communication Technologies, ICCCT 2014.
2. Fournier-Viger, P., Lin, J.C.-W., Duong, Q.-H., Dam, T.-L.: FHM+: Faster high-utility itemset mining using length upper-bound reduction, 29th International Conference on Industrial Engineering and Other Applications of Applied Intelligent Systems, IEA/AIE 2016.
3. Pillai, J., Vyas, O.P., Muyeba, M.: HURI - A novel algorithm for mining high utility rare itemsets, 2nd International Conference on Advances in Computing and Information Technology, ACITY 2012, pp. 531–540.
4. Jyothi Pillai, O.P.V., Muyeba, M.K.: A Fuzzy Algorithm for Mining High Utility Rare Itemsets – FHURI. In: Hope, D.M. (ed.) CEE. p. 11 ACEEE (A Computer division of IDES) (2014).
5. Gan, W., Lin, J.C.-W., Viger, P.F., Chao, H.-C.: More efficient algorithms for mining high-utility itemsets with multiple minimum utility thresholds, Lecture Notes in Computer Science (including subseries Lecture Notes in Artificial Intelligence and Lecture Notes in Bioinformatics) 2016, pp. 71–87.
6. Goyal, V., Dawar, S., Sureka, A.: High utility rare itemset mining over transaction databases, Lecture Notes in Computer Science (including subseries Lecture Notes in Artificial Intelligence and Lecture Notes in Bioinformatics) 2015, pp. 27–40.
7. Ryang, H., Yun, U., Ryu, K.H.: Discovering high utility itemsets with multiple minimum supports, Intelligent Data Analysis, 2014, pp. 1–533.
8. Chun-Wei Lin, Tzung-Pei Hong, Wen-Hsiang Lu,; An effective tree structure for mining high utility itemsets, Expert Systems with Applications, Vol 38, Issue 6, pp. 7419–7424.
9. Tsang S., Koh Y.S., Dobbie G. (2011) RP-Tree: Rare Pattern Tree Mining. In: Cuzzocrea A., Dayal U. (eds) Data Warehousing and Knowledge Discovery. DaWaK 2011. Lecture Notes in Computer Science, vol 6862. Springer, Berlin, Heidelberg.
10. Geetha M, R.J. D'Souza (2008): Discovery of Frequent Closed Itemsets using Reduced Pattern Count Tree, The international conference of Computer Science and Engineering, International Association of Engineers, Lecture Notes in Engineering and Computer Science, Vol.
11. Jiawei Han, Michelene Kamber, Jian Pei,: Data Mining: Concepts and Techniques.
12. Lichman, M. (2013). UCI Machine Learning Repository Irvine, CA: University of California, School of Information and Computer Science., http://archive.ics.uci.edu/ml.
13. Daqing Chen, Sai Liang Sain, and Kun Guo, Data mining for the online retail industry: A case study of RFM model-based customer segmentation using data mining, Journal of Database Marketing and Customer Strategy Management, Vol. 19, No. 3, pp. 197–208, 2012.

# Pre-Diagnostic Tool to Predict Obstructive Lung Diseases Using Iris Recognition System

**Atul Bansal, Ravinder Agarwal and R. K. Sharma**

**Abstract** In human beings Lungs are the essential respiratory organs. Their weakness affects respiration and lead to various obstructive lung diseases (OLD) such as bronchitis, asthma or even lung cancer. Predicting OLD at an earlier stage is better than diagnosing and curing them later. If it is determined that a human is prone to OLD, human may remain healthy by doing regular exercise, breathing deeply and essentially quitting smoking. The objective of this work is to develop an automated pre-diagnostic tool as an aid to the doctors. The proposed system does not diagnose, but predict OLD. A 2D Gabor filter and Support Vector Machine (SVM) based iris recognition system has been combined with iridology for the implementation of the proposed system. An eye image database, of 49 people suffering from OLD and 51 healthy people has been created. The overall maximum accuracy of 88.0% with a sample size of 100 is encouraging and reasonably demonstrates the effectiveness of the system.

**Keywords** Lung · Iridology · Gabor filters · SVM · Iris

A. Bansal (✉)
Department of Electronics and Communication Engineering,
GLA University, Mathura, UP, India
e-mail: atulbansal@rediffmail.com

R. Agarwal
EIED, Thapar University, Patiala, Punjab, India
e-mail: ravinder_eeed@thapar.edu

R. K. Sharma
CSED, Thapar University, Patiala, Punjab, India
e-mail: rksharma@thapar.edu

© Springer Nature Singapore Pte Ltd. 2019
B. K. Panigrahi et al. (eds.), *Smart Innovations in Communication and Computational Sciences*, Advances in Intelligent Systems and Computing 669, https://doi.org/10.1007/978-981-10-8968-8_7

# 1 Introduction

Iris is a thin circular diaphragm that lies between the cornea and lens of the eye. It is the colored area of an eye that surrounds dark pupil. An alternate medicine technique, that examines patterns, colors, and other characteristics of the iris to determine the health status of rest parts of the body is known as Iridology (also recognized as Iridodiagnosis) [1, 2]. Iridology does not diagnose any disease, it merely reveals the weakness in a particular part of the human body. Iris is like a map of the body. Left eye corresponds to left portion and right eye to right portion of human body. Iridology charts [3] divides iris into a number of sectors/zones where each sector/zone corresponds to specific parts of the body. Iridologists use these charts to compare the specific markings, such as a dark spot, circular ring, etc. of the subject with that of healthy human iris. Based upon these observations they determine which part of human body is weak and prone to suffer from disease much prior to actual symptoms.

Iris recognition, an accurate and reliable biometric system [4], identifies enrolled individual based upon iris features. Few studies [5–7] combining iris recognition algorithms and iridology to determine the status of health of an individual have been reported. Stearn et al. [1] used iridology to determine hearing loss and found an accuracy of 70%. Wibawa and Purnomo [5] employed iridology in conjunction with iris recognition system and obtained an accuracy of 94% in determining the broken tissues of pancreas. Ma and Li. [6] reported an accuracy of 86.4% and 84.9% for iris based diagnosis model to determine alimentary canal disease and nerve system respectively. Similarly, Lesmana et al. [7] obtained an accuracy of 83.3% for their model to determine pancreas disorder. As of now, computer aided diagnostic tools are becoming an important area of research in medical field [8]. Similarly, different researchers [9, 10] have proposed various methods and models related to lung diseases.

In this paper, an iridodiagnosis system, combining iridology and iris recognition technique, to predict OLD has been proposed and implemented. Due to atmospheric pollution and certain bad habits like smoking, lungs are prone to various diseases such as bronchitis, asthma or even lung cancer. These diseases are generally diagnosed using various pathological/clinical tests at very late stage. OLD can be predicted at a much earlier stage than actual symptoms. Spirometry is the widely used non-invasive test to assess how well lungs function and is helpful in diagnosing the OLD [11]. It measures how much air one inhales, how much one exhales and how quickly one can inhale and exhale. Smoking is one of the reasons for developing obstructive lung disease. The probability of smokers developing obstructive lung disease is higher when compared to non-smokers [12–15]. Quitting smoking and following healthy diets along with regular exercises can delay if not prevent the cancer. In this paper, experiments have been carried out on iris images to predict the OLD. 2D Gabor filters for feature extraction and Support Vector Machine (SVM) as a classifier has been used in the proposed model to predict OLD.

## 2 Materials and Methods

Five stages involved in iris recognition based diagnostic model are: iris image capture, iris image pre-processing, separating region of interest, feature extraction, disease classification. These stages are discussed in following subsections.

### 2.1 Iris Image Capture

To implement iris based pre-diagnostic model the most important thing is to capture the image of iris. Iris images have been captured using I-SCAN$^{TM}$-2 dual iris scanner of Cross Match Technologies Inc. [16]. Iris images from 100 subjects have been captured. Spirometry test has been used to diagnose the OLD in these subjects. This test uses two parameters, namely, Forced Expiratory Volume in 1 s (FEV1) and Forced Vital Capacity (FVC). If the ratio of these parameters ($\rho$ = FEV1/ FVC) is less than 0.8 for a subject, then one will have a high probability of suffering from obstructive lung disease. Spirometry test on 100 subjects was carried out and based upon spirometry test ratio, dataset was classified into healthy subjects and the subjects suffering from obstructive lung disease as shown in Table 1. Figure 1 shows the sample images captured by iris scanner.

Subjects filled a questionnaire related to the health details, such as whether cough with sputum, smoker's cough, shortness of breath, history of TB, blood in sputum. An informed written consent of each person and ethical approval from the Institutional ethics committee was taken for this research.

**Table 1** Information on the dataset

| Health status of subjects | $\rho$ | Total number of subjects | Number of smokers | Number of non-smokers |
|---|---|---|---|---|
| Suffering from obstructive lung disease | $\leq 0.7$ | 25 | 25 | 0 |
| | $0.7 < \rho < 0.8$ | 24 | 22 | 2 |
| Healthy | $\geq 0.8$ | 51 | 3 | 48 |

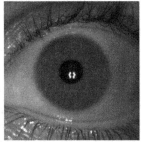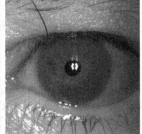

**Fig. 1** Sample images captured using iris scanner

## 2.2   Iris Image Pre-processing

Iris image pre-processing was the process to obtain suitable form of iris image such that significant features can be extracted. Firstly, iris region was isolated from an eye image by determining the boundaries of iris. Next process is to normalize the iris image to eliminate dimensional inconsistencies in iris image arising due to different image capturing conditions such as illumination, distance *etc.* In this process doughnut shape of iris was converted into a rectangular array of size $n \times m$ where $n$ represents radial resolution and $m$ represents angular resolution. Here, value of $n$ is 100 and that of $m$ is 500. Further, to adjust the contrast and poor illumination enhancement of normalized iris image is carried out. Iris localization was done using Circular Hough Transform (CHT) [17]. While, Daughman's rubber sheet model [4, 18] was implemented to normalize the iris image. Further, local histogram analysis and thresholding [19] were employed for image enhancement.

## 2.3   Separating Region of Interest

The health status of a particular organ of human body is reflected at the specific segment of iris. Lungs of human body were shown between 2 o'clock and 3 o'clock in left eye, 9 o'clock and 10 o'clock in right eye in the iridology chart [3]. To predict OLD, lungs region, i.e. region of interest (ROI) was segmented from the normalized iris image. In present work, the size of ROI considered was $100 \times 60$.

## 2.4   Feature Extraction

Iridologists compare the specific markings in the subject's iris at a particular location with healthy human iris. Based upon comparison results, iridologists determine the weakness of a particular organ. For an automated pre-diagnosis system, it becomes important to extract significant features from an iris. Different researchers had employed various feature extraction algorithms [6, 7] for automated iris diagnosis models. Here, iris texture features were extracted using $2D$ Gabor filters. The $2D$ Gabor filter in spatial domain is expressed as:

$$\left. \begin{array}{l} h(x, y) = h'(x, y) . \exp(j2\pi\omega) \\ h'(x, y) = \frac{1}{2\pi\sigma^2} \exp\left(-\frac{x^2 + y^2}{2\sigma^2}\right) \\ \omega = x . \cos\theta + y . \sin\theta \end{array} \right\} \tag{1}$$

where, $(x, y)$ is coordinates of spatial region, $\sigma$ is the standard deviation $\omega$ represents frequency component and $\theta$ is direction parameter of filter. Iris feature vector is created by combining 12 texture energy features computed along four different directions and with three different frequencies. Initially for each filter channel $(\omega_i, \theta_j)(i = 1, 2, 3)$ and $(j = 1, 2, 3, 4)$ the segmented lungs region of iris image $I(x, y)$ is convolved with Gabor filters $h(x, y)$ to obtain an image $S(x, y)$ in frequency domain. The energy for each channel was then computed as:

$$e(\omega_i, \theta_j) = \sqrt{\text{Re}(S)^2 + \text{Im}(S)^2} \tag{2}$$

where, $\text{Re}(S)$ and $\text{Im}(S)$ are the real and imaginary part of image $S(x, y)$. The feature vector so obtained is as shown in Eq. (3).

$$F = [e(\omega_1, \theta_1), e(\omega_1, \theta_2), e(\omega_1, \theta_3), e(\omega_1, \theta_4), \ldots,$$
$$e(\omega_3, \theta_1), e(\omega_3, \theta_2), e(\omega_3, \theta_3), e(\omega_3, \theta_4)] \tag{3}$$

## 2.5 Disease Classification

The obtained feature vector was used for classifying the subjects into healthy one or one prone to OLD using SVM classifier [20–23]. The SVM was proposed by Vapnik [20]. Nowadays, SVM is a popular classifier in the field of text classification, voice recognition, image recognition, pattern recognition *etc.* [21]. Burges [22] and Cristanini [23] provided in-depth information on SVM. It is a binary classifier and is based on the principal of structural risk minimization. In SVM, the main aim was to construct an optimal hyper plane that separates two classes. The hyper plane was constructed in such a way that the margin of separation between two classes is maximum [23, 24]. In this paper, a standard SVM was created for classifying the subjects into healthy one or one prone to obstructive lung diseases.

## 3 Results and Discussions

In the proposed iris diagnosis system by utilizing 2D Gabor filter for feature extraction and SVM for disease classification, were implemented using an image processing module of Matlab 7.1 on Intel Core2 Duo 1.80 GHz processor with 1 GB RAM. The experiments were validated using 10-fold cross validation technique. The whole dataset was divided randomly into ten equal groups. Here, nine groups were used for training the SVM and one group was used for testing. Accuracy of the system was measured for each iteration independently and overall accuracy of the system was the mean of accuracies obtained for individual iterations. Hence, there were ten iterations of training and testing the SVM. The accuracy of the proposed system was calculated by considering linear, polynomial and Gaussian RBF kernel functions for SVM. The maximum accuracy of the proposed method with these features was achieved when RBF kernel function was

**Table 2** Accuracy using
Gabor filter for subjects with
different ρ

| ρ | Number of subjects | Accuracy (in %) |
|---|---|---|
| ≤ 0.7 | 25 | 88.0 |
| 0.7 < ρ < 0.8 | 24 | 83.3 |
| ≥ 0.8 | 51 | 90.2 |

**Fig. 2** Distribution of
smokers and non-smokers in
subjects identified as healthy
and the subjects identified as
having OLD

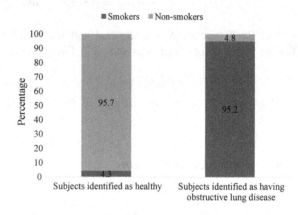

employed. This accuracy using 10-fold cross validation was 88.0%. Table 2 illustrates the accuracy of the proposed method for subjects with different spirometry test ratio.

Figure 2 depicts that 95.2% of the subjects identified as having obstructive lung disease were from the category of smokers and 95.7% of the subjects identified as healthy were from the category of non-smokers.

**Table 3** Comparison of accuracy of proposed obstructive lung disease prediction models with existing techniques

| Disease | Feature extraction technique | Classifier | No. of samples | Accuracy (in %) |
|---|---|---|---|---|
| Detecting broken tissues in pancreas [5] | Minimum filter | Visual inspection | 34 | 94.0 |
| Alimentary canal [6] | Gabor filter | SVM | 53 | 84.9 |
| Nerve system [6] | Gabor filter | SVM | 44 | 86.4 |
| Hearing loss [1] | Digital image enhancement | Visual inspection | 53 | 70.0 |
| Pancreas disorder [7] | GLCM | Neighborhood based modified back propagation (ANMBP) | 50 | 83.3 |
| Pulmonary diseases [25] | Fuzzy C-means clustering | Gray level analysis | 32 | 84.3 |
| Proposed model | Gabor filter | SVM | 100 | 88.0 |

Table 3 shows the comparison of proposed models to predict obstructive lung disease with existing iridodiagnosis models. The obtained accuracy with higher sample size demonstrates the effectiveness of the proposed system.

It has been observed that system's performance depends upon various factors [5]. The way samples of iris images were collected affects the system performance. External factors such as illumination, rotation of eyes, blinking of eyes and usage of iris scanner needs to be considered while acquiring iris images. Subject's unwillingness to discuss their habits and health conditions with researchers is another major aspect affecting the system performance.

# 4 Conclusion

A system to predict the OLD using iris images has been proposed and implemented. The proposed model combines the iris recognition system and iridology. Significant features from the iris have been extracted using Gabor filter. The SVM based classifier has been implemented for classifying subjects into healthy or one prone to OLD. An accuracy of 88.0% with a sample size of 100 is encouraging and reasonably demonstrates the effectiveness of system as compared to earlier iris diagnosis systems. Higher accuracy may be achieved considering more features and employing different classifiers such as Support Tensor Machines.

# 5 Ethical Statement

An informed written consent of each person and ethical approval from the Institutional Ethics Committee, Thapar University, Patiala, India, was taken for this research.

**Acknowledgements** The authors would like to acknowledge "GLA University, Mathura" for partially supporting this research. The authors would also like to acknowledge "Dr. Arun Bansal", "Dr. Poonam Agarwal" and all subjects who helped in developing the database.

# References

1. Stearn N, Swanepoel DW. Identifying hearing loss by means of iridology. African journal of traditional, complimentary and alternate medicines, 2007; 4: 205–10.
2. Jensen, B. The science and practice of Iridology. California Bernard Jensen Co. 1985; 1.
3. Jensen B. 'Iridology charts', http://www.bernardjensen.com/iridology-charts-c-38_42.html, accessed June 2011.

4. Daughman J. High confidence visual recognition of persons by a test of statistical independence. IEEE Transaction on Pattern Analysis and Machine Intelligence. 1993; 15: 1148–1161.

5. Wibawa AD, Purnomo M H. Early detection on the condition of pancreas organ as the cause of diabetes mellitus by real time iris image processing. IEEE Asia Pacific Conference on Circuits and Systems. 2006; 1008–10.

6. Ma L, Li N. Texture feature extraction and classification for iris diagnosis. International conference on medical biometrics, Lecture notes in computer science, Springer-Verlag. 2007; 168–75.

7. Lesmana IPD, Purnama IKE, Purnomo MH. Abnormal Condition Detection of Pancreatic Beta-Cells as the Cause of Diabetes Mellitus Based on Iris Image. International Conference on Instrumentation, Communication, Information Technology and Biomedical Engineering. 2011; 150–155.

8. Kumar A, Anand S. EEG Signal Processing for Monitoring Depth of Anesthesia. IETE Technical Review. 2006; vol 23(3), pp 179–186.

9. Smith, E, Stein P, Furst J, Raicu DS. Weak Segmentations and Ensemble Learning to Predict Semantic Ratings of Lung Nodules. Machine Learning and Applications (ICMLA), 12th International Conference on. 2013; vol.2, no., pp. 519–524.

10. Wang J, Valle MD, Goryawala M, Franquiz JM, Mcgoron AJ. Computer-assisted quantification of lung tumors in respiratory gated PET/CT images: phantom study. Medical Biological Engineering and Computing. 2010; 48:49–58.

11. Pierce R. 'Spirometry: an essential clinical measurement' Australian family physician, 2005, vol. 37, no. 7, pp. 535–539.

12. Lundback B. et al. 'Not 15 but 50% of smokers develop COPD?-Report from obstructive lung disease Northern Sweden studies", Respiratory Medicine, 2003, vol. 97, pp. 115–122.

13. Vestbo J. 'Definition and Overview', Global Strategy for the Diagnosis, Management, and Prevention of Chronic Obstructive Pulmonary Disease. Global Initiative for Chronic Obstructive Lung Disease, 2013, pp. 1–7.

14. Yuh-Chin and Huang T. 'A clinical guide to occupational and environmental lung diseases', Humana Press, 2012, pp. 266.

15. Min J. Y., Min K. B., Cho S. and Paek D. "Combined effect of cigarette smoking and sulphur dioxide on lung functions in Koreans" Journal of Toxicology and Environmental Health, Part-A, 2008, vol. 71, pp. 301–303.

16. I-SCAN-2 Dual iris Scanner. http://www.crossmatch.com/i-scan-2.php. Accessed 16 July 2011.

17. Wildes R, Asmuth J, Green G, Hsu S, Kolczynski R, Matey J, McBride S. A system for automated iris recognition. IEEE Workshop on Applications of Computer Vision, Sarasota, FL. 1994; 121–28.

18. Daughman J. How iris recognition works? IEEE Transaction on Circuits and Systems for Video Technology. 2004; 14: 21–30.

19. Zhu Yong, Tan Tieniu, Wang Yunhong. Biometric personal identification based on iris patterns. Proceedings of the IEEE international conference on pattern recognition. 2000; 2801–2804.

20. Cortes C, Vapnik V. Support vector networks-Machine Learning. Kluwer Academic Publishers, Boston. 1995; 273–97.

21. Kim KA, Chai JY, Yoo TK, Kim SK, Chung K, Kim DW. Mortality prediction of rats in acute hemorrhagic shock using machine-learning techniques. Medical Biological Engineering and Computing. 2013; 51:1059–1067.

22. Burges CJC. A Tutorial on Support Vector Machines for Pattern Recognition. Kluwer Academic Publishers, Boston. 1998.

23. Cristianini N, Shawe TD. An Introduction to Support Vector Machines and other kernel-based Learning Methods" Cambridge University press, Cambridge.
24. Haykin S. Neural Networks-A comprehensive foundation. Pearson Education, 2nd ed. 2004.
25. Sivasankar K., Sujaritha M., Pasupathi P. and Muthukumar, S. "FCM based iris image analysis for tissue imbalance stage identification", in Proc. of International Conference on Emerging Trends in Science, Engineering and Technology (INCOSET), 13–14 Dec. 2012, pp. 210–215.

# HANFIS: A New Fast and Robust Approach for Face Recognition and Facial Image Classification

R. Senthilkumar and R. K. Gnanamurthy

**Abstract** The purpose of this paperwork is to improve the recognition rate and reducing the recognition time required for face recognition application. The method involved in this work is feature extraction which is done using texture-based Haralick feature extraction method and facial image classification is achieved by using modified ANFIS classifier. This method is simply called as HANFIS. The four standard face databases ORL, Yale, Surveillance, and FERET are tested by the HANFIS method along with other existing standard texture, approach, and documentation-based face recognition methods. The results obtained clearly show that the proposed approach gives almost 90–100% face recognition accuracy, and it also requires less recognition time (i.e., faster than existing approaches).

**Keywords** ANFIS · Classifier · Bag of visual words · Face recognition
Haralick features · KNN · Naïve Bayes · Recognition rate
SVM · 2DPCA

## 1 Introduction

There are two main problems to be considered in the present facial image recognition using appearance-based [1–5], texture-based [6–10], and documentation-based [11–14] methods: improvement in the recognition rate and reduction in the time for recognition. The Haralick feature extraction method [6] is a texture-based feature extraction method. In this method, first a gray-level co-occurrence matrix [9, 10]

R. Senthilkumar (✉)
Department of Electronics and Communication Engineering,
Institute of Road and Transport Technology, Erode, Tamil Nadu, India
e-mail: rsenthil_1976@yahoo.com

R. K. Gnanamurthy
Department of Electronics and Communication Engineering,
P.P.G. Institute of Technology, Coimbatore, Tamil Nadu, India
e-mail: rkgnanam@yahoo.co.in

© Springer Nature Singapore Pte Ltd. 2019
B. K. Panigrahi et al. (eds.), *Smart Innovations in Communication
and Computational Sciences*, Advances in Intelligent Systems
and Computing 669, https://doi.org/10.1007/978-981-10-8968-8_8

is developed and from this matrix 22 statistical features are extracted. The classification of faces is done with help of either Euclidean distance [15]-based or Naïve Bayes [16] classifier.

Jian Yang [17] developed an approach-based face recognition called two-dimensional principal component analysis (2DPCA). Jian Yang tested three face databases, AR face database [18], Yale [19], and ORL [20]. The maximum classification rate achieved in his work is 96% for ORL otherwise called as AT&T, 96.1% for AR and 84.24%. The kNN classifier was used in his work. Again the 2DPCA feature extraction performance is analyzed in 2014 [21], and Senthil face database [22] is used instead of AR face database and to increase the performance of 2DPCA method, Senthilkumar et al. [23] proposed a novel approach for facial image recognition called as Duplicating Facial Images Based on Correlation Study (DFBCS). This method improves the recognition rate up to 98% for Yale face database without resizing the face images.

The Naïve Bayes classifier first introduced for face detection. Later, it has been combined with PCA and LDA [24] and used for face recognition. The combined local feature selection for face recognition based on naïve Bayesian classification [25] gives 78.75% recognition accuracy. The support vector machine (SVM) [26–28] classifier is the best method for both linear and nonlinear separable data. T.H. Le and L. Bui [29] tested 2DPCA with SVM classifiers [30] for face recognition. The 2DPCA+SVM classifier has given 97.3% recognition rate for ORL face database.

The bag of words (BOW) first introduced by Csurka [11] for feature extraction and later it has been adapted for computer vision applications [12, 14]. In this method, input face images are split it into many sub-images. For each and every sub-image, a codebook is created. This codebook is called as bag of words. The features are retrieved from the codebook entry which are then given to SVM classifiers for facial image recognition [31].

The aim of our work is to increase the recognition rate and to reduce the recognition time required for the test faces for all cases, i.e., face image may be too large, pose variant, illumination variant, occluded or low correlated. For that, we have proposed an approach which combines Haralick statistical methods and advanced ANFIS classifier [32, 33]. The Haralick method applied to face images to extract its textual statistical features and modified ANFIS classifier [34] is used to classify those face images based on textual content. The combined new approach is called as Haralick features with ANFIS classifier (HANFIS).

The results produced by the proposed method is the best when compared to the existing methods such as Haralick+Naïve Bayes (texture-based method), 2DPCA+kNN (appearance-based method), 2DPCA+Naïve Bayes (appearance-based method), 2DPCA+SVM (appearance-based method), and BOW+SVM (documentation-based method). Our method gives almost 100% recognition accuracy for all the four standard face databases. The proposed approach also reduces the recognition time required for test face image recognition.

This entire paperwork is arranged as follows: Sect. 2 discusses the steps involved in existing methods. The proposed method and the corresponding algorithm for feature extraction and classification are explained in Sect. 3. Section 4 discusses the experimental results and related enhancement of our work. Section 5 draws conclusion and related enhancement of our work.

## 2 Existing Methods

### 2.1 Haralick Feature Extraction and Naïve Bayes Classifier

In this method, first a gray-level co-occurrence matrix (GLCM) is built from the 2D gray face image matrix. The gray matrix size depends on the number of gray levels represented by the image. This GLCM is the basic building block for all the 22 features derived by Haralick and others. These features are fed to the Naïve Bayes classifier for classification. Figure 1 shows the block diagram representation of this method.

### 2.2 2DPCA Feature Extraction and K-Nearest Neighbor Classifier

The 2DPCA method depends on the 2D covariance matrix which is directly evaluated from the 2D face image matrix. Then the eigenvectors are evaluated from this covariance matrix. These eigenvectors are called as the principal components because they are the basic building block for the reconstruction of original test face image. The classification is done with the help of k-nearest neighbor classifier. Figure 2 shows the block diagram representation of this method.

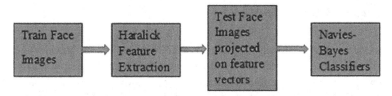

**Fig. 1** Haralick feature extraction and Naïve Bayes classifier

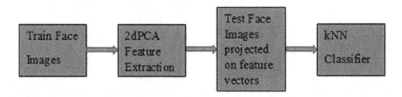

**Fig. 2** 2DPCA feature extraction and kNN classifier

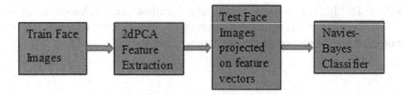

**Fig. 3** 2DPCA feature extraction and Naïve Bayes classifier

## 2.3 2DPCA Feature Extraction and Naïve Bayes Classifier

The principal component vectors are evaluated from the 2D covariance matrix, and feature vectors are found out by projecting the 2D test face images on the first few selected principal components. The projected vectors and associated class labels are fed to the Naïve Bayes classifier. Figure 3 shows block diagram representation of this method.

## 2.4 2DPCA Feature Extraction and SVM Classifier

The principal components are evaluated as discussed above. The principal components of the train set face images are SVM trained. Then a SVM classifier (linear or nonlinear) is used to classify the test set face images based on the support vectors obtained from the train set. This method is represented in block diagram as shown in Fig. 4.

## 2.5 Bag of Visual Words

The first step in this approach is to extract scale-invariant feature transform (SIFT) [35] local feature vectors from the set of face images. Then they are put into a single set. Then a clustering algorithm (k means) is applied over the set of local feature vectors to find coordinates of centroid. An ID is assigned to each centroid.

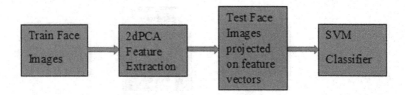

**Fig. 4** 2DPCA feature extraction and SVM classifier

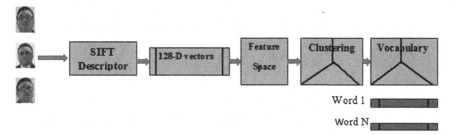

**Fig. 5** Bag of visual words diagrammatic representation

The vocabulary will be formed from these centroids. The global feature vector called as histogram that counts number of times each centroid occurred in each face image is found out. These global feature vectors are called as bag of words which are the face image labels. Finally, the face image labels are fed to SVM classifier as represented in block diagrammatic form in Fig. 5.

## 3 Proposed Method

### 3.1 HANFIS Architecture

In our proposed method HANFIS, the first 22 Haralick features are extracted. The extracted features are applied to advanced neuro-fuzzy inference system (ANFIS) architecture. A slight modification has been done in ANFIS classifier final stage. The HANFIS architecture is shown in Fig. 6 and associated pseudo-code is enlisted in Fig. 7. For each subject, only one face image feature vectors are selected for ANFIS training and for remaining faces of the same subject, their feature vectors are directly applied to ANFIS classifier. There is no need to train their feature vectors. This simplifies the classification and reduces the time consumed for training and recognition. The genfis() function is used to generate fuzzy inference system structure from data, anfis() is the function for Sugeno-type fuzzy inference system, and the evalfis() function perform fuzzy inference calculations.

### 3.2 Haralick Features

Given an image 'A,' of size 'nxm,' the co-occurrence, matrix 'p' can be defined as (1):

$$p_k(i,j) = \sum_{r=1}^{m} \cdot \sum_{s=1}^{n} \begin{cases} 1, & \text{if} \quad A(r,s) = i \text{ and } A(r+\Delta x, s+\Delta y) = j \\ 0, & \text{otherwise} \end{cases} \quad (1)$$

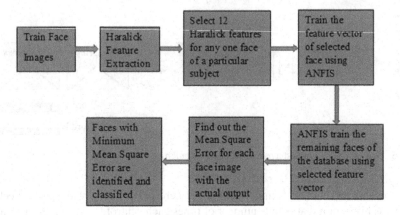

**Fig. 6** HANFIS architecture–Block diagram representation

```
Haralick_feature =readexcel ('FaceDatabase.xls');
y = Haralick_feature {1 to 12}; //trainData
C = Num_of_classes;
N = Numb_of_subjects;
L = CxN; //Total no. of faces in entire database
y = norm(y); //Normalize the features
define memType; //define Fuzzy membership type
define memFun; //define Fuzzy membership function
IN_FIS = genfis(y,memType,memFun);
[TrainFIS,TestFIS} = anfis
(y,IN_FIS,epoch,testData);
Out1 =evalfis([1 to 12],TrainFIS);
Out2 =evalfis([1 to 12],TestFIS);
MSE = sqrt(((Out1-Out2)^2))/Length(Out1));
Recog = 0;
If MSE< min_MSE
 Recog = Recog+1;
End
Recog_Accuracy = (Recog/L)x100;
```

**Fig. 7** HANFIS Algorithm–Pseudo-code

In '$p_k$' matrix, element (i, j) is the total events of numbers 'i' and 'j' that their distance from each other is ($\Delta x$, $\Delta y$). The '$\Delta x$' is the small distance along row direction, and '$\Delta y$' is the small distance along column direction. The 'k' indicates the four directions 0°, 45°, 90°, and 135°. The basis of Haralick features is the gray-level co-occurrence matrix (GLCM) is given in (2). This matrix is a square matrix of size 'NgxNg.' The 'Ng' indicates the number of gray levels in the face image. The matrix element values depend on the number pixel adjacencies. The final probability values in the matrix indicate that the number of times the pixel value 'i' will be adjacent to the pixel value 'j.' The adjacency in each of four directions in a 2D square pixel face image (horizontal, vertical, left and right diagonals) can be considered and calculated. Out of 22, only 12 Haralick features are listed from Eqs. (3) to (18).

$$GC = \begin{bmatrix} p(1,1) & p(1,2) & \ldots\ldots & p(1,Ng) \\ p(2,1) & p(2,2) & \ldots\ldots & p(2,Ng) \\ \ldots & \ldots & \ldots\ldots & \ldots\ldots \\ p(Ng,1) & p(Ng,2) & & p(Ng,Ng) \end{bmatrix} \qquad (2)$$

1. Homogeneity:

$$H = \sum_i \sum_j \frac{1}{1+(i-j)^2} \cdot p(i,j) \qquad (3)$$

2. Dissimilarity:

$$DS = \sum_i \sum_j |i-j| \cdot p(i,j) \qquad (4)$$

3. Energy:

$$E = \sum_i \sum_j p(i,j)^2 \qquad (5)$$

4. Contrast:

$$CT = \sum_i \sum_j (i-j)^2 \cdot p(i,j) \qquad (6)$$

5. Symmetry:

$$S = |p(i,j) - p(j,i)| \qquad (7)$$

6. Entropy:

$$EH = -\sum_i \sum_j p(i,j) \cdot \log(p(i,j)) \qquad (8)$$

7. Correlation:

$$\gamma = \frac{\sum_i \sum_j (i-\mu_x) \cdot \left(j-\mu_y\right) \cdot p(i,j)}{\sigma_x \sigma_y} \qquad (9)$$

where

$$\mu_x = \sum_i \sum_j i \cdot p(i,j) \qquad (10)$$

$$\mu_y = \sum_i \sum_j j \cdot p(i,j) \qquad (11)$$

$$\sigma_x = \sum_i \sum_j (i-\mu_x)^2 \cdot p(i,j) \qquad (12)$$

$$\sigma_y = \sum_i \sum_j \left(j-\mu_y\right)^2 \cdot p(i,j) \qquad (13)$$

8. Inverse Difference Moment:

$$IDM = \sum_i \sum_j \frac{1}{1+(i-j)^2} \cdot p(i,j) \qquad (14)$$

9. Cluster Shade:

$$CS = \sum_i \sum_j \left(i+j-\mu_x-\mu_y\right)^3 \cdot p(i,j) \qquad (15)$$

10. Cluster Prominence:

$$CP = \sum_i \sum_j \left(i + j - \mu_x - \mu_y\right)^4 \cdot p(i, j) \tag{16}$$

11. Maximum Probability:

$$P_m = \underset{i,j}{MAX} \cdot p(i, j) \tag{17}$$

12. Angular Second Moment:

$$ASM = \sum_i \sum_j p(i, j)^2 \tag{18}$$

## 3.3 ANFIS Classifier

In our application, Sugeno-type fuzzy inference system (FIS) is used. This ANFIS uses a hybrid learning algorithm to identify the parameters of Sugeno-type fuzzy inference systems. It is a combination of the least squares method and the back-propagation gradient descent method in order to train the FIS membership function parameters to emulate a given training dataset. The Sugeno-type ANFIS has the following properties:

- The system order must be first or zero.
- Single output obtained using defuzzification weighted average method. The output membership functions must be the same type.
- The number of output membership functions is equal to the number of rules.
- Each rule must have unity weight.

The rules and functionality of ANFIS five layers are detailed below:

**Rules**:

Rule 1: If 'x' is $A_1$ and 'y' is $B_1$, then $f_1 = P_1 x + Q_1 y + r_1$

Rule 2: If 'x' is A2 and 'y' is $B_2$, then $f_2 = P_2 x + Q_2 y + r_2$

**Layer 1**: Every node 'i' in this layer 1 is an adaptive node with a node function:

$$O_{1,i} = \mu_{A_i}(x), \quad \text{for} \quad i = 1, 2$$
$$O_{1,i} = \mu_{Bi-2}(y), \quad \text{for} \quad i = 3, 4 \tag{19}$$

Membership function,

$$\mu_A(x) = \frac{1}{1 + \left|\frac{x - C_i}{a_i}\right|^{2b}} \tag{20}$$

where $\{a_i, b_i, c_i\}$ is the parameter set. The parameters in this layer are referred to as premise parameters.

**Layer 2**: Every node in this layer is a fixed node, whose output is the product of all the incoming signals:

$$O_{2,i} = W_i = \mu_{A_i}(x)\mu_{B_i}(y), \quad i = 1, 2 \tag{21}$$

The each node output represents the firing strength of a rule.

**Layer 3**: Every node in this layer is a fixed node, labeled 'N.' The outputs of this layer are called normalized firing strengths. The '$i$th' node calculates the ratio of the '$i$th' rules firing strength,

$$O_{3,i} = \overline{W_i} = \frac{W_i}{W_1 + W_2}, \quad i = 1, 2 \tag{22}$$

**Layer 4**: Every node 'i' in this layer is an adaptive node with a node function,

$$O_{4,i} = \overline{W_i} \cdot f_i = \overline{W_i}(p_i x + q_i y + r_i) \tag{23}$$

where '$\overline{W_i}$' is a normalized firing strength from layer '3,' $\{p_i, q_i, r_i\}$ is the parameter set of this node. The parameters of this layer are referred to as consequent parameters.

**Layer 5**: Single node in this layer is a fixed node labeled '$\sum$,' which computes the overall output as the summation of all incoming signals:

$$\text{overall} \quad \text{output} = O_{5,1} = \sum_i \overline{W_i} \cdot f_i = \frac{\sum_i W_i f_i}{\sum_{j=i} W_i} \tag{24}$$

There are two passes in the hybrid learning procedure for ANFIS: forward pass and backward pass. For premise parameters, the forward pass is fixed and backward pass is gradient descent. For consequent parameters, forward pass is least squares estimator and backward pass is fixed. The signals are node output for forward pass and error signals for backward pass.

# 4 Experimental Results and Discussion

For our experiment, four standard face databases ORL otherwise called as AT&T, Yale, Surveillance [36], and FERET [37] are considered for testing. For training and testing of Haralick+Naïve Bayes method, 2DPCA+kNN, 2DPCA+Naïve Bayes, and 2DPCA+SVM the following strategy is adopted. 50% of faces of each subject are trained and 50% remaining faces are tested. For example, the ORL database consists of 40 subjects each with 10 faces per subject, totally 400 faces in its entire database. For this face database, 5 faces per subject are trained and remaining 5 faces are considered for testing. The Yale database consists of 15 subjects each with 11 faces per subject, totally 165 faces in its entire face database. For this database similar to ORL, 5 used for training and 5 used for testing and the 11th face is left out.

For big face databases like Surveillance face database, a test probe consists of 10 subjects and 10 faces per subject. In the FERET face database, a test probe consists of 100 subjects and 5 faces per subject, and a total of 500 faces are tested. For training and testing our approach, the following strategy is adopted. Only one face of each subject is trained, and remaining faces are included in test set. This will automatically reduce the feature extraction time for each subject in the proposed method. That is why our approach must be faster than the other techniques. Since our method is based on predefined membership function, it requires less recognition time as compared to other existing methods.

Table 1 lists the recognition time required in seconds for different face recognition techniques. The recognition time for Haralick+Naïve Bayes method calculated for three different cases: 12 features, 22 features, selected five features. The recognition time for other methods 2DPCA+kNN, 2DPCA+Naïve Bayes, and 2DPCA+SVM are calculated for three different feature sizes such as Cx1, Cx2, and Cx3. The 2DPCA+kNN technique and 2DPCA+SVM techniques require large recognition time compared to other methods. Our method is tested for three different mean square errors: 1e-8, 1e-7.9, and 1e-7.5. The HANFIS method requires less than 1% recognition time compared to other methods which is shown in Fig. 8. Figure 8a is for ORL, Fig. 8b is for Surveillance, and Fig. 8c for Yale face databases. The recognition time required for FERET face database is not shown in figure, but it is enlisted in Table 1.

Table 2 lists the recognition accuracy in percentage for different face recognition methods tested for four databases. The recognition accuracy of BOVW method is

**Table 1** Comparison of recognition time in seconds of different face recognition techniques (intel core i3 processor, 4 GB RAM)

| Feature extraction method | Face databases | | Recognition time in seconds | | | |
|---|---|---|---|---|---|---|
| | | | ORL (AT&T) | Surv | Yale | FERET |
| *Texture-based FR* Haralick feature extraction Naïve Bayes classifier | Feature dimension | First 12 | 0.039 | 0.0132 | 0.016 | 0.109 |
| | | All 22 | 0.042 | 0.0134 | 0.020 | 0.115 |
| | | Selected five (contrast, correlation, energy, entropy, homogeneity) | 0.038 | 0.0117 | 0.017 | 0.099 |
| *Appearance-based FR* 2DPCA kNN classifier | Feature dimension | CxK = 1 | 6.171 | 0.415 | 1.879 | 54.72 |
| | | CxK = 2 | 7.362 | 0.621 | 2.575 | 79.79 |
| | | CxK = 3 | 7.643 | 1.000 | 4.371 | 74.04 |
| *Appearance-based FR* 2DPCA Naïve Bayes classifier | Feature dimension | CxK = 1 | 0.294 | 1.572 | 0.453 | 3.734 |
| | | CxK = 2 | 0.301 | 0.434 | 0.431 | 4.366 |
| | | CxK = 3 | 0.322 | 0.972 | 0.981 | 3.319 |
| *Appearance-based FR* 2DPCA SVM classifier | Feature dimension | CxK = 1 | 4.854 | 1.383 | 4.541 | 35.52 |
| | | CxK = 2 | 7.322 | 0.933 | 2.272 | 33.44 |
| | | CxK = 3 | 7.343 | 0.821 | 3.682 | 33.91 |
| *Documentation-based FR* Bag of visual words | Feature dimension | 500 | 0.269 | 0.266 | 0.4118 | 0.6957 |
| *Haralick feature extraction ANFIS classifier (HANFIS)* | Feature dimension (First 12) | MSE 1e-8 | 0.049 | 0.0272 | 0.0381 | 0.0251 |
| | | MSE 1e-7.9 | 0.048 | 0.0275 | 0.0376 | 0.0319 |
| | | MSE 1e-7.5 | 0.043 | 0.0283 | 0.0375 | 0.0337 |

comparable with HANFIS method, but it requires larger recognition time compared to the proposed method. The Haralick+Naïve Bayes classifier is a worst classification method compared to other four methods. This method not only gives poor recognition rate but also consumes large recognition time. The 2DPCA+kNN, 2DPCA+Naïve Bayes, and 2DPCA+SVM methods produce moderate recognition accuracy. The 2DPCA+kNN require larger recognition time for FERET and ORL face databases. This is because they are big databases compared to other face databases discussed here.

Figure 9a–d shows the performance comparison of different face recognition methods with the proposed method HANFIS. Figure 9a shows the plot for AT&T (ORL) face database, Fig. 9b is the plot for Surveillance database, Fig. 9c is the

**Fig. 8  a** Pie chart shows
HANFIS recognition time in
seconds for ORL database.
**b** Pie chart shows HANFIS
recognition time in seconds
for Surveillance database.
**c** Pie chart shows HANFIS
recognition time in seconds
for Yale database

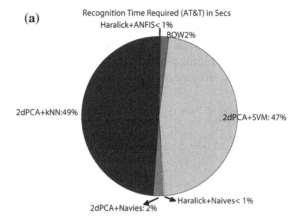

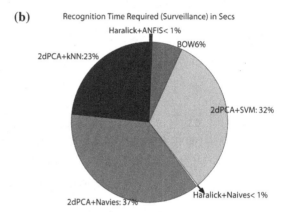

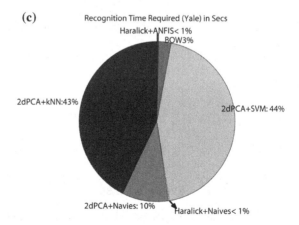

plot for Yale face database, and Fig. 9d shows the plot for FERET face database. The dark line (HANFIS) in Fig. 9a reaches highest recognition rate, and red line is the lowest recognition rate achieved in Haralick+Naïve Bayes method.

Figure 9a–d is the plot between feature vector and mean square error. Figure 10 compares the recognition rate obtained for our HANFIS method for three different

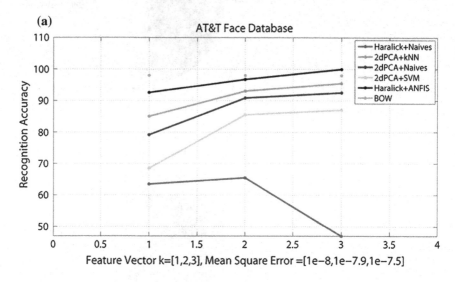

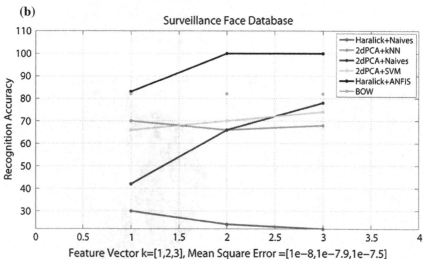

**Fig. 9** **a** Comparison of HANFIS method recognition rate for ORL face database. **b** Comparison of HANFIS method recognition rate for Surveillance face database. **c** Comparison of HANFIS method recognition accuracy for Yale database. **d** Comparison of HANFIS method recognition rate for FERET face database

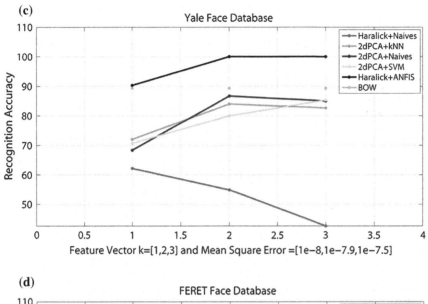

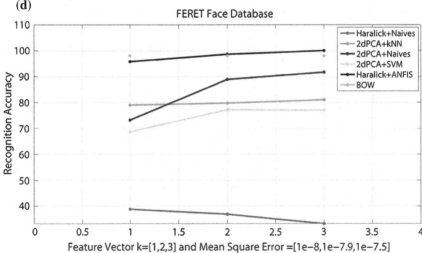

**Fig. 9** (continued)

MSEs. From the result, it is clear that as the MSE decreases, the recognition rate drastically increases. In this case, for MSE <1e-7.5, maximum recognition rate is achieved. For MSE <1e-8, minimum recognition rate is achieved.

**Table 2** Performance comparison of recognition accuracy in percentage of different face recognition techniques

| Feature extraction method | Face databases | | Recognition accuracy in percentage | | | |
|---|---|---|---|---|---|---|
| | | | ORL (AT&T) | Surv | Yale | FERET |
| *Texture-based FR* Haralick feature extraction and Naïve Bayes classifier | Feature dimension | First 12 | 63.50 | 30.00 | 62.19 | 38.87 |
| | | All 22 | 65.50 | 24.00 | 54.87 | 36.84 |
| | | Selected five (contrast, correlation, energy, entropy, homogeneity) | 47.00 | 22.00 | 42.68 | 33.19 |
| *Appearance-based FR* 2DPCA kNN Classifier | Feature dimension | CxK, K = 1 | 85.00 | 70.00 | 72.00 | 79.04 |
| | | CxK, K = 2 | 93.00 | 66.00 | 84.00 | 79.79 |
| | | CxK, K = 3 | 95.50 | 68.00 | 82.66 | 81.06 |
| *Appearance-based FR* 2DPCA Naïve Bayes classifier | Feature dimension | CxK, K = 1 | 79.17 | 42.00 | 68.33 | 73.23 |
| | | CxK, K = 2 | 90.83 | 66.00 | 86.67 | 88.89 |
| | | CxK, K = 3 | 92.50 | 78.00 | 85.00 | 91.67 |
| *Appearance-based FR* 2DPCA SVM classifier | Feature dimension | CxK, K = 1 | 68.50 | 66.00 | 70.66 | 68.68 |
| | | CxK, K = 2 | 85.50 | 70.00 | 80.00 | 77.27 |
| | | CxK, K = 3 | 87.00 | 74.00 | 85.33 | 77.02 |
| *Documentation-based FR* Bag of visual words | Feature dimension | 1 × 500 | 98.00 | 82.00 | 89.33 | 97.9798 |
| *Haralick feature extraction and ANFIS classifier (HANFIS)* | Feature dimension First 12 parameters | MSE1 1e-08 | 92.25 | 83.00 | 90.3 | 95.75 |
| | | MSE2 1e-07.9 | 96.75 | 100 | 100 | 98.58 |
| | | MSE3 1e-07.5 | 100 | 100 | 100 | 100 |

C is the number of columns in the 2D face image matrix

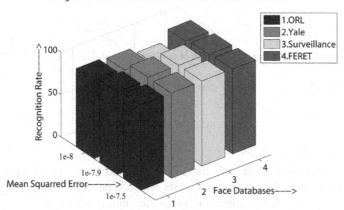

**Fig. 10** Comparison of recognition rate of HANFIS approach for different mean square error and for different face databases

# 5 Conclusion and Future Work

From the experimental results, it is clear that our proposed approach makes two primary advantages: it reduces the recognition time required and improves the recognition accuracy as the MSE reduces. The HANFIS is a fast method, because it requires lesser recognition tine. The main drawback in our method is the Haralick feature extraction time is too large compared to others.

As the MSE reduces, the recognition rate increases. The mean square error plays a vital role in determining the recognition rate. Since the feature extraction time is large, our future work involves the usage of genetic algorithm for feature selection. This GA-based HANFIS will produce far better results compared to other methods. It reduces not only the recognition time but also the feature extraction time.

# References

1. M.A.Turk and A.P. Pentland, "Face recognition using eigenfaces", in Proc. IEEE Conf. on Computer Vision and Pattern Recognition. (1991), 586–591.
2. M. Turk and A. Pentland. 1991. Eigenfaces for recognition. J. Cognitive Neurosci. 3, 1(1991), 71–86.
3. S. Swets and J. Wen, "Using Discriminant Eigen features for Image Retrieval", IEEE Trans. On Pattern Analysis and Machine Intelligence. 18, 8, July 1996, pp. 831–836.
4. S. Bartlett, J.R. Movellan, and T.J. Sejnowski, "Face Recognition by Independent Component Analysis", IEEE Trans. Neural Networks. 13, 6, June 2002, pp. 1450–1464.
5. M.H. Yang, "Kernel Eigen faces vs. Kernel Fisherfaces: Face Recognition Using Kernel Methods", In Proc. 5th IEEE Int'l Conf. Automatic Face and Gesture Recognition (AFGR'02), May 2002, pp. 215–220.
6. R.M. Haralick, K. Shanmugam and I. Dinstein, "Textural Features for Image Classification", IEEE Transactions on Systems, Man, and Cybernetics, vol. 3, 6, Nov'1973, pp. 610–621.
7. L. Soh, C. Tsatsoulis, "Texture Analysis of SAR Sea Ice Imagery Using Gray Level Co-Occurrence Matrices", IEEE Transactions on Geoscience and Remote Sensing, vol 37, 2, March 1999, pp. 780–795.
8. D.J. Clausi, "An analysis of co-occurrence texture statistics as a function of grey level quantization", Can. J. Remote Sensing, vol. 28, 1, pp. 45–62, 2002.
9. A. HazratiBishak, Z. Ghandriz, T. Taheri, "Face Recognition using Co-occurrence Matrix of Local Average Binary Pattern", Journal of Selected Areas in Telecommunication, pp. 15–19, April 2012.
10. P. Mohanaiah, P. Sathyanarayana and L. Gurukumar, "Texture Feature Extraction Using GLCM Approach", International Journal of Scientific and Research Publications", vol. 5, 5, May 2013, pp. 1–5.
11. G. Csurka, C.R. Dance, L. Fan, J. Willamowski and C. Bray, "Visual categorization with bags of keypoints", In Workshop on Statistical Learning in Computer Vision, ECCZV, pp. 1–22.
12. Z. Li, J. Imai, and M. Kaneko, "Robust face recognition using block-based bag of words", The 20th International Conference on Pattern Recognition, 2010, pp. 1285–1288.
13. Y.S. Wu, H.S. Liu, G.H. Ju, T.W. Lee and Y.L. Chiu, "Using the visual words based on affine-sift descriptors for face recognition", In Signal Information Processing Association Annual Summit and Conference, 2012, pp. 1–5.
14. R. Shekhar and C. Jawahar, "Word image retrieval using bag of visual words", In 10th IAPR International Workshop on Document Analysis Systems, 2012, pp. 297–301.

15. B. Kishore, R. VijayaArjunan, R. Saha and S. Selvan, "Using Haralick Features for the Distance Measure Classification of Digital Mammograms", International Journal of Computer Applications, 2014, pp. 17–21.
16. T.V. Pham, M. Worring and W.M. Smeulders, "Face detection by aggregated Bayesian network classifiers", Pattern Recognition Letters, vol. 23, 2002, pp. 451–461.
17. J. Yang, D. Zhang, A.F. Frangi and J. Yang, "Two-Dimensional PCA: A New Approach to Appearance-Based Face Representation and Recognition", IEEE Transactions on Pattern Analysis and Machine Intelligence, vol. 26, no. 1, Jan'2004, pp. 131–137.
18. A.M. Martinez and R. Benavente, "The AR Face Database", CVC Technical Report, no. 24, June 1998.
19. Yale Face Database, Computer Vision Laboratory, Computer Science Engineering Department of UCSD. http://vision.ucsd.edu/datasets/.
20. The ORL Database of Faces. Our Database of Face available at http://www.cl.cam.ac.uk/research/dtg/attarchive/facedatabase.html. 1994.
21. R. Senthilkumar and R.K. Gnanamurthy, "A Comparative Study of 2DPCA Face Recognition Method with Other Statistically Based Face Recognition Methods", J. Inst. Eng. India Ser. B (September 2016) 97(3):425–430. https://doi.org/10.1007/s40031-015-0212-6.
22. Senthil Face Database, Senthilkumar face database (ver1.0). http://www.face-rec.org/databases/.
23. R. Senthilkumar and R.K. Gnanamurthy, "A New Approach in Face Recognition: Duplicating Facial Images Based on Correlation Study", in: ACM Proceedings, The 11th International Knowledge Management in Organizations (KMO 2016) Conference on the changing face of Knowledge Management Impacting Society, July 2016, pp. 1–6. http://dx.doi.org/10.1145/2925995.2926032.
24. D. Kumar, R. Singh, A. Kumar and N. Sharma, "An Adaptive Method of PCA for Minimization of Classification Error Using Naïve Bayes Classifier", Procedia Computer Science, vol. 70, 2015, pp. 9–15.
25. W. Quarda, H. Trichili, A.M. Alimi and B. Solaiman, "Combined local features selection for face recognition based on Naïve Bayesian classification", the IEEE International Conference on Hybrid Intelligent Systems, Dec'2013.
26. B.E. Boser, I.M. Guyon, and V.N. Vapnik, "A training algorithm for optimal margin classifiers", In 5th Annual ACM Workshop on COLT, 1992, pp. 144–152.
27. C. Cortes and V. Vapnik, "Support-vector networks," Machine learning, vol. 20, pp. 273–297, 1995.
28. P. Pallabi and T. Bhavani, "Face Recognition using Multiple Classifiers", in the Proceedings of the 18th IEEE International Conference on Tools with Artificial Intelligence, 2006, pp. 1–8.
29. T.H. Le and L. Bui, "Face Recognition Based on SVM and 2DPCA", International Journal on Signal Processing, Image Processing and Pattern Recognition, vol. 4, no. 3, Sep'2011, pp. 85–94.
30. L.S. Oliveira, A.L. Koerich and M. Mansano, "2D Principal Component Analysis for Face and Facial-Expression Recognition", the IEEE Magazine on Computing In Science & Engineering, Mar'2011, pp. 2–6.
31. M.F. Karaaba, O. Surinta, L.R.B. Schomaker and M.A. Wiering, "Robust Face Identification with Small Sample Sizes using Bag of Words", In the Proceedings of the 11th Joint Conference on Computer Vision, Imaging and Computer Graphics Theory and Applications, vol. 4, July 2016, pp. 582–589.
32. J.S.R. Jang, "Fuzzy Modeling Using Generalized Neural Networks and Kalman Filter Algorithm", Proc. of the Ninth National Conf. on Artificial Intelligence, July 1991, pp: 762–767.
33. J.S.R. Jang, "ANFIS: Adaptive-Network-based-Fuzzy Inference Systems", IEEE Transactions on Systems, Man and Cybernetics, vol. 23, no. 3, May 1993, pp. 665–685.
34. P. Sukumar, R.K. Gnanamurthy, "Computer Aided Detection of Cervical Cancer Using Pap Smear Images Based on Adaptive Neuro Fuzzy Inference System Classifier", in J. Medical Imaging and Health Informatics, vol. 6, no. 2, April 2016, pp. 312–319.
35. M. Grgic, K. Delac, S. Grgic, S, "SCface–surveillance cameras face database", J. Multimedia. Tools Appl., Vol. 51, 3, 863–879, 2011.

36. P.J. Phillips, H. Moon, P.J. Rauss, S. Rizvi, "The FERET evaluation methodology for face recognition algorithms", IEEE Trans. On Pattern Anal. Mach. Intell., vol. 22, no. 10, Oct'2000, pp. 1090–1104.
37. D.G. Lowe, "Distinctive Image Features from Scale-Invariant Keypoints", International Journal of Computer Vision, vol. 60, no. 2, 2004, pp. 91–110.

# An Adaptive Java Tutorials Using HMM-Based Approach

Aditya Khamparia and Babita Pandey

**Abstract** In the present scenario, higher education is one of the most important requirements. To facilitate the higher education, E-Learning systems are best substitute of tradition education system. It outstrips the distance and time constraints. E-Learning can be used not only for collaborative learning, but also for autonomous, individual learning. E-Learning can also provide another concept that is called learning by teaching. In this paper, we have proposed a Web-based tool named as Adaptive Java Tutorials that has used Hidden Markov Model to make E-Learning more adaptive. The development of such system is a challenging task that could support both learning by teaching and learning environment. Adaptive Java Tutorials are an attempt to develop a Web tool that could quench the intentions of both learning and learning by teaching systems. The teacher can learn from the reviews provided by learner. The learner's experience is used to train the Adaptive Java Tutorials. Hidden Markov Model is used to make the system adaptive. When a concept is browsed by the learner, it will train the system based on the reviews provided by the learner. It will prioritize concepts that are best reviewed or most visited.

**Keywords** E-Learning · Learning by teaching · Adaptive Java Tutorials HMM

A. Khamparia (✉)
Department of Computer Science and Engineering, Lovely Professional University, Jalandhar 144411, India
e-mail: aditya.khamparia88@gmail.com

B. Pandey
Department of Computer Applications, Lovely Professional University, Jalandhar 144411, India

© Springer Nature Singapore Pte Ltd. 2019
B. K. Panigrahi et al. (eds.), *Smart Innovations in Communication and Computational Sciences*, Advances in Intelligent Systems and Computing 669, https://doi.org/10.1007/978-981-10-8968-8_9

101

# 1   Introduction

E-Learning is one of the modernistic subjects in the today's digital world. E-Learning facilitates the adaptive approach of learning. E-Learning is providing educational concepts accessible to each and every person at any time. In the traditional education system, the education was provided as one-way process. The teaching is centered on the instructor only. But in E-Learning system, the education is provided according to intention of the learner. Learner is the main core of E-Learning. E-Learning is learners' dominant discipline. The learner can select any mode of learning according to the priorities, interest. The user can also evaluate its learning process by means of quiz and can provide the feedback about the study material at the same time. The student can rate the data for improving the quality of the data.

E-Learning provides the concept of learning, i.e., learning by teaching. In this approach, the instructor provides the online content for study. The learner can pick any topic of their choice. The learner can rate the topic. The rating can help the instructor to improve the quality of content. Adaptive Java Tutorial has been developed to gratify the concept of learning by teaching. Adaptive Tutorial carters the study material on the basis of the user requirements. The user can select the study can elect a topic to study after login only that topic is provided to the user. The rest of the topics are not visible to the user. On the basis of topic elected, the student can be provided suggestions to study the next topic. The topic suggestion is provided on the basis of the topic elected by student. The next topic suggestion is calculated by using parameters in the Hidden Markov Model.

In the Adaptive Java Tutorials, there are three major modules: admin module, instructor module, and learner module. The admin is responsible for managing the learner and instructor module. The instructor can add the topic under the domain. The instructor also entertains the learners' review regarding particular learning material. The learning material is revised on the basis of the student reviews. The learner module is provided with the learning material. The learner can grasp the knowledge from the material and can rate the data as best, good, average, bad, and worst. The learner's feedback is used to review the data and improve its quality. The fore-mentioned modules are implemented by using Hidden Markov Model.

# 2   Related Works

HMM had been proposed in the series of papers by Leonard E. Baum in mid-1960. HMM had been successfully used in the various fields like human identification using gait, human action recognition from time sequential images, facial expression identification from videos, speech recognition, E-Learning, etc. Hogyeong Jeong had summarized that HMM can be used in metacognitive prompting in the learning environment to determine the students' pattern activities [1].

According to Gautam Biswas, the teaching agents can be used to promote the learning environment by collaborating learning by teaching with self-regulation mentoring. It can be used to train the teaching agent via learning by teaching task [2].

John Wagster had devised a system that can be trained by using learning by teaching with metacognitive support. The teaching prepared the topics by integrating the structured knowledge and then trains the teacher [3].

Morteza Saberi Anari had reported that artificial intelligence can be used to improve and train the electronic education system by using set of actions performed by the user [4].

Lee, M. G had highlighted the Web-based instruction system that could be used as an alternative education aspect to compass the various difficulties of the traditional education system [5].

Kelly and Tangney had summarized that each learner has their own learning style. The learner's grasping also varies from learner to learner. They have emphasized on adaptive E-Learning profile that can gratify the learning behavior of learner [6].

# 3 Hidden Markov Model

Hidden Markov Model is a statistical and stochastic Markov model that speculates Markov model along hidden states. Leonard E. Baum had introduced HMM in the series of statistical papers in the mid-1960. It is represented by using dynamic Bayesian network. It expresses output dependent on the probability of the state rather than state itself, unlike Markov model. Each output has probability associated with it. The transitions among states are governed by set of probabilities called transition probabilities. The states are undercover in this model; hence, it is called Hidden Markov Model. HMM use two layers: one observation layer and one hidden layer. The observation layer specifies the probability of the state, and the hidden layer is the Markov process that is used to find out transition in the model. There are two types of Hidden Markov Model, i.e., Supervised Hidden Markov Model and Unsupervised Hidden Markov Model. In Supervised Hidden Markov Model, the trainer has complete knowledge about the sequence generated by patterns. The transition emission and emission probabilities are also known. In Unsupervised Hidden Markov Model, the trainer updates the HMM parameter based on the new samples. The transition estimation and emission probabilities are not known.

Hidden Markov Model is represented as quintuple of $(\pi, A, B)$ represented by set of hidden states $Q = \{q_1, q_2, q_3, ..., q_{|Q|}\}$, a set of observations $O = \{o_1, o_2, o_3, ..., o_{|O|}\}$, a set of transition probabilities $A = \{a_{ij} = P(q_j^{t+1}|q_i^t)\}$, and series of output (emission probabilities) $B = \{b_{ik} = P(o_k|q_i)\}$. In other words, $q_i^t$ denotes the event that state is $q_i$ at time t. $a_{ij}$ is the probability that next state is $q_j$ and given current state is $q_i$; $b_{ik}$ is the probability that output is $o_k$ given current state is $q_i$. The initial state probabilities are denoted by $\pi = \{\pi_i = P(q_i^1)\}$ where i lies between 1 and Q.

## 4  Proposed Approach

From the above literature survey, it has been clarified that E-Learning provides the student with the study material according to their interest, profile, and attributes.

The Hidden Markov Model is used in this context because this sphere is parametric in nature but the parameters hidden from HMM have been implemented with the E-Learning, to predict the likelihood of the study material. It also implies that outcomes are not known and probability of that happening can be concluded from the current parameters. The model is stochastic, it is random process which is hidden from the learner, and its result can be calculated by stochastic processes that conclude reaction of considerations.

The proposed methodology focuses on three major sections: In the first section, data is collected and organized in the topics in the Web tool. This section corresponds to expert system that has domain knowledge. In the second section, the HMM parameters are initialized for each and every topic in the module, and in the third section, unknown parameters were computed using Baum Welch Algorithm. The fourth section of an HMM is to find the most likely series of states given an observed series of observations (Viterbi Algorithm).

1. **Add topics**: The preliminary step is to add the topic and subtopics in the Web tool. The topics are added according to the blueprint of the domain created using the ontology. The data is added regarding every topic to clarify the concepts. The data is classified as definition, syntax, and example (if exists).
2. **Assign probabilities to each topic**: The HMM consists of three states: Start, Relevant (R), and Irrelevant (I). The next step is to add the basic probabilities to each topic. There are two states considered for the topic named as R (0) and I (1). The set of observation sequence (0, 1, 2, 3) represents the way in which topics can be read. Zero corresponds fundamentals, 1 corresponds OOPS, 2 corresponds GUI, and 3 corresponds Web programming.

   The HMM parameters are initial state probabilities $\pi = \{\pi_1, \pi_2\}$, state transition probabilities $A = [a_{00}, a_{01}, a_{10}, a_{11}]$, relevant state emission probabilities $B_R = [b_{11}, b_{12}, b_{13}, ..., b_{1N}]$, and irrelevant state emission probabilities $B_I = [b_{01}, b_{02}, ..., b_{0N}]$ as shown in Fig. 1.

   There are four possible associations according to which probabilities are provided: RR, RI, IR, II. The following initial probabilities are provided to topic.

$$\begin{array}{cc} & \begin{array}{cc} R & I \end{array} \\ A = \begin{array}{c} R \\ I \end{array} & \begin{bmatrix} 0.9 & 0.1 \\ 0.2 & 0.8 \end{bmatrix} \end{array} \tag{1}$$

We use two states $Q = \{0, 1\}$ to denote that the user is reading relevant (0) or irrelevant (1) learning material. To define A and B, we use the Baum–Welch algorithm (a.k.a forward–backward algorithm). We call it HMMs Training. The algorithm has two main steps, forward and backward procedures.

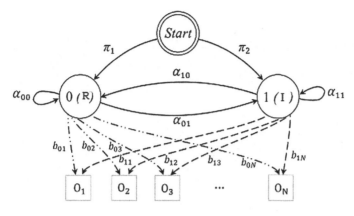

**Fig. 1** Proposed HMM

$$B = \begin{array}{c} \\ R \\ I \end{array} \begin{array}{cccc} 01 & 02 & 03 & 04 \\ \left[ \begin{array}{cccc} 0.5 & 0.3 & 0.1 & 0.07 \\ 0.03 & 0.02 & 0.2 & 0.4 \end{array} \right] \end{array} \tag{2}$$

$$\begin{array}{cc} R & I \\ \pi = [0.7 & 0.3] \end{array} \tag{3}$$

$$O = (0, 1, 2, 3) \tag{4}$$

3. **Re-estimation of probabilities using Baum–Welch Algorithm**: Let $\lambda = (\pi, A, B)$ with random initial conditions. The algorithm updates the parameters of $\lambda$ iteratively until convergence, following the next procedures. The re-estimation is done on the basis of Baum–Welch Algorithm. It calculates the forward probability ($\alpha$) as well as backward probability ($\beta$), and then both are summed to calculate new probability.

   In forward procedure, $\alpha_i^t = P(O^1, ..., O^t, Q^t = q_{i|} \lambda)$, the computation of $\alpha_i^t$ can be done recursively as $\alpha_i^1 = \pi_i b_i(O^1)$ and $\alpha_j^{t+1} = b_j(O^{t+1})$. $\alpha_i^t \alpha_{ij}$.

   In backward procedure, $\beta_i^t$ is the probability of ending partial observation sequence $O^{t+1}, ..., O^t$. $\beta_i^T = 1$ and $\beta_i^t = \beta_j^{t+1}$. $\alpha_{ij} b_j(O^{t+1})$.

3.1 **Update data**: On the basis of user rating and the re-estimated probability values, the quality of data is revised to impart best data to user. The best and quality content is carter to the learner. The feedback fosters the learning process by rating the quality of data. The instructor aims at providing quality and easy content to student that can help student to learn the concepts quickly and easily as well as help instructor to set the good examples among learner and enhance their knowledge by devoting more efforts on the content to make it quite easy and simple.

$$\gamma_i^t = P(Q^t = q_i | O, \lambda) = \frac{\alpha_i^t \beta_i^t}{\sum_{j=1}^{N} \alpha_j^t \beta_j^t}$$

$$\xi_{ij}^t = P\left(Q^t = q_i, Q^{t+1} = q_j | O, \lambda\right) = \frac{\alpha_i^t \alpha_{ij} \beta_j^{t+1} b_j(O^{t+1})}{\sum_{i=1}^{N} \sum_{j=1}^{N} \alpha_i^t \alpha_{ij} \beta_j^{t+1} b_j(O^{t+1})}$$

$$\pi_i = \gamma_i^1$$

$$\alpha_{ij} = \frac{\sum_{t=1}^{T-1} \xi_{ij}^t}{\sum_{t=1}^{T-1} \gamma_i^t}$$

$$b_{ik} = \frac{\sum_{t=1}^{T} \delta(O^t, o_k)\gamma_i^t}{\sum_{t=1}^{T} \gamma_i^t}$$

4. **Finding the optimal state sequence using Viterbi Algorithm**: One of the most common queries of an *HMM* is to find the most likely series of states given an observed series of observations. In our case, we can find the state sequence (e.g., IRIR) that most likely happens given the observation sequence. Here, observation 1 is fundamentals, 2 is OOPS, 3 is GUI, and 4 corresponds to Web programming as shown in Fig. 2.

If the value of the P (O| $\lambda$) increases, then re-estimation of probabilities is done again. The alpha, beta, gamma values are calculated again using above-mentioned formulas. The highest values denote that what will be the probability of deciding that based on given observations whether the topic is relevant or irrelevant for user.

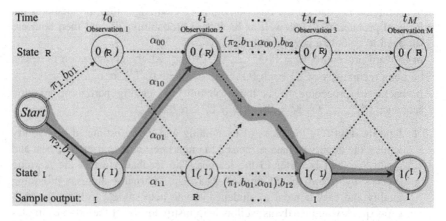

**Fig. 2** State sequence by Viterbi Algorithm

## 5 Implementation

Adaptive Java Tutorial is a Web-based tool that is implemented by using the PHP as shown in Fig. 3. The index page has two login forms, i.e., one for the learner's login and the other for the instructor's login. When the instructor logged in, he/she can upload the topics and subtopics in the domain, and the instructor can review the feedback provided by the learner. When the user logged in as learner, the learner is asked about the topics he/she wants to study. Then, those topics are displayed on the Web tool. The user can study the topic and can rate it. The data is arranged under headings named as: purpose, syntax, and example. At the end, option to rate the data is provided. As soon, the user rates the topic, clicks on the submit button, according to the aforementioned algorithm, and then receives a notification to update the data. The user can select at most four topics to study. If the user selects only one module, the rest of the parameter in the algorithm is set to nil. The user can read out any subtopic under the main topic. The user sequence is monitored by the Web tool. This sequence is used to train the system. The student is the centric entity in this system.

The user can pick any topic of its choice. There is no restriction to follow a sequence, and ultimately, it leads to save time of the user. This feature contrasts the traditional learning environment where user has to follow the proper sequence to reach particular topic. E-Learning can also carter the content in no time, hence widely acceptable in today's fast world. The third module is admin module. The admin module is used to manage the instructor and the learner. The admin can add and delete the instructor as well as learner. The admin is responsible to monitor activities performed by the learner and instructor. The admin can manage the accounts of both categories of user. The admin is able to suspend the learner and instructor from the tool on misconduct.

**Fig. 3** Index page of Adaptive Java Tutorial

## 5.1 Empirical Study

The purpose of this experimental research is to examine whether Adaptive Java Tutorials are able to reduce cognitive load, level of motivation, and degree of technology acceptance. To evaluate the Knowledge Analyzer and Adaptive Java Tutorials, a test has been conducted before and after the implementation of Web tool. We focused on investigating the effect of using tool on the cognitive load, level of motivation, and degree of technology acceptance. Pretest and posttest are used as research tools. The pretest is conducted to evaluate the prior knowledge of the learner. It was designed by head of the department. Sixty students are divided into two groups, namely experimental group and control group. The students of the experimental group are taught with the adaptive system, and control systems are taught traditionally. The students are evaluated before experiment and after experiment with the help of pretest, posttest, and questionnaire (Fig. 4).

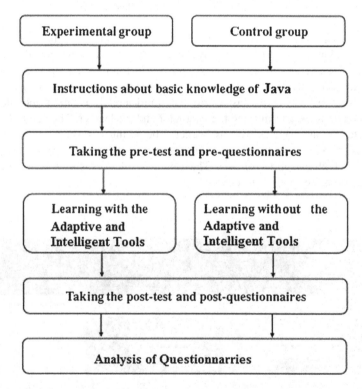

**Fig. 4** Process of experimental design

**Table 1** Pre-t-test result of the learning motivation pre-post questionnaire ratings of the two groups

| Group | N | Mean | S.D | t-test |
|-------|---|------|-----|--------|
| Control | 30 | 3.83 | 0.98 | 1.34 |
| Experimental | 30 | 4.3 | 0.95 | |

**Table 2** Pre-t-test result of the cognitive load pre-post questionnaire ratings of the two groups

| Group | N | Mean | S.D | t-test |
|-------|---|------|-----|--------|
| Control | 30 | 5.33 | 0.96 | 0.0012 |
| Experimental | 30 | 5.09 | 0.94 | |

**Table 3** Pre-t-test result of the technology acceptance pre-post questionnaire ratings of the two groups

| Group | N | Mean | S.D | t-test |
|-------|---|------|-----|--------|
| Control | 30 | 5.63 | 0.89 | 0.052 |
| Experimental | 30 | 5.33 | 0.85 | |

Above table depicts the results of the pretest conducted to measure the cognitive skills, learning motivation, and technology acceptance degree of the students. From the above tables, it is concluded that our experiment is able to reduce the cognitive skills and enhance motivation and technology can be accepted easily. The mean and standard deviation of experimental and control groups in Table 1 are 4.3, 0.95 and 3.83, 0.98, respectively. The mean and standard deviation of experimental and control groups in Table 2 are 5.33, 0.96 and 5.09, 0.94, respectively. The mean and standard deviation of experimental and control groups in Table 3 are 5.63, 0.89 and 5.33, 0.85, respectively. The t-test is used here to find out the dependency between the two classes. The t-test is less than p at 5% of the significance level. It concludes that the level of motivation is enhanced by fore-mentioned tools (Figs. 5 and 6).

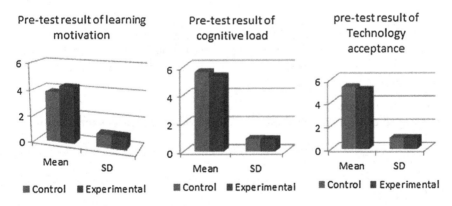

**Fig. 5** Graphical representation of pretest results

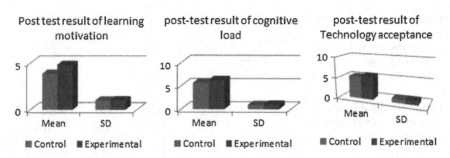

**Fig. 6** Graphical representation of posttest results

## 6  Conclusions

HMM can be used to implement the learning by teaching Web tool. It helps in calculating the likelihood of the topics as well as it helps to find out what shall be the probability of navigation in particular manner. It caters the two main aspects E-Learning and learning by teaching. It fosters the learning process by considering learner as a centric entity and delivers the lecture demanded by them in an efficient manner. It aims at making the learning process as best as possible. It also suggests next topics to be studied by the user. Another aspect focuses on the learning by teaching aspect. It helps the instructor in self-regulating and self-monitoring to groom their knowledge. It instructs the student as well as learns itself by teaching. It contributes a lot in education-imparting process. It tends to mimic the behavior of human instructor. This helps the instructor in understanding the concepts of Java domain as well conveying the same concepts to the students in an effective way. It provides the abstraction in imparting the concepts in the structured and well-recognized compact format.

*Ethics Approval and consent to participate*

All procedures performed in studies involving human participants were in accordance with the ethical standards of the institutional and/or national research committee and with the 1964 Helsinki declaration and its later amendments or comparable ethical standards.

*Consent for Publication*

Informed consent was obtained from all individual participants included in the study.

## References

1. A. Fok, H.S. Wong, and Y. S. Chen, "Hidden Markov model based characterization of content access patterns in an E-Learning environment", *IEEE International conference on Multimedia and Expo*, pp. 201–204, 2005.

2. Biswas, G., Leelawong, K., Schwartz, D., Vye, N., TAG-V: Learning By Teaching: A New Agent Paradigm for Educational Software. In: Applied Artificial Intelligence. AAI, vol. 19, pp. 363– 392 (2005).
3. Wagster, J., Tan, J., Wu, Y., Biswas, G., Schwartz, D: Do Learning by Teaching Environ-ments with Metacognitive Support Help Students Develop Better Learning Behaviors? The 29th Annual Meeting of the Cognitive Science Society, Nashville, TN, pp. 695–700. (2007).
4. M.S. Anari, S Anari, "Intelligent E-Learning Systems using Student Behavior Prediction", *Journal of basic applied sciences*, pp. 12017–12023, 2012.
5. Yaghobi Mousavi, Saeed, Pazuki, Iraj, E-Learning and its impact on educational system, Islamic AzadUniversity Press, South Tehran Branch, 2008.
6. Kelly, D. and B. Tangney., "Adapting to intelligence profile in an adaptive educational system." Interacting with computers: 85–409, 2006.

# KMST+: A K-Means++-Based Minimum Spanning Tree Algorithm

**Sabhijiit Singh Sandhu, B. K. Tripathy and Shivansh Jagga**

**Abstract** There are a number of algorithms that have been proposed in the graph theory literature to compute the minimum spanning tree of a given graph. These include the famous Prim's and Kruskal's algorithm, among others. The main drawback of these algorithms is their greedy nature, which means they cannot be applied to large datasets. In 2015, Zhong et al. proposed a fast MST (FMST) algorithm framework that uses K-means to find the MST with a reduced complexity of $O(N^{1.5})$. In this paper, we have introduced an improved version of the FMST algorithm by using K-means++ to further increase the efficiency of the FMST algorithm. The use of K-means++ instead of K-means results in lesser complexity, faster convergence, and more accurate results during the clustering step which improves the accuracy of the MST formed.

**Keywords** FMST · MST · K-means++ · Binary heap

## 1 Introduction

The concept of minimum spanning tree was introduced by Otakar Boruvka [10] in 1926. New algorithms for solving the same were later proposed, most notable of which were given by Prim [11] (in 1957) and Kruskal [9] (in 1956). MSTs have found extensive use in both the literature for solving complex graph theory problems such as the travelling salesman problem and in real-world applications such as telecommunications for laying cables.

S. S. Sandhu (✉) · B. K. Tripathy · S. Jagga
School of Computing Science and Engineering, Vellore Institute of Technology,
Vellore 632014, Tamil Nadu, India
e-mail: sabhijiit@gmail.com

B. K. Tripathy
e-mail: tripathybk@vit.ac.in

S. Jagga
e-mail: shivansh.jagga2013@vit.ac.in

© Springer Nature Singapore Pte Ltd. 2019
B. K. Panigrahi et al. (eds.), *Smart Innovations in Communication
and Computational Sciences*, Advances in Intelligent Systems
and Computing 669, https://doi.org/10.1007/978-981-10-8968-8_10

MSTs find extensive use in various fields such as image segmentation, manifold learning, clustering, classification, as they can be used to approximately estimate the inherent structure of a dataset. Despite their wide applications, the drawback these algorithms face is their high complexity, which is of the order $O(N^2)$. This reduces their scalability to large-sized datasets. To overcome this, a recently proposed framework, the FMST by Zhong et al. [13], suggests the use of K-means and a divide-and-conquer-based approach to split the dataset into smaller parts to approach the construction of the MST. This reduces the complexity to $O(N^{1.5})$, allowing its application to much larger datasets.

Clustering is one of the principal problems in the field of data science and is central to machine learning. Clustering algorithms can be classified into numerous types depending on the cluster model used, with there being approximately as many as hundred different algorithms that have been proposed till date. Still, the K-means algorithm, proposed by Lloyd in 1957, remains the most popular clustering method. Its popularity stems from the simplicity it offers. The procedure presents an elementary method to classify a given dataset into "K" clusters. The idea is to choose k initial centroids at random and then to take the remaining data points and assign them to the nearest centroid. The centroids are then recomputed on the basis of the current cluster assignment. This procedure is repeated till the residual sum of squares function (objective function) is not minimized and the solution does not change for two consecutive rounds.

The major drawbacks the K-means algorithm faces are its lack of efficiency and quality along with its likelihood of converging to a local optimum, rather than the global optimum which turn out to be counterintuitive. Nevertheless, K-means remains unmatched in terms of both speed and simplicity, and as a result, further work has focused on finding a better method of initialization which increases the performance in terms of quality as well reduces the time taken for convergence.

By augmenting K-means with an improved initialization technique, Arthur et al. [1] in 2007 proposed the K-means++ algorithm which was $O(\log K)$ competitive with the optimal solution and often produced quite sizeable improvements, both in accuracy and speed of K-means. The idea here is to choose the first centroid at random; however, for choosing the remaining "K-1" centroids, the preference is given to the point that lies farthest from the previously selected centers. This algorithm was meant to exploit the fact that a good clustering is always spread out.

## 1.1  Our Contribution

The FMST framework given in [13] computes an approximate MST, and even though it is close to the true MST, it is something that can be improved upon. To make an attempt at the same, we propose using the K-means++ algorithm instead of the K-means so as to increase the accuracy of clustering and hence get a more accurate MST. The paper also shows the results of an implementation of the FMST framework—using binary heaps with adjacency lists where the clustering step uses

K-means++ on real-world datasets, and finally, the experimental results are compared with the K-means-based FMST.

Our key observations in the experiments are:

- The K-means++ implementation is considerably faster than the K-means-based implementation of the FMST algorithm.
- The cost of the minimum spanning tree (i.e., sum of weights of edges) is considerably closer to that of the one obtained by an exact clustering algorithm such as Kruskal's algorithm.

## 2 Related Work

The first work related to construction of minimum spanning trees was the Borvuka's algorithm [10]. The algorithm comprises of a sequence of "Boruvka steps," which are as follows. All the vertices are initialized as individual sets, and while there are more than one sets remaining, for each component, the closest weight edge joining it to another set is found and if not already added to MST, it is done so. Boruvka's algorithm has a complexity of O(|E| log |V|) (where E is the number of edges and V is set of vertices) because each Boruvka step takes linear time and the number of vertices is reduced by at least half in each step.

Among the classic algorithms, Prim's [11] and Kruskal's [9] algorithms are also present. Prim's grows the MST one edge at a time by adding the least weight edge connecting a new vertex to the tree, till no vertices are left. Both Prim's and Kruskal's algorithms also have a complexity of O(|E| log |V|).

There have been some fast MST algorithms also that were proposed. For a sparse graph, algorithms with a complexity of O(|E| log log |V|) were proposed by Yao [12] and Cheriton and Tarjan [3]. Fredman and Tarjan [4] posited using the Fibonacci heap to implement a priority queue for constructing an exact MST, thereby reducing the computational complexity to O(|E| $\beta$ (E, V)) where $\beta(E, V)$ =min{i log$^{(i)}$ V <= E/V}. Gabow et al. [5] further incorporated the idea of packets into the previous work and reduced the complexity to O(|E| log $\beta$ (E, V)). Most recently, Zhong et al. [13] proposed the FMST algorithm which is the center of our paper's discussion.

In clustering, Stuart Lloyd introduced the K-means algorithm in 1957. There have been a number of papers that describe O(1 + $\varepsilon$)-competitive algorithms for the K-means problem that are essentially unrelated to Lloyd's method. Unfortunately, all these methods have their drawbacks and hence are not at all viable in practice. Some are highly exponential in "K", while others are too slow in practice. The K-means++ algorithm that was O(log K) competitive was independently developed by Arthur and Vassilvitskii [1] in 2007.

# 3 Algorithm

In this section, we present our proposed algorithm for finding the MST of a graph. We begin by describing the FMST algorithm from [13] followed by a description of the K-means++ and binary heap algorithms (Sect. 3.1). Then, we present our algorithm, the KMST+, for constructing a minimum spanning tree (Sect. 3.2). Finally, we show experimental results of the performance of our algorithm on some real-world datasets and compare it with that of [13] (Sect. 3.3).

## 3.1 Background

The FMST algorithm employs a divide-and-conquer method to produce an approximate minimum spanning tree with a reduced complexity of $O(N^{1.5})$. It does so in two stages. The first stage, the divide-and-conquer stage, uses K-means to split up the dataset into $\sqrt{N}$ clusters. Then, an exact MST algorithm (such as Prim's or Kruskal's) is applied to each cluster and the $\sqrt{N}$ MSTs obtained as a result are then connected using a specified criterion to form the first approximate MST. In the refinement stage, which is the second stage, clusters produced in the first stage form $\sqrt{N} - 1$ neighboring pairs, and the dataset is then repartitioned into as many clusters. This is done to compensate for boundary effects by partitioning the boundary of the given pair into clusters. With these, $\sqrt{N} - 1$ clusters, a second new approximate MST is generated. A final approximate MST is then generated from the graph formed by combining the first two approximate MSTs.

The refinement stage is done to compensate for any boundary effects that may arise, wherein data points close to the boundaries are prone to be misconnected. Also, we chose $\sqrt{N}$ clusters for two reasons. First is to satisfy the locality property when constructing an MST. Second is that for $K = \sqrt{N}$, the overall time complexity of [13] is minimized.

The K-means++ algorithm was obtained when Arthur and Vassilvitskii modified the initialization step of the K-means algorithm. Instead of random initialization, here, they choose the initial centroids in a more systematic way. The first centroid is chosen at random, and the current choice of centers is then used to stochastically bias the choice of the next center. This is followed by Llyod's iteration to perform the clustering. The algorithm was shown to be $O(\log k)$ competitive with the optimal algorithm.

For implementing the Kruskal's algorithm using binary heaps, we use an adjacency list representation along with a Min Heap. Here, we maintain two disjoint subsets of vertices—one set has a list of vertices that are already included in the MST, while the other set contains vertices not yet included. With adjacency list representation, BFS takes $O(|V| + |E|)$ time to traverse all the vertices of a given graph. The idea is to perform traversal using breadth-first search and to store the vertices that are not yet included in a Min Heap. The priority queue is implemented using a Min Heap to get the least weighted edge of the cut. Min Heap is used because it has a time complexity of $O(\text{Log } |V|)$ for operations like extracting minimum element and decreasing key value. As a result, the overall complexity of performing the MST operation becomes $O(|E| \log |V|)$.

## 3.2 Synopsis of the Proposed Algorithm

Our algorithm is largely based on the framework proposed in [13]; however, it differs in the fact that we use K-means++ to perform clustering. The motivation is to reduce the number of unnecessary comparisons as much as possible, and clustering is performed initially to fragment the dataset into smaller parts so as to achieve the same.

The divide-and-conquer-based algorithm is as follows:

1. Divide-and-conquer stage

   a. Divide step: Apply K-means++ to partition dataset of N points into $\sqrt{N}$ clusters.
   b. Conquer step: Construct an MST for each subset separately using the binary heap-based Prim's algorithm.
   c. Combine step: The individual MSTs are then combined using a connection criterion to give a primary approximate MST.

2. Refinement stage

   a. Partitioning is again carried out but this time, with a focus on the boundaries of the clusters constructed in the previous stage.
   b. The conquer and combine steps are used to construct a secondary approximate MST.
   c. The two approximate MSTs are then merged to obtain a new more accurate MST from the resulting graph using the binary heap version of Prim's algorithm.

## 3.3 Using K-Means++ for Partitioning

We use K-means++ for partitioning to preserve the locality property of the edges in the MST as it will partition locally related data points into the same clusters. The only requirement of the algorithm is that the number of clusters is to be specified.

The algorithm is:

1. The first centroid is chosen uniformly at random from the dataset.
2. For each data point x, compute the distance from it to the nearest center that has already been chosen.
3. Choose a new data point as the next new centroid. This is done using weighted probability distribution where the probability of a point x being chosen is $(D(x)^2/\Sigma_{x \in X} D(x)^2)$. This weighting is called "$D^2$ weighting".
4. Repeat steps 2 and 3 till all centroids have been initialized.
5. Proceed with the standard K-means algorithm to perform clustering.

## 3.4 Constructing MST Using Binary Heap-Based Implementation of Kruskal's Algorithm

Since the application of Kruskal's algorithm using binary heaps takes $O(|E| \log |V|)$ time to construct the MST, it has been used in the algorithm to ensure maximum possible efficiency.

The algorithm is as follows:

1. A simple connected graph is taken as input.
2. Partition the set of vertices into two.
3. Use the weights of the edges to construct a minimum heap.
4. Remove the root node (i.e., least weight) of the heap and extract its corresponding vertices (V).
5. Add the edge (E) in the MST if doing so does not create a loop.
6. Reconstruct the heap.
7. Repeat steps 3, 4, and 5 until $|E| = |V| - 1$.

## 3.5 Combining MSTs of the Subsets

As given in [13], using an MST to determine the neighboring subsets is faster than using brute force to achieve the same. Also, an intuitive connection criterion was proposed to connect adjacent clusters for a low computational cost, though with the drawback of not always finding the optimal connections. This, however, is remedied using the refinement stage.

## 4 Experiments

In this section, we present our experimental analysis showing the evaluation of KMST+ and a comparison of its performance to the FMST algorithm and the Kruskal's algorithm.

The algorithms were run on a PC with the Intel Core i5-3230 M CPU 2.60 GHz with a 6 GB memory running a Windows 8 OS. The code for the algorithm was written in python and executed using Spyder IDE for python given in the Anaconda python distribution.

### 4.1 Datasets

We use the t4.8 k dataset for performing the experimental analysis. It is a synthetic dataset that was designed for testing the CHAMELEON algorithm [8] and is described as follows: The t4.8 k dataset has six clusters of different size, shape, and orientation, as well as random noise points and special artifacts such as streaks running across clusters and 8000 data points. The dataset is available from [7], and the image is available at [6] (Fig. 1).

### 4.2 Performance Metric

We use two metrics to evaluate the performance. They are as follows:

- **COST**: The cost of the MST is computed. This is the sum of the weights of the edges included in the minimum spanning tree. The lower the cost, the better, and the aim is to get it as close to the cost of the MST from the exact Kruskal's algorithm.
- **EXECUTION TIME**: The time taken to return the computed MST. This calculates the time from clustering to construction of the final approximate MST.

Table 1 gives the costs and execution times of the minimum spanning trees produced by an exact MST algorithm (Kruskal's algorithm) for varying values of N from the t4.8 k dataset (Fig. 2).

The following tables, Tables 2, 3, and 4, show a comparative analysis of the performance of the proposed KMST+ algorithm with the FMST framework [13] for different values of K with varying N.

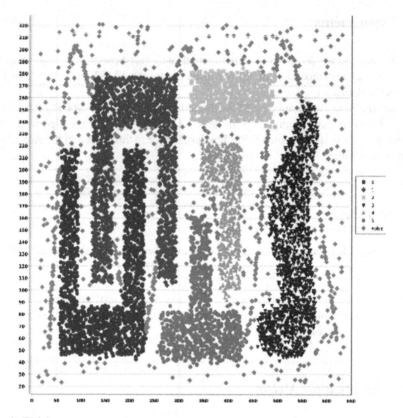

**Fig. 1** T4.8 k

The two algorithms have been compared with respect to the time taken to compute the minimum spanning tree for a given set of values of K and N (Figs. 3 and 4).

## 4.3 Result Analysis

It is observed that the proposed algorithm succeeds in fine-tuning the results by returning better costs in lesser time as is evident from Tables 2, 3, and 4. Both the time taken and cost of MST formed are lesser in most cases. We also see that the cost of MSTs produced is very near to the exact MST costs obtained from Kruskal's algorithm, while the times taken are much less.

**Table 1** Results for Kruskal's algorithm on t4.8 k

| Kruskal's algorithm | | N = 100 | N = 200 | N = 300 | N = 400 | N = 500 | N = 600 | N = 700 | N = 800 | N = 900 | N = 1000 |
|---|---|---|---|---|---|---|---|---|---|---|---|
| | Cost | 2307.2 | 3244.5 | 3911.5 | 4464.1 | 5068.9 | 5523.2 | 5993.8 | 6407.4 | 6713.1 | 7040.6 |
| | Time | 1.3404 | 10.603 | 36.305 | 91.14 | 180.23 | 305.94 | 486.25 | 727.92 | 1031.9 | 1306.7 |

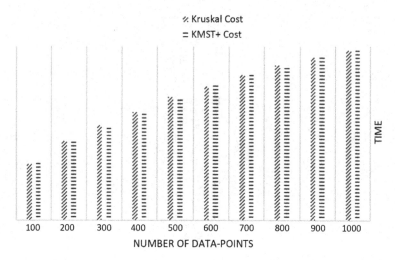

**Fig. 2** Cost-wise comparison of Kruskal's algorithm versus KMST+ algorithm

It can be noticed that our algorithm is outperformed by [13] in some instances; however, the occurrence of such is minimal. This happens because K-means++ is O(log k) competitive with the optimal method and does not guarantee the better results every time.

Also, with increase in value of K, the cost can be seen to increase. This is because t4.8 k ideally has six clusters and a different value of K leads to misrepresentation. However, the execution time is reduced drastically with K > 6.

## 4.4 Discussion on Effect of Size of N

Better initialization allows K-means++ not only to return better results, but also to converge faster and as a result, to have lesser run time. However, due to its sequential nature of initialization, as the size of data increases, so does the run time. Consequently, the complexity increases.

Also, as N increases, the gap between execution times can mostly be seen to widen, with KMST+ increasingly outperforming [13].

**Table 2** Results for t4.8 k with K = 4 and varying N

| | | N = 100 | N = 200 | N = 300 | N = 400 | N = 500 | N = 600 | N = 700 | N = 800 | N = 900 | N = 1000 | N = 1500 |
|------|------|---------|---------|---------|---------|---------|---------|---------|---------|---------|----------|----------|
| KMST+ | Cost | 2307.2 | 3244.5 | 3914.3 | 4467.9 | 5072.8 | 5523.2 | 6000.8 | 6407.4 | 6713.1 | 7047.3 | 8731.8 |
| | Time | 0.535 | 2.44 | 7.799 | 22.33 | 38.69 | 78.78 | 127.77 | 181.76 | 263.51 | 281.49 | 1291.3 |
| FMST | Cost | 2340.1 | 3262.8 | 3914.5 | 4509.8 | 5068.9 | 5551.5 | 6000.8 | 6414.3 | 6720.1 | 7040.6 | 8731.8 |
| | Time | 0.6287 | 2.9099 | 8.8455 | 23.985 | 37.83 | 80.634 | 130.05 | 190.25 | 272.93 | 301.21 | 1286.5 |

**Table 3** Results for t4.8 k with K = 8 and varying N

| | | N = 100 | N = 200 | N = 300 | N = 400 | N = 500 | N = 600 | N = 700 | N = 800 | N = 900 | N = 1000 | N = 1500 |
|---|---|---|---|---|---|---|---|---|---|---|---|---|
| KMST+ | Cost | 2307.7 | 3278.4 | 3919.4 | 4471.9 | 5085.4 | 5534.3 | 6098.8 | 6411.9 | 6843.4 | 7112.8 | 8775.3 |
| | Time | 0.3896 | 1.1994 | 2.7327 | 5.0768 | 10.153 | 15.101 | 23.254 | 33.638 | 53.074 | 66.314 | 212.01 |
| FMST | Cost | 2337.5 | 3300.9 | 3976.7 | 4533.8 | 5091.7 | 5664.6 | 6107.1 | 6411.9 | 6769.5 | 7101.2 | 8797.8 |
| | Time | 0.443 | 0.98 | 2.88 | 5.9261 | 9.3981 | 17.898 | 24.231 | 35.077 | 47.096 | 70.227 | 231.65 |

**Table 4** Results for t4.8 k with K = 12 and varying N

| | | N = 100 | N = 200 | N = 300 | N = 400 | N = 500 | N = 600 | N = 700 | N = 800 | N = 900 | N = 1000 | N = 1500 |
|---|---|---|---|---|---|---|---|---|---|---|---|---|
| KMST+ | Cost | 2307.7 | 3273.1 | 3945.6 | 4502.5 | 5086.8 | 5590.7 | 6023.4 | 6461.2 | 6738.1 | 7094.7 | 8755.5 |
| | Time | 0.4127 | 0.8914 | 1.8017 | 3.3469 | 6.2955 | 8.6953 | 13.762 | 21.582 | 22.897 | 33.276 | 108.12 |
| FMST | Cost | 2385.1 | 3309.2 | 3953.3 | 4536.6 | 5144.1 | 5683.9 | 6088.2 | 6457.8 | 6744.4 | 7132.3 | 8831.8 |
| | Time | 0.4374 | 1.0006 | 1.9942 | 3.5256 | 5.7656 | 10.053 | 13.377 | 22.083 | 24.494 | 36.772 | 106.28 |

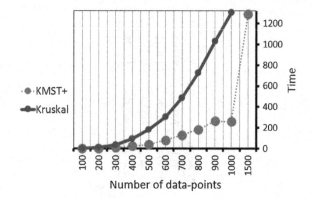

**Fig. 3** Execution time comparison: KMST+ versus Kruskal for K = 4

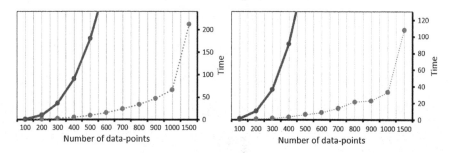

**Fig. 4** Execution time comparison: KMST+ versus Kruskal for K = 8 and 12, respectively

## 5    Conclusion

In this paper, we have proposed the KMST+ algorithm which replaces the K-means algorithm with K-means++ in the divide-and-conquer step of [13]. The assumption was that the improved initialization of the K-means++ algorithm will result in better performance of the FMST framework. It was expected to see an improvement in both the run time and the accuracy of the result by an approximate O(log k) competitive complexity. Using cost and execution time as performance metrics, the algorithm was tested on a real-world dataset and experimental results support the prediction.

## 6  Future Work

A problem faced by the K-means++ algorithm is its lack of scalability. As it employs a sequential technique for initialization, the algorithm is not scalable to large datasets. As future work, scalable K-means++ [2], also called parallel K-means (‖K-means), can be used instead. Empirical evaluation of the ‖K-means algorithm on massive datasets has shown it outperforms K-means++ for both sequential and parallel configurations.

## References

1. Arthur, D., Vassilvitskii, S.: k-means++: The advantages of careful seeding. In SODA, pages 1027–1035 (2007).
2. Bahmani, B., Moseley, B., Vattani, A., Kumar, R., Vassilvitskii, S.: Scalable k-means++, *Proceedings of the VLDB Endowment* 5.7, 622–633(2012).
3. Cheriton, D., Tarjan, R.E.: Finding minimum spanning trees, SIAM J. Comput. 5 24–742 (1976).
4. Fredman, M.L., Tarjan, R.E.: Fibonacci heaps and their uses in improved network optimization algorithms, J. ACM, 34, 596–615(1987).
5. Gabow, H.N., Galil, Z., Spencer, T.H., Tarjan, R.E.: Efficient algorithms for finding minimum spanning trees in undirected and directed graphs, Combinatorica, 6 109–122(1986).
6. https://github.com/deric/clustering-benchmark.
7. https://github.com/deric/clustering-benchmark/tree/master/src/main/resources/datasets/artificial.
8. Karypis, G., Han, E.H., Kumar, V.: CHAMELEON: a hierarchical clustering algorithm using dynamic modeling, IEEE Trans. Comput. 32 68–75(1999).
9. Kruskal, J. B.: On the shortest spanning subtree of a graph and the travelling salesman problem, Proc. Amer. Math. Soc. 7, 48–50 (1956).
10. Nesetril, J., Milkova, E., Nesetrilov, H.: Otakar Boruvka on minimum spanning tree problem – Translation of both the 1926 papers, comments, history, Discrete Mathematics, 233, 3–36 (2001).
11. Prim, R. C.: The shortest connecting network and some generalization, Bell Systems Tech. J. 36, 1389–1401(1957),
12. Yao, A.C.: An O(E log log V) algorithm for finding minimum spanning trees, Inform. Process. Lett. 4, 21–23(1975).
13. Zhong, C., Malinen, M., Miao, D. and Fränti, P.: A fast minimum spanning tree algorithm based on K-means. *Information Sciences* 295, 1–17(2015).

# Resurgence of Deep Learning: Genesis of Word Embedding

Vimal Kumar Soni, Dinesh Gopalani and M. C. Govil

**Abstract**  As the complexity in the structure of natural language increases, the input, output, and processing for a computer system become more challenging. Development of computational techniques and models for automatic analysis and representation of such natural languages is known as natural language processing (NLP). The base unit of any natural language is a word, and its representation is a challenging task as decoding its actual semantic role is vital for any NLP application. One of the most popular computation models is artificial neural network (ANN). However, with the birth of deep learning, a new era has started in computational linguistic research as representation of words has been redefined in terms of word embeddings which capture words semantics in the form of real-valued vectors. This paper presents lifespan of ANN from discovery of first artificial neuron to current era of deep learning. Further, it follows the journey of word embeddings, analyzes their generation methods along with their objective functions, and concludes with current research gaps.

**Keywords**  Artificial neural network · Word representation · Word embedding

## 1  Introduction

Natural language processing (NLP) is an area that deals with computational techniques and models for automatic analysis and representation of human language/ natural language (NL) in textual or speech forms. The textual form is more formal and rule bounded than speech. In speech communication, humans perceive added information like pause, tone, intonation which is completely unavailable in text.

V. K. Soni (✉) · D. Gopalani
Department of Computer Science and Engineering, Malaviya National Institute
of Technology, Jaipur, India
e-mail: ruvimals@gmail.com

M. C. Govil
Department of Computer Science and Engineering, National Institute of Technology,
Ravangla, Sikkim, India

© Springer Nature Singapore Pte Ltd. 2019
B. K. Panigrahi et al. (eds.), *Smart Innovations in Communication
and Computational Sciences*, Advances in Intelligent Systems
and Computing 669, https://doi.org/10.1007/978-981-10-8968-8_11

Absence of such vital information leads to complexities and ambiguities in text processing. However, these issues can be resolved up to some extent by supplying some meta information at the time of text processing [1].

Every natural language has its own set of words that gets continuously updated. Several new words are coined and added to it, a number of words are imported from different sources (from other languages), some of the words become obsolete due to various reasons, and meanings of many words change or get updated with due course of time [2].

A corpus in NLP is a huge collection of text, i.e., sentences in text file separated by new line. However, the NLP techniques are mostly dependent on the manually labeled corpus or annotated corpora. A labeled corpus is one where part of speech for each word is specified. Such annotated corpus is not widely available; therefore, it increases the overhead for every researcher to retrieve the unstructured data from the Internet and using machine learning techniques to pre-process like data cleaning and restructuring for their particular research purpose. Due to free and easy access, huge amount of unstructured text in the disguise of information is uploaded to Internet every day. Taking leverage of this, plain corpus can easily be developed for commonly used natural languages.

Deep learning is inspired from brain and is future of ANN. ANN contains very few hidden layers while deep learning uses multiple hidden layers. Deep learning is considered the best approach to learn patterns from unstructured data. It has one simple rule: more input (unstructured) data will lead to more feature extraction and thus result in better learning.

A word is minimal unit that represents semantic information either individually or by contributing some part in a phrase/sentence. All classical NLP problems like word-sense disambiguation, named entity recognition, machine translation can be better handled if the semantic information encoded in the word/phrase is recaptured exactly as per the context present at the time of encoding the message. Thus, word is the key unit to process and its position with other words represents different semantic aspects.

Considering aforementioned facts, vectors were created to represent words. Initially, these word vectors were containing binary values (one-hot representation); however, with the genesis of distributed representation of words, these vectors become real valued. This paper is an attempt to present expedition of deep learning starting from inception of ANN which gave birth to word embedding. It also analyzes different techniques to generate word embeddings.

## 2  Motivation for Conducting this Survey

Word embeddings have gained huge popularity since its beginning as it is outperforming other methods for various NLP applications. There are number of techniques to generate word embeddings based on different approaches which are mentioned in Table 1.

**Table 1** Different approaches to compute word vectors

| S. no. | Approach | Technique |
| --- | --- | --- |
| 1 | Hard clustering | Brown clustering [3] |
| 2 | Soft clustering | Latent semantic indexing [4] |
| 3 | Mathematical | Noise contrastive estimation [5] |
| | | Principal component analysis [6] |
| | | GloVe [7] |
| 4 | Neural network based | Bengio et al. [8] |
| | | Collobert and Weston [9] |
| | | word2vec [10, 11] |
| | | fastText [12] |

Brown clustering is an agglomerative clustering algorithm to generate word vectors by clustering them based on their occurrence with adjacent words [3]. LSA and LSI are basically count-based models while mathematical model such as GloVe uses word co-occurrence matrix. However, a number of researchers have reported that neural network-based word embeddings are qualitative and perform better than embeddings generated by other approaches [13, 14]. Therefore, the focus of this paper is on neural word embeddings. Before in-depth analysis of these neural word embeddings, a glimpse of resurgence of deep learning from the birth of ANN has been presented in Sect. 3.

## 3 Expedition of ANN and Deep Learning

An artificial neural network is a model containing number of computing units with intermediate weighted connections that is actually inspired by biological neurons. In other terms, it can be considered as a programmed imitation of biological neurons for a specific task. Table 2 summarizes the journey of ANN from the proposal of first artificial neuron to current deep learning era.

The very first simplified model of a neuron was attempted in 1943 by McCulloch and Pitts [15] which was used to classify the set of inputs into two classes. The next breakthrough in ANN was achieved by Rosenblatt in 1958 by designing perceptron [17]. A perceptron is more powerful as it can have different weights at different inputs unlike McCulloch-Pitts neuron, which has identical weights at all inputs. In 1962, Widrow and Hoff gave a rule that specified learning algorithm to update the weights based on difference between desired output and actual output of neuron [18]. This rule got famous as Widrow and Hoff learning rule named after its originators.

The very first deep learning algorithm with number of hidden layers was published by Ivakhnenko and Lapa in 1965. Each layer in the network select the finest features using statistical methods and forward them to next layer. 1982 is a

**Table 2** Journey of ANN and deep learning

| S. no. | Year | Progress in ANN | Remarks |
|---|---|---|---|
| 1 | 1943 | McCullogh-Pitts neuron [15] | To solve OR, AND kind of simple logical operations |
| 2 | 1949 | Hebbs Rule [16] | Algorithm to update weights of connections of neuron |
| 3 | 1958 | Perceptron-Rosenblatt [17] | Specified learning rule |
| 4 | 1962 | Widrow and Hoff [18] | Specified learning algorithm for the updation of weights |
| 5 | 1967 | First deep learning algorithm by Ivakhnenko and Lapa [19, 20] | Learning algorithm for supervised deep feedforward multilayer perceptrons |
| 6 | 1980 | Neocognitron by Kunihiko Fukushima [21] | Convolutional neural network |
| 7 | 1982 | Hopfield Neural Networks by John Hopfield [22] | Content addressable memory system |
| 8 | 1982 | Self-Organizing Maps by Kohonen and Teuvo [23] | • Unlike ANNs error-correction learning, SOM applies competitive learning • Use a neighborhood function to maintain the topological properties of input space |
| 9 | 1986 | Backpropagation by Rumelhart [24] | Minimize the error by propagating the error from final layers to previous layers |
| 9 | 1986 | Jordan's Recurrent n/w [25] | Basis of recurrent neural network |
| 10 | 1988 | Radial basis function n/w by Broomhead and Lowe [26] | Radial basis function as activation function |
| 11 | 1995 | Support Vector Machine by Vapnik et al. [27] | Used for classification and regression analysis |
| 12 | 1997 | Long Short-Term Memory Network by Schmidhuber and Hochreiter [28] | Vanishing gradient problem solved |
| 13 | 2006 | Deep Belief Network by Hinton and Salakhutdinov [29] | Resurgence of deep learning |

remarkable year in ANNs journey because of two significant researches. First by John Hopfield when he proposed Hopfield Neural Networks to be used as content addressable memory system [22]. The next substantial research achievement of that year was from Kohonen and Teuvo in the form of Self-Organizing Maps [23]. A major research contribution was made in 1986 which established ANN as the most favorite computation technique among researchers. The famous Backpropagation NN was introduced by Rumelhart which was first to adjust weights to minimize the error by propagating it from final layers to previous layers [24]. The foundation of Recurrent Neural Network was laid in 1986 when Jordan's recurrent network was published

[25]. Broomhead and Lowe contributed radial basis function network which uses radial basis function as activation function in 1988 [26]. Support Vector Machine was the next big gun of ANN after Backpropagation algorithm. It was proposed by Vapnik et al. [27] in 1995. Schmidhuber and Hochreiter proposed Long Short-Term Memory Network in 1997 to solve Vanishing gradient problem [28]. Resurgence of deep learning is considered in 2006 when Hinton and Salakhutdinov proposed Deep Belief Network [29].

The deep learning comprises huge matrix multiplication along with other operations. These operations can be executed parallel on Graphics Processing Unit (GPU) because a GPU is designed for vector and matrix operations. Therefore, it can execute deep learning algorithms significantly faster than a CPU. One of the key reasons of deep learning performing so well and getting so widespread among researchers is the computation power of GPU is increasing exponentially after 2006 making GPU ten times more powerful than CPU.

## 4 Distributed Representation of Words

One-hot representation is used to characterize a word however this type of representations suffers from data sparsity problems due to the high dimensionality of the word count vectors [5]. In one-hot representation, every word is represented as a sparse vector of the size of the vocabulary and only one bit is turned one which means only one dimension is on [30]. This can be seen as all the lemmatized words are assigned a unique ID in vocabulary (i.e., position of word in vocabulary). The one-hot vector of an ID is a vector that contains only single 1 corresponding to the position associated with that ID and with 0 at all other positions. Suppose, there are two documents: D1 and D2, as shown below:

D1: AIIMS is situated in Delhi.
D2: AIIMS Delhi is a government hospital.
Therefore, the vocabulary contains following eight words:
{AIIMS, is, situated, in, Delhi, a, government, hospital}

Table 3 lists all the words of vocabulary along with their one-hot vectors; while the last two rows of it, represent documents D1 and D2, respectively.

The foremost shortcoming with this representation is that it can not estimate whether a word is semantically close to or far from other word. Distributed Representation produces word vectors that are real valued, dense, and low dimensional which are not only computationally efficient but also capture semantics.

Bengio et al. [8] devised the term "Distributed Representation of Words" in their proposal of a neural probabilistic language model in 2003. This is a feedforward neural network which predicts next word in sequence using one hidden layer. $T$ is total number of training words in the training corpus and with the context of previous $n - 1$ words, i.e., $w_{(t-1)}, w_{(t-2)}, \ldots, w_{(t-n+1)}$, the model tries to maximize the objective function as specified in Eq. 1.

**Table 3** One-hot representation

|            | AIIMS | is | situated | in | Delhi | a | government | hospital |
|------------|-------|-----|----------|-----|-------|-----|------------|----------|
| a          | 0     | 0   | 0        | 0   | 0     | 1   | 0          | 0        |
| government | 0     | 0   | 0        | 0   | 0     | 0   | 1          | 0        |
| in         | 0     | 0   | 0        | 1   | 0     | 0   | 0          | 0        |
| is         | 0     | 1   | 0        | 0   | 0     | 0   | 0          | 0        |
| hospital   | 0     | 0   | 0        | 0   | 0     | 0   | 0          | 1        |
| Delhi      | 0     | 0   | 0        | 0   | 1     | 0   | 0          | 0        |
| AIIMS      | 1     | 0   | 0        | 0   | 0     | 0   | 0          | 0        |
| situated   | 0     | 0   | 1        | 0   | 0     | 0   | 0          | 0        |
| D1         | 1     | 1   | 1        | 1   | 1     | 0   | 0          | 0        |
| D2         | 1     | 1   | 0        | 0   | 1     | 1   | 1          | 1        |

$$L = \frac{1}{T} \sum_t logf(w_{(t)}, w_{(t-1)}, \ldots, w_{(t-n+1)}; \theta) + R(\theta) \tag{1}$$

where $f(w_{(t)}, w_{(t-1)}, \ldots, w_{(t-n+1)})$ is output of neural network and $R(\theta)$ is a regularization term.

In this, each training sentence updates the model about an exponential number of semantically close sentences. The model automatically learns about each word's distributed representation and the probability function for sequences of words, expressed in terms of these representations. The model consists of following layers: input layer or embedding layer that generates word embeddings by multiplying an index vector with a word-embedding matrix; projection layer along with some hidden layers and Softmax layer or output layer that produces a probability distribution over words in vocabulary as shown in Fig. 1.

Because of limited computing power, training over large corpora was not possible which resulted in a delay of almost 5 years to cover next step. In 2008, Collobert and Weston [9] proved that word embeddings trained on adequately large corpora contain syntactic and semantic information.

They removed the Softmax bottleneck as Softmax functions computing cost grows rapidly with the increasing size of vocabulary. The network was trained in such a manner that it will generate higher score for a correct word sequence than for an incorrect one. A pair-wise ranking criterion mentioned as ranking cost is defined in Eq. 2.

$$\sum_{s \in S} \sum_{w \in V} Max(0, 1 - f(s) + f(s^w)) \tag{2}$$

where $s$ is a sentence windows of text and $S$ is set of all $s$. Vocabulary is represented by $V, f(.)$ denotes the Neural Network, and $s^w$ is modified $s$ in which the middle word

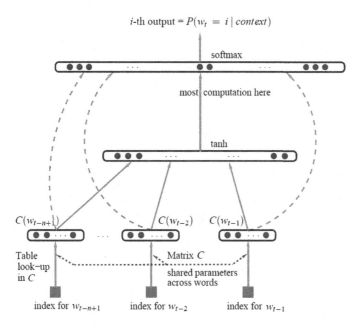

**Fig. 1** Neural architecture: $f(i, w_{t-1}, \ldots, w_{t-n+1}) = g(i, C(w_{t-1}), \ldots, C(w_{t-n+1}))$ where $g$ is the neural network and $C(i)$ is the $i$th word feature vector [8]

has been replaced by the word $w$. This replacement was to make language model to categorize a two-class classification task, i.e., whether the context is relevant or not after replacing the word $w$ in the input window. Figure 2 represents the architecture of this model.

In 2013, Mikolov et al. [10] presented two new models for learning distributed representation of words with lesser computation cost than existing methods. The first model is known as continuous-bag-of-words (CBOW) and is shown in Fig. 3. Previous models focused only on neighboring words in the window of target word but CBOW model takes five words occurring before and five words occurring next to the target word. Equation 3 mentions its objective function.

$$\frac{1}{T} \sum_t logP(w_t|w_{t-n}, \ldots, w_{t-1}, w_{t+1}, \ldots, w_{t+n}) \tag{3}$$

The second model is skip gram, which exactly mirrors the CBOW model. Skip-gram model predicts neighbor/context words where single word is given as input while CBOW predicts one word for given $n$ words as shown in Fig. 4. It performs addition of the logarithmic probabilities of adjacent $n$ words appeared before and after the target word. Equation 4 mentions its objective function.

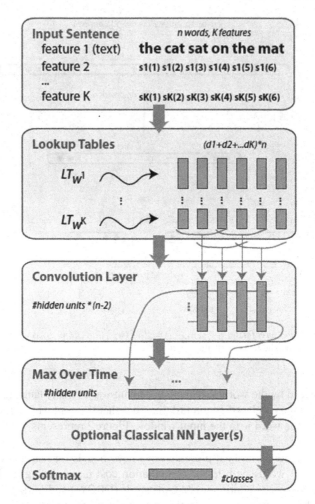

**Fig. 2** Collobert and Weston model [9]

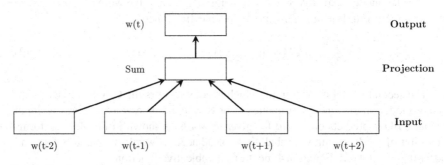

**Fig. 3** Continuous-bag-of-words model [10]

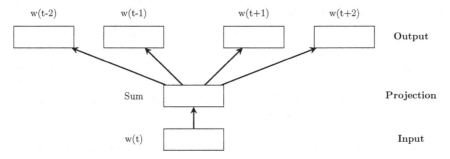

**Fig. 4** Skip-gram model [11]

$$\frac{1}{T} \sum_{t=1}^{T} \sum_{-c \leq j \leq c, j \neq 0} logP(w_{t+j}|w_t) \tag{4}$$

Instead of using full Softmax, Hierarchical Softmax was used as it evaluates only $log_2(W)$ nodes, to generate probability distribution while full Softmax process all $W$ nodes. On the track of Collobert and Weston whose model was trained by classifying correct phrase above noise, Mikolov also introduced negative sampling in the skip-gram model. To maintain the balance between rare words and frequently occurring words, subsampling of frequent words was also performed.

Facebook fastText is an extension of skip-gram model incorporating morphological information to generate high-quality word embedding. With the help of a new logistic loss function $l : xlog(1 + e^{(-x)})$, the objective function is stated in Eq. 5.

$$\sum_{t=1}^{T} \sum_{c \in C_t} l(s(w_t, w_c)) + \sum_{n \in N_{t,c}} l(-s(w_t, n)) \tag{5}$$

A natural parameterization for the score function between a word $w_t$ and a context word $w_c$ is to take the scalar product between word and context embeddings $s(w_t, w_c) = u_{(w_t)}^T v_{w_c}$ where $u_{w_t}^T$ and $v_{w_c}$ are vectors in $R^d$ [12]. Unlike the skip-gram model, fastText considers internal structure of words. For a given word w, a set of n-grams appearing in w is denoted by $G_w \subset \{1, ..., G\}$. Each n-gram $g$ is associated with a vector $z_g$, and a word is represented by the summation of the vector representations of its n-grams. To define this subword model, they proposed a scoring function, here denoted as Eq. 6.

$$s(w, c) = \sum_{g \subset G_w} z_g^T v_c \tag{6}$$

This subword model empowers fastText to train across its morphological variations thus producing better vectors and proved in text classification task [31].

# 5   Conclusion and Future Work

This survey paper is an attempt to provide an insight on the journey of deep learning and word embedding. The whole journey from the discovery of first artificial neuron to rebirth of deep learning is briefed in this paper. Expanding the neural word embedding, we have discussed different models along with their objective functions. An extensive study on these models proved that the fastText is most advanced and powerful model among the existing ones.

However, there is scope of improvement in word embedding using word dependencies and phrase structure. Generating embeddings for ambiguous words is another research challenge as single-word vector is not sufficient to represent all contexts of a word. Researchers are also trying to generate and improve word vectors using multilingual corpora. Hence researchers can focus on these issues.

# References

1.  Collobert, R., Weston, J., Bottou, L., Karlen, M., Kavukcuoglu, K., & Kuksa: Natural Language Processing (almost) from Scratch. Journal of Machine Learning Research, 12, pp. 2493–2537, (2011)
2.  C. Lala and S. B. Cohen: The Visualization of Change in Word Meaning over Time using Temporal Word Embeddings. CoRR, vol. abs/1410.4 (2014)
3.  P. F. Brown, P. V Desouza, R. L. Mercer, V. J. Della Pietra, and J. C. Lai: Class-based n-gram models of natural language. Comput. Linguist., vol. 18, no. 4, pp. 467–479 (1992)
4.  S. Deerwester, S. T. Dumais, G. W. Furnas, T. K. Landauer, and R. Harshman: Indexing by latent semantic analysis. Journal of the American Society for Information Science, vol. 41, pp. 391–407 (1990)
5.  A. Mnih: Learning word embeddings efficiently with noise-contrastive estimation. Nips, pp. 2265–2273 (2013)
6.  Lebret R, Collobert R.: Word emdeddings through hellinger PCA. arXiv preprint arXiv:1312.5542 (2013)
7.  Pennington J, Socher R, Manning CD. Glove: Global Vectors for Word Representation. EMNLP, Vol. 14, pp. 1532–1543 (2014)
8.  Y. Bengio, R. Ducharme, P. Vincent, and C. Janvin: A Neural Probabilistic Language Model. J. Mach. Learn. Res., vol. 3, pp. 1137–1155 (2003)
9.  R. Collobert and J. Weston: A Unified Architecture for Natural Language Processing: Deep Neural Networks with Multitask Learning. Architecture, vol. 20, no. 1, pp. 160–167 (2008)
10. T. Mikolov, G. Corrado, K. Chen, and J. Dean: Efficient Estimation of Word Representations in Vector Space. Proc. Int. Conf. Learn. Represent. (ICLR 2013), pp. 112 (2013)
11. T. Mikolov, K. Chen, G. Corrado, and J. Dean: Distributed Representations of Words and Phrases and their Compositionality. Nips, pp. 1–9 (2013)
12. P. Bojanowski, E. Grave, A. Joulin, and T. Mikolov: Enriching word vectors with subword information. arXiv Prepr. arXiv:1607.04606 (2016)
13. F. Hill, K. Cho, S. Jean, C. Devin, and Y. Bengio: Not All Neural Embeddings are Born Equal. arXiv, p. 4, (2014)
14. M. Baroni, G. Dinu, and G. Kruszewski: Dont count, predict! A systematic comparison of context-counting vs. context-predicting semantic vectors. in Proceedings of the 52nd Annual Meeting of the Association for Computational Linguistics (Volume 1: Long Papers), pp. 238–247 (2014)

15. W. S. McCulloch and W. Pitts: A logical calculus of the ideas immanent in nervous activity. Bull. Math. Biophys., vol. 5, no. 4, pp. 115–133, (1943)
16. Hebb, Donald Olding.: The organization of behavior: A neuropsychological approach. John Wiley & Sons, (1949)
17. F. Rosenblatt: The perceptron: A probabilistic model for information storage and organization in the brain. Psychol. Rev., vol. 65, no. 6, p. 386, (1958)
18. B. Widrow and M. E. Hoff: Associative Storage and Retrieval of Digital Information in Networks of Adaptive Neurons. in Biological Prototypes and Synthetic Systems, Springer, p. 160, (1962)
19. Ivakhnenko, A.G. and Lapa, V.G.: Cybernetic predicting devices. (No. TR-EE66-5). Purdue Univ Lafayette Ind School of Electrical Engineering. (1966)
20. Ivakhnenko, A.G. and Lapa, V.G.: Cybernetics and forecasting techniques. (1967)
21. Mermelstein, P. and Eden, M.: Experiments on computer recognition of connected handwritten words. Information and Control, 7(2), pp. 255–270. (1964)
22. J. J. Hopfield: Neural networks and physical systems with emergent collective computational abilities. Proc. Natl. Acad. Sci., vol. 79, no. 8, pp. 2554–2558, (1982)
23. T. Kohonen: Self-organized formation of topologically correct feature maps. Biol. Cybern., vol. 43, no. 1, pp. 5969, (1982)
24. D. E. Rumelhart, G. E. Hinton, and R. J. Williams: Learning internal representation by back propagation. Parallel Distrib. Process. Explor. Microstruct. Cogn., vol. 1, (1986)
25. M. I. Jordan: Serial order: a parallel distributed approach (ICS Report 8604). San Diego: University of California. Inst. Cogn. Sci., (1986)
26. D. S. Broomhead and D. Lowe: Radial basis functions, multi-variable functional interpolation and adaptive networks. R. SIGNALS RADAR Establ. MALVERN (UNITED KINGDOM), vol. No. RSRE-M, (1988)
27. C. Cortes and V. Vapnik: Support-vector networks. Mach. Learn., vol. 20, no. 3, pp. 273–297, (1995)
28. S. Hochreiter and J. Schmidhuber: Long short-term memory. Neural Comput., vol. 9, no. 8, pp. 1735–1780, (1997)
29. G. E. Hinton and R. R. Salakhutdinov: Reducing the dimensionality of data with neural networks. Science (80.), vol. 313, no. 5786, pp. 504–507 (2006)
30. J. Guo, W. Che, H. Wang, and T. Liu: Revisiting Embedding Features for Simple Semi-supervised Learning. ACL pp. 110–120 (2014)
31. A. Joulin, E. Grave, P. Bojanowski, and T. Mikolov: Bag of tricks for efficient text classification. arXiv Prepr. arXiv:1607.01759 (2016)

# Financial Planning Recommendation System Using Content-Based Collaborative and Demographic Filtering

Nymphia Pereira and Satishkumar L. Varma

**Abstract** The one stop to all problems is the Internet. But finding relevant information is difficult. The interest of the user lies in different forms of information content such as images, text, audio, or videos. The recommendation system is a process of information filtering that helps users to find better products, financial plans, and other related information by personalizing the suggestions. There are different recommendations techniques such as collaborative filtering, demographic recommendation, knowledge-based recommendations, content-based recommendation, and utility-based recommendation system. These techniques fail to eliminate the drawbacks such as data sparsity, new user cold start problem, new item cold start problem, overspecialization, and shilling attacks. In today's generation, saving income is very important. In this work, a recommendation system for financial planning is proposed. Here, the idea is to modify the recommendation process to improve the recommendations in the best possible way. The above-mentioned drawbacks are eliminated using hybrid approach. In the hybrid approach, the techniques of collaborative filtering, i.e., user–user and item–item similarity along with demographic filtering, are combined. The experimental result is evaluated using performance metrics precision and recall. An ROC curve is used for evaluating the system.

**Keywords** Recommendation system · Demographic filtering · Item–item collaborative filtering · User–user collaborative filtering · Markov chain

N. Pereira (✉) · S. L. Varma
Department of Information Technology, PCE, New Panvel, India
e-mail: nymphiapereira@gmail.com; npereira@mes.ac.in

S. L. Varma
e-mail: vsat2k@yahoo.com

© Springer Nature Singapore Pte Ltd. 2019  141
B. K. Panigrahi et al. (eds.), *Smart Innovations in Communication and Computational Sciences*, Advances in Intelligent Systems and Computing 669, https://doi.org/10.1007/978-981-10-8968-8_12

# 1 Introduction

Recommendation system is a process of information filtering for providing personalized suggestions to the user. A good recommending system benefits not only to the end users but also to those who are developing the systems for better businesses and gaining profit. Recommender system can be applied to a variety of applications. Most of the online systems such as Amazon, Movielens, and Flipkart are being widely dependent on it [1]. Its task is to automatically suggest items to the users with minimum efforts of the user from a large collection of items.

In order to do this, various recommendation techniques (Fig. 1) are developed to fulfill the needs. In this work, techniques under personalized recommendation systems are studied. Various techniques under personalized recommendation systems commonly suffer from data sparsity, that is, insufficient amount of data, scalability issues which cannot scale after a certain limit, and cold start problems which if a new user comes fail to recommend relevant items or information [2]. Overspecialization issues as sufficient amount of user ratings are not available and shilling attacks commonly known as push and nuke attacks [3]. Therefore, combining two techniques collaborative filtering and demographic named hybrid is adopted to overcome these drawbacks.

This paper is organized into 5 sections. Section 1 gives an overview of the paper. It describes the use of recommendation system in financial planning sector. Section 2 describes the related work done by different authors. In Section 3, we propose a hybrid approach to improve the recommendation system as well as recommend a best financial plan to the user. Section 4 gives a comparative analysis and evaluation done on combined ranking techniques. The last section of this paper summarizes the results. It reveals that a lot of improvement can still be done in the field of recommendation systems for financial planning.

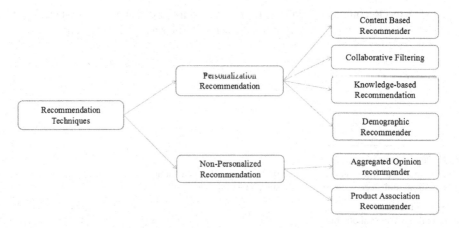

**Fig. 1** Recommendation techniques [4, 5]

## 2 Related Work

Recommendation systems have become the center of attraction during the past few years with increase in e-businesses. A lot of research work has been carried out but still the system fails to overcome few of the issues mentioned above. Therefore, below literature shows few papers overcoming not all but few problems.

Pasquale Lops et al. discusses the major issues related to content-based recommendation systems such as overspecialization. The basic concepts of content-based recommender systems and their advantages and limitations are presented. User profiles are created [6]. It overcomes overspecialization and serendipity problem. It does not discuss new user cold start problem [4].

Robin Burke explains various hybrid recommendation techniques that can be used to improve the recommendation systems. It combines two or more techniques together to overcome the limitation of the other. It uses various hybrid methods such as cascading, mixed, augmented, which can be used in recommendation systems based on the application for better accuracy and results [7].

Realization of individualized recommendation system on book sales uses association rules, data preprocessing, data inputs, and simple techniques to provide individualized recommendation system for book sales. The results of the Web site based on the simple techniques have shown great performance. But it does not address the issue of new user cold start problem and data sparsity [8].

Hybrid approach plays an important role in collaborative filtering. It combines user–user similarity and item–item similarity. Authors talk about a new hybrid approach for solving the problem of finding the ratings of unrated items. Two major challenges of recommender systems, accuracy and sparsity of data, are addressed in this proposed system [9].

A new approach to book recommendation system by combining features of content-based filtering, collaborative filtering, and association rule mining is proposed by A. S. Tewari et al. [10]. The paper reveals the various parameters like content and quality of the book by doing collaborative filtering of ratings by other buyers. Issues such as new user cold start problem and overspecialization are not addressed in this work.

The Web-based personalized hybrid book recommendation system [11] discusses using content-based filtering, collaborative filtering, and demographics recommendation system. It uses hybrid approach [3] to overcome the new user cold start problem and new item cold start problem. But it uses tools to do so which cannot extend after a certain limit. Therefore, it does not overcome scalability issues.

Travel-based recommendation system is developed and presented in [12, 13]. It develops a model to extract the topics conditioned on both the tourists and the intrinsic features. The system combines demographic and content-based filtering together in order to overcome the new user cold start problem as shown in Fig. 2. The input of this is given to collaborative filtering which overcomes the drawback of new item cold start problem. It uses mobile application. It is very efficient.

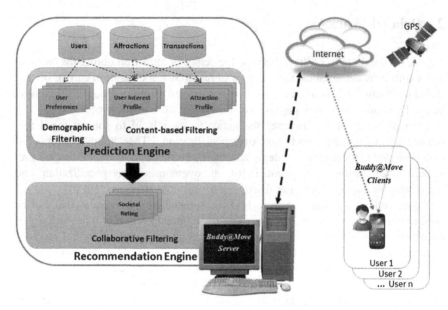

**Fig. 2** Existing architecture [12]

But it does not consider ontology features and external climatic conditions as demographic factors. Also, it uses predefined data which is limited in context to travel-based recommendation systems.

## 3 Proposed System

The proposed work is to develop a financial planning application. This application provides appropriate recommendations for investments to the users. A hybrid approach is used for this purpose. A hybrid approach is a combination of content-based filtering, collaborative filtering, and demographic filtering. To improve the quality of recommendation, two ranking techniques are used. One is TF-IDF ranking and the other is Markov chain [14] aggregate ranking function. This helps in better prediction of the result. Also, the proposed system overcomes overspecialization [4], new user cold start problem, new item cold start problem [15], and data sparsity faced by other recommendation systems.

### 3.1 System Architecture

A recommender system has many components. The proposed architecture consists of five steps: data preprocessing, content-based filtering, rule technique, using

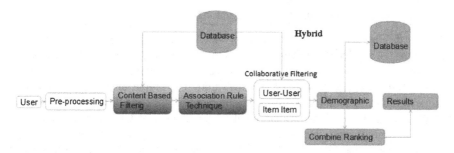

**Fig. 3** Architecture of recommendation system for financial planning

hybrid approach, i.e., collaborative filtering and demographic filtering, and ranking. The enhancement in the system is shown by the shaded portion in Fig. 3. Different methods are used other than the existing methods in each technique.

Step 1: Data Preprocessing

Information regarding financial planning is collected from various company datasets. Preprocessing has to be done on the datasets [9]. Preprocessing involves eliminating insignificant data that would affect the results and eliminating noise that would have direct influence over the calculation of the output.

Step 2: Content-Based Filtering

Classification is an important step in the recommendation process. Classification technique is used to find similarity between the past buying histories of the user. Different classification techniques exist. Here, C4.5 algorithm is used as it is better compared to Naive Bayes technique earlier used in recommendation technique. Naive Bayes works well with small amount of data [10]. Hence, C4.5 algorithm helps to find similarity between the users by building a tree-like structure. It calculates the information gain as shown in Eq. 1.

$$Info\,(D) = -\sum_{i=1} Pi\log_2(Pi) \tag{1}$$

The main advantage of using this algorithm is that, it overcomes the drawback of new user cold start problem.

Step 3: Association Rule Technique

In large datasets, association rule technique helps to find interesting relations between variables. Based on users input, association rule mining (FP Growth) technique is applied to find similar items. These rules allow to discover interestingness between items that frequently appear together. Here, the user is asked as to where he could like to invest. Based on this an FP tree is generated. For example, if (age = "21–25," salary = 1–1.9999 lacs, occupation = engineer), then (option = lic, health).

Step 4: Hybrid Filtering and Combine Ranking

This is the last step of recommendation system. Here, the output of FP growth technique is applied to collaborative filtering to find item–item and user–user similarity [16]. Item–item similarity achieved by using association rule. User–user similarity is achieved by using C4.5 algorithm. The output of these two collaborative filtering techniques is given to demographic filtering features. TF-IDF is applied to find the most similar neighbors and rank them. Rank aggregation function, Markov chain is applied to the result of user–user ranking and item–item ranking to give the best ranking result in order to recommend. Markov chain is given by

$$\text{Markov Chain} = (0.5 * N + N_{BA})/(N + C_{BA}) \tag{2}$$

where N—scale to convince a probability. $N_{BA}$ is nothing but number of times item–item IDF value beat user–user IDF value and $C_{BA}$ is number of times they were compared. This helps us in knowing which particular age individuals are more interested in financial planning as well as the system can learn and predict the best result. The performance metrics ROC curve is used to represent the system. Top-ranked plans are recommended to the users.

# 4 Results and Discussion

In this section, evaluation method is given which is used for testing our approach. We also provide the details of the dataset that was used for testing the system. The quality of a recommender system can be evaluated by comparing recommendations to a test set of known user profiles. The current system is measured using predictive accuracy metrics, which emphasizes on providing good financial planning solution as well as improving the recommendation process.

## 4.1 Experimental Setup

Description of two datasets used in the experiment is shown in Table 1. First dataset consists of 600 financial plans with four categories, and the second dataset consists of investments which people have previously purchased. This dataset

**Table 1** Dataset description

| Policy type | LIC | Health | Liquid fund | SIP | Total records |
|---|---|---|---|---|---|
| Dataset1 | 150 | 200 | 150 | 100 | 600 |
| Dataset2 | 3000 | 1500 | 1500 | 1000 | 7000 |

consists of 7000 entries. People from all age groups from 21 to 60 are considered in order to recommend. Precision and recall are used to measure the quality of recommendation system. Definition of precision and recall is given in Eqs. 3 and 4

**Table 2** Matrix for evaluating precision and recall for item–item and user–user

| Cutoff | Recall | | Precision | | |
|---|---|---|---|---|---|
| | tp/(tp + fn) | tn/(tn + fp) | tp/(tp + fp) | tn/(tn + fn) | |
| | Sensitivity | Specificity | Predictive | Predict-Neg | 1-Specificity |
| *TF-IDF item–item* | | | | | |
| 0 | 85.71 | 42.86 | 93.75 | 23.08 | 57.14 |
| 10 | 82.86 | 50.00 | 90.63 | 33.33 | 50.00 |
| 20 | 83.78 | 50.00 | 89.86 | 36.84 | 50.00 |
| 30 | 81.58 | 57.14 | 91.18 | 36.36 | 42.86 |
| 40 | 82.19 | 58.82 | 89.55 | 43.48 | 41.18 |
| 50 | 81.95 | 55.56 | 88.24 | 50.00 | 44.44 |
| 60 | 81.76 | 62.50 | 91.04 | 62.50 | 37.50 |
| 70 | 80.80 | 66.67 | 92.54 | 62.50 | 33.33 |
| 80 | 0 | | | | 0 |
| 90 | 0 | | | | 0 |
| *TF-IDF user–user* | | | | | |
| 0 | 87.84 | 42.86 | 94.20 | 25.00 | 57.14 |
| 10 | 85.71 | 44.44 | 92.96 | 26.67 | 55.56 |
| 20 | 84.15 | 50.00 | 94.52 | 23.53 | 50.00 |
| 30 | 83.54 | 54.55 | 92.96 | 31.58 | 45.45 |
| 40 | 83.56 | 58.33 | 92.42 | 36.84 | 41.67 |
| 50 | 82.89 | 62.50 | 95.45 | 27.78 | 37.50 |
| 60 | 82.28 | 66.67 | 95.59 | 30.00 | 33.33 |
| 70 | 80.77 | 71.43 | 96.92 | 25.00 | 28.57 |
| 80 | 0 | | | | 0 |
| 90 | 0 | | | | 0 |
| *Markov chain* | | | | | |
| 0 | 96.59 | 42.86 | 95.51 | 50.00 | 57.14 |
| 10 | 94.51 | 50.00 | 93.48 | 54.55 | 50.00 |
| 20 | 93.26 | 50.00 | 93.26 | 50.00 | 50.00 |
| 30 | 92.05 | 53.85 | 93.10 | 50.00 | 46.15 |
| 40 | 93.26 | 58.33 | 94.32 | 53.85 | 41.67 |
| 50 | 91.95 | 66.67 | 94.12 | 58.82 | 33.33 |
| 60 | 91.11 | 70.00 | 96.47 | 46.67 | 30.00 |
| 70 | 83.15 | 84.62 | 97.37 | 42.31 | 15.38 |
| 80 | 0 | | | | 0 |
| 90 | 0 | | | | 0 |

$$Precision = \frac{No\ of\ relevant\ financial\ plans\ retrieved}{No\ of\ retrieved\ plans} \qquad (3)$$

$$Recall = \frac{No\ of\ relevant\ financial\ plans\ retrieved}{No\ of\ relevant\ plans} \qquad (4)$$

## 4.2 Experimental Results

The performance of the system is computed using precision and recall for each class label, and the individual performance on class labels is analyzed as shown in Table 2. For evaluating precision and recall of the system, 150 sample data are considered for each class of ages 21–60. TF-IDF calculated for item–item similarity and user–user similarity is shown in section (A and B). The result of TF-IDF is given to Markov chain. Evaluation of Markov chain is shown in section (C).

The recommendation system is evaluated to check whether the recommended financial plans are correct or not. Most recommendation solution either provides relevant answer or based on similarity of the user query. Let us consider an example.

For a small sample of age group 21–25, the term frequency (TF) and inverse document frequency (IDF) is shown in Fig. 4. The x-axis shows the items like LIC, Health, SIP, and liquid fund (LF) purchased by the users and their corresponding frequencies is shown on y-axis. In Fig. 5, the result of TF-IDF is given to aggregate rank function, Markov chain. So we can understand that of particular age group how many users have accessed the system and how many items have been purchased.

The evaluation of the system is done using receiver operating characteristic (ROC) curve. The advantage of using this technique is, at various thresholds a binary classifier system is compared. The curve consists of true positive rate

**Fig. 4** TF-IDF for age group 21–25

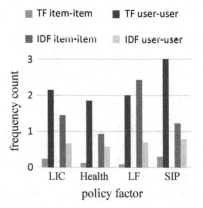

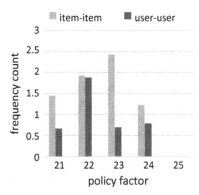

**Fig. 5** Markov chain for age group 21–25

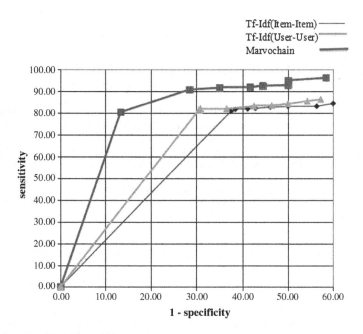

**Fig. 6** Results of combine ranking used

(sensitivity) on y-axis and false positive rate (fall out) on x-axis. The evaluation of precision and recall is shown in Table 2. Fig. 6 shows the graph of ROC curve based on TF-IDF ranking for item–item and user–user filtering and Markov chain.

Based on the analysis, we can say that Markov chain ranking gives the best result. From the graph, it can be concluded that the accuracy of the system is 96%.

# 5 Conclusion

In this work, a recommendation system using various data mining techniques for financial planning is implemented. The user is recommended best financial plans based on his/her age, salary, and expenses. The purpose of building a system for financial planning is that a user can make appropriate decisions based on his age and salary as how well he can protect his/her income and save for the future. The performance metrics precision and recall are used to evaluate the system. ROC curve is used to represent the state of the system. The system can be used for different online and offline applications. As the system provides suggestions only for individuals, it can be enhanced further for family planning by taking in more information from the user such as whether the user has previous loans, marital status, etc.

# References

1. RVVSV Prasad and V Valli Kumari, "A categorical review of recommender sys- tems", International Journal of Distributed and Parallel Systems (IJDPS) Vol. 3, No. 5, September 2012.
2. Jyoti Gupta and Jayant Gadge, "A Framework for a Recommendation System Based On Collaborative Filtering and Demographics", International Conference on Circuits, Systems, Communication and Information Technology Applications (CSCITA), pp 300–304, 2014.
3. Xiang Li, Min Gao, Wenge Rong, Qingyu Xiong and Junhao Wen, "Shilling Attacks Analysis in Collaborative Filtering Based Web Service Recommendation Systems", IEEE International Conference on Web Services (ICWS), pp 538–545, 2016.
4. Pasquale Lops, Marco de Gemmis and Giovanni Semeraro, "Content-based Recommender Systems: State of the Art and Trends", Springer Science Business Media, LLC, pp 73–100, 2011.
5. Anil Poriya, Neev Patel, Tanvi Bhagat and Rekha Sharma, Ph. D, "Non-Personalized Recommender Systems and User-based Collaborative RecommenderSystems", International Journal of Applied Information Systems (IJAIS)- ISSN:2249-0868 Foundation of Computer Science, Volume 6, No. 9, March 2014.
6. Bahram Amini, Roliana Ibrahim and Mohd Shahizan Othman, "Discovering the impact of knowledge in recommender systems: A comparative study", International Journal of Computer Science & Engineering Survey (IJCSES) Vol. 2, No. 3, August 2011.
7. Robin Burke, "Hybrid Recommender Systems: Survey and Experiments", User Modelling and User-Adapted Interactions, Volume 12, Issue 4, Nov 2002, pp 331–370.
8. Gilbert Badaro, Hazem Hajj, Wassim El-Hajj and Lama Nachman, "A Hybrid ap-proach with collaborative filtering for recommender systems", 9th International Wire-less Communications and Mobile Computing Conference (IWCMC), 2013.
9. Luo Zhenghua, "Realization of Individualized Recommendation System on Books Sale", IEEE International Conference on Management of e-Commerce and e-Government, pp 10–13, 2012.
10. Tewari A.S., Kumar A. and Barman A.G., "Book recommendation system based on combine features of content based filtering, collaborative filtering and association rule mining", International Advance Computing Conference (IACC), pp 500–503, 2014.

11. Salil Kanetkar, Akshay Nayak, Sridhar Swamy and Gresha Bhatia, "Web-based Personalized Hybrid Book Recommendation System", IEEE International Conference on Advances in Engineering and Technology Research (ICAETR - 2014).
12. Shini Renjith and Anjali C, "A Personalized Mobile Travel Recommender System using Hybrid Algorithm", 2014 First International Conference on Computational Systems and Communications (ICCSC), December 2014.
13. Qi Liu, Yong Ge, Zhongmou Li, Enhong Chen and Hui Xiong, "Personalized Travel Package Recommendation", IEEE 11th International Conference on Data Mining (ICDM), pp 407–416, Dec. 2011.
14. Qingyan Yang, Ju Fan, Jianyong Wang and Lizhu Zhou, "Personalizing Web Page Recommendation via Collaborative Filtering and Topic-Aware Markov Model," IEEE 10th International Conference on Data Mining (ICDM), pp 1145–1150, Dec. 2010.
15. Jyoti Gupta and Jayant Gadge, "Performance Analysis of Recommendation System Based On Collaborative Filtering and Demographics", International Conference on Communication, Information and Computing Technology (ICCICT), Jan 2015.
16. Ruisheng Zhang, Qi-dong Liu, Chun Gui, Jia-Xuan Wei and Huiyi Ma, "Collab-orative Filtering for Recommender Systems", Second International Conference on Advanced Cloud and Big Data (CBD), 2014.

# Entity Relation Extraction for Indigenous Medical Text

**J. Betina Antony, G. S. Mahalakshmi, V. Priyadarshini and V. Sivagami**

**Abstract** Understanding the relation between entities in a vast context is a basic yet tedious task in the field of information extraction and machine translation. Entities are the prime concept in any domain, and all the information are constructed around this prime concept. The information that is to be understood is nothing but how the entity prevails in its arena and how it is supported by other entities. The task of apprehending these vital details is the backbone behind relation extraction. In this work, we present a two stage information extraction systems from English traditional medicine research articles. In the first stage, the system recognizes entities in the context by random forest-based learning which performed relatively well with an f-score of 88%. The second stage is the extraction of relation between entities using pattern recognition.

**Keywords** Entity recognition · Relation extraction · Traditional medicine
Feature extraction

## 1 Introduction

Traditional medicines or indigenous medicines are medicines obtained from Mother Nature. They form the backbone for a medical system developed from traditional knowledge that existed centuries ago. These information are gaining popularity in

J. Betina Antony (✉) · G. S. Mahalakshmi · V. Priyadarshini · V. Sivagami
Department of Computer Science and Engineering, College of Engineering Guindy,
Anna University, Chennai 600025, Tamil Nadu, India
e-mail: betinaantony@gmail.com

G. S. Mahalakshmi
e-mail: gsmaha@annauniv.edu

V. Priyadarshini
e-mail: priyastar23897@gmail.com

V. Sivagami
e-mail: sivagami285@gmail.com

© Springer Nature Singapore Pte Ltd. 2019
B. K. Panigrahi et al. (eds.), *Smart Innovations in Communication and Computational Sciences*, Advances in Intelligent Systems and Computing 669, https://doi.org/10.1007/978-981-10-8968-8_13

153

the recent days as their results have lasting effect than English medicines. Many such information have been translated and registered as research articles. Our work focuses on extracting and presenting these information. The details extracted can be used for Ontology building, knowledge graph construction, machine translation, etc.

In this work, we concentrate on extracting entities and their corresponding relation tuples. We focus on building our own entity recognizers rather than using off-the-shelf methods as the categories of entities are different for traditional medicines when compared to English biomedicine. We then start on a simple pattern recognition system for identifying relation statements and training them.

## 2 Related Works

Entity recognition and relation extraction are text mining techniques that have been in practice for more than two decades. Though many algorithms have been tested and worn out for the field of biomedicine, there always exists a gap to be filled. There are two ways of approaching these two techniques. One is pipelining method where one is carried out first and their result is fed as input to the next. The next is mutual learning or joint learning where both techniques are simultaneously trained.

Named-entity recognition is the method of identifying entity names that form the backbone for many text mining applications. It is carried out by dictionary-based [1], rule-based [2], statistical [3], and machine learning algorithms and has been analyzed by ample number of experiments. Some of the commonly employed methods are conditional random fields [4], Markov models [5–7], support vector machine (SVM) [8], decision tree methods [9], and deep learning [10, 11].

Relation extraction too went through a series of evolution and their method of implementation varied among different applications. Initial methods focused on feature building with syntactic and semantic features [12, 13] playing a key role in identifying relations between entities. To improve accuracy of the system and to overcome the overheads placed by the features, more of statistical and machine learning approaches were roped in. In particular, artificial neural networks [14, 15] were a popular choice for identifying and classifying relations. However, their level of accuracy still needs improvement.

## 3 System Description

There are four basic phases in the system. They are preprocessing phase, feature extraction phase, entity extraction phase, and relation recognition phase. Each phase has many intermediate modules which are explained in the following sections (Fig. 1).

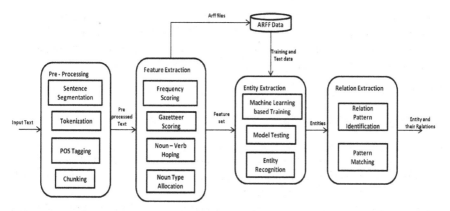

**Fig. 1** Overall flow diagram for entity relation extraction

## 3.1 Pre-Processing Phase

Preprocessing phase has three modules, namely sentence segmentation, word tokenization, and parts of speech tagging. The input of this module is biomedical-based text file. The file contains paragraphs of data pertaining to biomedical information that do not have any structure or pattern to it. The output of this phase is POS tagged chunks of words with noun and verb chunks.

- **Sentence Segmentation**

The whole text document is given as input to the system, and it is split into each separate sentences using punctuation such as ".", ";" or ":". Thus, we get text document segregated sentence-wise, and these sentences are stored into a list form of data structure to be processed later.

- **Word Tokenization**

The list of sentences are then converted into separate words as individual words form the building units for further processing. The result is again maintained in a list.

- **POS Tagging**

The output from the word tokenize module is given as input to this module. A POS tag gives the sense of a word in the given context and plays a crucial part in designing the role of word. We use NLTK toolkit to identify POS tags for the tokenized words.

- **Chunking Module**

Chunking is the process of grouping a set of the words into chunks such that each chunk forms a meaningful phrase or unit. In our work, these chunks are grouped in a specified grammatical form so that such formatted words can be used later for entity recognition and relation extraction phase. Chunking involves specifying grouping grammar such as (NN) * (NNP) ? VB, extracting combination of words that occur in this format and storing it in an array. The noun chunks determine entities, and the verb chunks determine relation between entities.

## 3.2 Feature Extraction

For the entity recognition process, we designed our own named-entity recognition module using certain features. There are two reasons for designing our own NER module instead of using other open-source Biomed NER tools available online. Firstly, our dataset contains information about indigenous plants and herbs, and their ailing uses to the field of medicine. The available tools are not trained to this particular domain. Even if they can be trained though being a tedious process, the second reason for not training them is the categories are mostly focused on English biomedicine. The categories are mainly DNA patterns, protein model, etc., which do not occur in our dataset. Hence, we designed a simple NER module trained using WEKA [16] where the feature set includes frequency of words, gazetteer score, distance between noun and verb, verb type (action verb or other), and verb location (preceding or succeeding).

The frequency scoring involves finding the frequency of a word in a text and normalizing it based on the total number of words in the file. That way the value remains constant for all words across documents. The next feature is a lookup score calculated by checking with a dictionary containing list of biomedical words.[1] The dictionary contains a list of 98,100 words. Since entities are mostly nouns, we take into account the location of nouns in a context. Hence, the distance of nouns from the nearest verb and their position is taken as location features. Also the type of the nearest verb is specified as either action or other verb. Examples for action verbs are treats, cures, heals, etc. With these information, the system is trained to differentiate a *Drug*, a *Disease*, or a *Misc* word.

## 3.3 Entity Recognition

In the entity recognition phase, the feature set from the training corpus is trained using a machine learning algorithm. As an experiment, we ran the data through 7 different classifiers out of which random forest algorithm gave the best result. Random forest is an ensemble machine learning technique that gives the mode value of a number of decision trees run on the training data. Since the feature set is highly variant with low bias, general decision tree develops an overfitting problem.

---

[1]https://github.com/Glutanimate/wordlist-medicalterms-en.

This is averaged out by the random forest. The trained model is tested against a set of new data that gave promising result. The model is then used to identify entities belonging to disease and drug categories.

## 3.4 Relation Extraction

In this phase, biomedical entities which are related are extracted. We employ basic pattern recognition method the pattern here meaning format of sentences. This phase consists of two modules, relation format recognition and relation extraction. The type of sentences that form candidates for relations are obtained from the corpus fed for entity recognition. The pattern we mention mainly symbolizes the action verbs and number and position of entities surrounding it.

Some words are able to express the relations between named entities (here it is biomedical terms) and some are not. Thus, it would be useful to know the words that inherently express these relations this is done in the first module, i.e., relation format recognition module. The extraction of those sentences containing the cue words, and their evaluation is done in second module, i.e., relation extraction module.

- **Relation format recognition**

In this module, a pattern is given to identify the sentence containing related keywords. The input file is processed, and the sentence containing the pattern is found by matching the pattern with the input and referring the reference dataset. This process leaves all other sentences and recognizes only the particular sentence which has the pattern. Some of the patterns and their occurrences are shown in Table 1.

```
String = re.findall(r'\s*([^..]*.used to treat[^..]*[.])',
input)
```

The above code snippet explains the relation pattern recognition method. It finds all the occurrence of the sentences which contains the pattern "used to treat" from the input dataset. They get stored in a string from which further processing can be done. The following gives an example of a context in which this case would apply.

**Table 1** Examples for relation patterns

| Pattern | Example occurrences |
| --- | --- |
| X treats Y | Ginger treats cold |
| X such as Y | …medicine such as insulin… |
| X controlled by Y | …fever is controlled by basil… |
| X used to cure Y | Gokshura is used to cure hypertension… |

**Table 2** Entities and their relation

| Subject | Pattern | Object |
|---------|---------|--------|
| Gokshura | used to treat | hypertension |
| Rasona | used to treat | pressure |

> *Indians used to take foods on daily basis which have medicinal properties for about last 5000 years. As for example we take citrus lemon, chilli, black pepper, haldi, tulsi, ginger, garlic, etc. Terminaliaarjuna produces dose-dependent sensitivity in anaesthetized dogs. Action of mechanism for this particular herb is that it acts like a beta-blocker and is a powerful antioxidant, liver protectant and contains cardio-protective, hypolipidemic, anti-angina, and anti-atheroma properties. Gokshura is used to treat many diseases including hypertension. It has shown to be diuretic and an ACE inhibitor. Rasona is used to treat pressure.*

For the above example, the entities which are related to each other and recognized by the pattern "used to treat" are shown in Table 2.

- **Relation extraction**

In this module, the sentences with the cue words are retrieved. These sentences are then categories based on the type of entity and type of verb chunks. Thus, the relation between entities is retrieved. Some of the cue words used are shown in Fig. 2. Learning-based relation extraction is a work in progress.

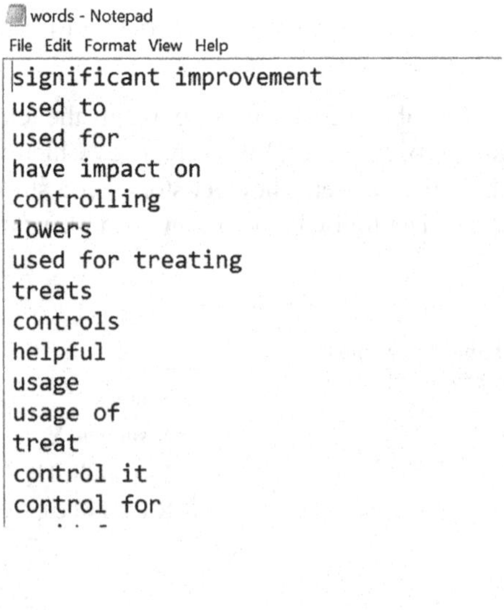

**Fig. 2** Sample action words for pattern recognition

```
words - Notepad
File Edit Format View Help
significant improvement
used to
used for
have impact on
controlling
lowers
used for treating
treats
controls
helpful
usage
usage of
treat
control it
control for
```

# 4 Results and Discussion

## 4.1 Dataset

The work was carried out for documents containing information about traditional medicines collected from AYUSH and Indian Journal of Traditional Knowledge. The dataset included more than 48,000 words and approximately 3500 sentences. For training the model for entity recognition, the feature set for 515 instances was used. These instances are candidate entities. The model was then tested for 20 documents containing about 600 sentences with 4423 instances. For pattern identification, a total of 3125 research papers were used from two different biomedical journals. In addition to this, a wordlist containing 91,000 biomedical terms were used as lookup base.

## 4.2 Results and Discussion

The evaluation part is divided into two stages. One is for the entity recognition phase, and the other is for relation extraction. For entity recognition, the training was done using five features such as frequency score, gazetteer score, verb type, hop distance, and location. The test was run for different algorithms, and their results was registered (Table 3). Out of the seven models, random forest model was found to give the best result. This situation persisted even when the number of instances was altered.

The model was then evaluated for 20 documents. These documents had unstructured and unfiltered text. The test set goes through the same preprocessing stages and feature extraction stages. Out of the whole text, only the candidate terms were run through the model built. The system gave an average f-measure of 0.714 (Table 4). The accuracy of the model can be improved by adding additional indicative features.

For the relation extraction phase, a collection of cue words were collected for different actions. The evaluation was carried out for *healing/curing* action which had a set of 41 cue words described in Table 5. A sentence which contains the

**Table 3** Evaluation results from different algorithms

| Classifier | F-measure | Correctly classified instances (%) |
|---|---|---|
| Naïve Bayes | 0.683 | 74.5631 |
| Naïve Bayes Multinominal | 0.650 | 75.534 |
| Decision Stump | 0.697 | 77.8641 |
| REP Tree | 0.804 | 81.5534 |
| Random Tree | 0.865 | 86.6019 |
| Random Forest | 0.881 | 88.3495 |

**Table 4** Evaluation results for entity recognition

| TEST dataset no. | No. of instance | Correctly classified instances (%) | Precision | Recall | F-score |
|---|---|---|---|---|---|
| 1 | 4 | 75 | 0.583 | 0.75 | 0.65 |
| 2 | 5 | 100 | 1 | 1 | 1 |
| 3 | 60 | 74.85 | 0.615 | 0.689 | 0.649 |
| 4 | 110 | 79.5666 | 0.583 | 0.75 | 0.65 |
| 5 | 40 | 69.1333 | 0.715 | 0.748 | 0.658 |
| 6 | 100 | 79.9383 | 0.82 | 0.796 | 0.748 |
| 7 | 124 | 69.3182 | 0.824 | 0.799 | 0.754 |
| 8 | 52 | 75.2135 | 0.481 | 0.693 | 0.568 |
| 9 | 352 | 75.2841 | 0.795 | 0.753 | 0.694 |
| 10 | 43 | 72.4512 | 0.694 | 0.746 | 0.683 |
| 11 | 210 | 74.6512 | 0.571 | 0.755 | 0.65 |
| 12 | 234 | 79.5666 | 0.805 | 0.816 | 0.804 |
| 13 | 312 | 79.9383 | 0.864 | 0.866 | 0.865 |
| 14 | 370 | 69.3182 | 0.881 | 0.833 | 0.881 |
| 15 | 411 | 75.2841 | 0.72 | 0.781 | 0.733 |
| 16 | 443 | 88.3495 | 0.57 | 0.753 | 0.649 |
| 17 | 521 | 86.6019 | 0.571 | 0.755 | 0.65 |
| 18 | 461 | 74.5631 | 0.693 | 0.779 | 0.697 |
| 19 | 451 | 75.3398 | 0.5 | 0.753 | 0.649 |
| 20 | 120 | 81.5534 | 0.571 | 0.755 | 0.65 |

**Table 5** Evaluation results for relation extraction by pattern matching

| Dataset no. | No. of pattern matched sentences | Precision |
|---|---|---|
| 1 | 5 | 0.4 |
| 2 | 7 | 0.143 |
| 3 | 2 | 1 |
| 4 | 4 | 0.25 |
| 5 | 5 | 0.6 |
| 6 | 8 | 0.25 |
| 7 | 3 | 0.333 |
| 8 | 11 | 0.272 |
| 9 | 4 | 0.5 |
| 10 | 6 | 0.167 |
| 11 | 8 | 0.375 |
| 12 | 7 | 0.286 |
| 13 | 3 | 0.667 |

(continued)

**Table 5** (continued)

| Dataset no. | No. of pattern matched sentences | Precision |
|---|---|---|
| 14 | 10 | 0.2 |
| 15 | 6 | 0.5 |
| 16 | 8 | 0.375 |
| 17 | 9 | 0.333 |
| 18 | 1 | 1 |
| 19 | 5 | 0.4 |
| 20 | 12 | 0.333 |

pattern may not have the keywords related to each other. Such sentences are also get recognized by using the pattern recognition method which is a drawback of this method. Also, this method requires hand building patterns for each relation that keeps growing which is tedious to write and maintain.

# 5 Conclusion

Any information extraction system is built on the two strong basements of entity center and their relation structure. In our work, we try to focus on these two aspects, to retrieve structural information from a noisy bland collection of data. The system was able to identify entities with an accuracy of about 88%. Their accuracy maybe improved with additional features and categories though alteration of ML algorithms may not have an effect on it. The relation extraction on the other hand requires ample updating as the pattern matching is still in the basic levels of operations, and the patterns lack variety and values. Also we assume that the amount of processing can be highly reduced if a better statistical or learning algorithm is used.

**Acknowledgements** This research is funded by **DST-INSPIRE Fellowship** under the governance of The **Department of Science and Technology**, India.

# References

1. Hirschman, L., Morgan, A.A. and Yeh, A.S., Rutabaga by any other name: extracting biological names. Journal of Biomedical Informatics, 35(4), pp. 247–259 (2002).
2. Gaizauskas, R., Humphreys, K., Cunningham, H. and Wilks, Y., November. University of Sheffield: description of the LaSIE system as used for MUC-6. In Proceedings of the 6th conference on Message understanding, pp. 207–220. Association for Computational Linguistics (1995).
3. Da Silva, J.F., Kozareva, Z. and Lopes, J.G.P., May. Cluster Analysis and Classification of Named Entities. In LREC (2004).

4. McCallum, A. and Li, W., May. Early results for named entity recognition with conditional random fields, feature induction and web-enhanced lexicons. In Proceedings of the seventh conference on Natural language learning at HLT-NAACL 2003-Volume 4, pp. 188–191. Association for Computational Linguistics (2003).
5. Jansche, M., August. Named entity extraction with conditional markov models and classifiers. In proceedings of the 6th conference on Natural language learning-Volume 20, pp. 1–4. Association for Computational Linguistics (2002).
6. Shen, D., Zhang, J., Zhou, G., Su, J. and Tan, C.L., July. Effective adaptation of a hidden markov model-based named entity recognizer for biomedical domain. In Proceedings of the ACL 2003 workshop on Natural language processing in biomedicine-Volume 13, pp. 49–56. Association for Computational Linguistics (2003).
7. Zhang, J., Shen, D., Zhou, G., Su, J. and Tan, C.L., Enhancing HMM-based biomedical named entity recognition by studying special phenomena. Journal of biomedical informatics, 37(6), pp. 411–422 (2004).
8. Isozaki, H. and Kazawa, H., August. Efficient support vector classifiers for named entity recognition. In Proceedings of the 19th international conference on Computational linguistics-Volume 1, pp. 1–7. Association for Computational Linguistics (2002).
9. Sekine, S. and Nobata, C., May. Definition, Dictionaries and Tagger for Extended Named Entity Hierarchy. In LREC, pp. 1977–1980 (2004).
10. Dong, X., Qian, L., Guan, Y., Huang, L., Yu, Q. and Yang, J., August. A multiclass classification method based on deep learning for named entity recognition in electronic medical records. In Scientific Data Summit (NYSDS), IEEE New York, pp. 1–10 (2016).
11. Sun, Y., Li, L., Xie, Z., Xie, Q., Li, X. and Xu, G., 2017, March. Co-training an Improved Recurrent Neural Network with Probability Statistic Models for Named Entity Recognition. In International Conference on Database Systems for Advanced Applications, pp. 545–555. Springer, Cham (2017).
12. Kambhatla, N., July. Combining lexical, syntactic, and semantic features with maximum entropy models for extracting relations. In Proceedings of the ACL 2004 on Interactive poster and demonstration sessions, p. 22. Association for Computational Linguistics (2004).
13. Rink, B. and Harabagiu, S., July. Utd: Classifying semantic relations by combining lexical and semantic resources. In Proceedings of the 5th International Workshop on Semantic Evaluation, pp. 256–259. Association for Computational Linguistics (2010).
14. Zeng, D., Liu, K., Lai, S., Zhou, G. and Zhao, J., August. Relation Classification via Convolutional Deep Neural Network. In COLING, pp. 2335–2344 (2014).
15. Santos, C.N.D., Xiang, B. and Zhou, B., Classifying relations by ranking with convolutional neural networks. arXiv preprint arXiv:1504.06580 (2015).
16. Mark Hall, Eibe Frank, Geoffrey Holmes, Bernhard Pfahringer, Peter Reutemann, and Ian H. Witten. The WEKA Data Mining Software: An Update. SIGKDD Explorations, 11(1), (2009).

# Effort Estimation for Mobile Applications Using Use Case Point (UCP)

Anureet Kaur and Kulwant Kaur

**Abstract** Estimating the software size and effort helps in early prediction of uncertainties in software development. Determining the size helps in ascertaining the effort, cost, and schedule for entire project. In context of mobile software, the existing techniques for software effort estimations could be adapted as such or with modifications. There are many software size and effort estimation metrics. In this paper, Use Case Point (UCP) metric is aimed for estimating size and effort for mobile application. Five android mobile applications are considered as a case study, and difference in actual effort and estimated effort is evaluated. Modified UCP has been also proposed to improve the results by considering mobile-specific characteristics.

**Keywords** Effort · Estimation · Mobile application · Software metric
Use Case Point (UCP)

## 1 Introduction

The rapid growth of mobile technology has attracted many researchers to encompass their studies in any area related to mobile domain. Mobile applications are the software that runs on mobile devices. With the increase in number of mobile devices, the mobile users will create an increasing demand for all sorts of applications that run on mobile devices. Conferring to Cisco Visual Networking Index [1], the traffic of data over mobile for year 2015–2020, is forecasted to have more than 5.5 billion mobile users, which would be 70% of world population. Also, Agreeing to App Annie [2], there will be massive growth in the app market by

A. Kaur (✉)
I.K. Gujral Punjab Technical University, Kapurthala, India
e-mail: anumahal@gmail.com

K. Kaur
Apeejay Institute of Management Technical Campus, Jalandhar, India
e-mail: kulwantkaur@apjimtc.org

© Springer Nature Singapore Pte Ltd. 2019
B. K. Panigrahi et al. (eds.), *Smart Innovations in Communication and Computational Sciences*, Advances in Intelligent Systems and Computing 669, https://doi.org/10.1007/978-981-10-8968-8_14

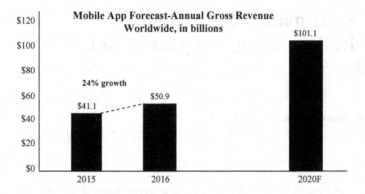

**Fig. 1** Annual growth forecast of mobile applications by App Annie [2]

2020. Not only that, Google Play downloads are expected to triple by 2020, when the global app market is expected to break the $100 billion barrier. Figure 1 depicts the annual growth forecast of mobile applications. For development of mobile application, the existing models used in software development are adapted. Also, the software size estimation methodologies used for software development like as Line of Code, Feature Point Metric, COCOMO models, Use Case Points, Function Point Analysis [3–5] are adapted to mobile application size estimation [6–8]. So the existing effort estimation Use Case Point (UCP) can be adapted for estimating effort in mobile applications. In this paper, Sect. 2 reviews existing literature related to Use Case Point (UCP). In Sect. 3, the methodology adopted for calculating estimated effort is summarized and a case study on five android mobile applications for implementing Use Case Point is taken and results are evaluated. Section 4 proposes a new methodology for estimating size for mobile applications. Finally, Sect. 5 concludes the paper.

## 2   Existing Literature

Till now, many researchers have reported about Use Case Point (UCP) and several case studies have also been done using Use Case Point method. Use Case Point is a method which is founded on use case model that is further based on functional range of the software [9]. Robiolo, G. and Orosco R. [10] highlighted drawbacks of function points listing need to know all the requirements well known in advance. Such limitations are improved using Use Case Point but still with a limitation having implemented only on projects with common environmental circumstances. Schneider, G et al. [11] also compared Use Case Point to be better effort estimation measure but concluded saying that it lacks certain standards for total control. Damodaran M. and Washington A [12] also showed that there is still lack of proper standards in Use Case Point and if these limitations are removed then UCP can be considered as an efficient method for size and effort estimation. Carroll, E. R. [13] changed the original UCP by modifying factors affecting the calculations and then applied the modified version on 200 projects. Ali Bou Nassif et al. [14] have proposed a new Use Case Model based on regression alongside fuzzy model for effort estimation. Ali Bou Nassif et al. [15] suggested using artificial neural network (ANN) for estimating effort founded on the Use Case Point (UCP) model. J. Smith [16] provided a model to calculate Line of Code (LOC) along with use cases. Jonathan Lee et al. [17] offered a modified Use Case Point metric for calculating the unadjusted Use Case Point-based relations and goals with the fuzzy membership functions.

## 3   Existing Methodology

This section briefly explains the steps taken for calculating effort estimation by means of Use Case Point method [18]. The steps are as follows:

1. Compute Unadjusted Use Case Weight (UUCW): In this step, number of use cases are recognized and categorized as simple, average, and complex founded on no. of operations contained in each use case as shown in Table 1.

$$UUCW = (\text{Total Simple Use Cases} \times 5) + (\text{Total Average Use Case} \times 10) \\ + (\text{Total Complex Use Cases} \times 15) \tag{1}$$

2. Compute Unadjusted Actor Weight (UAW): In this step, all the actors in all the use cases are recognized and categorized as simple, average, and complex and are assigned weights accordingly as shown in Table 2.

$$UAW = (\text{Total Simple actors} \times 1) + (\text{Total Average actors} \times 2) \\ + (\text{Total Complex actors} \times 3) \tag{2}$$

3. Compute the Technical Complexity Factor (TCF): In this step, 13 Technical Factors are considered and assigned weights in the range beginning 0 (No Impact) to 5 (Strong Impact) as shown in Table 3. Each factor's given value from range is then multiplied by the corresponding weight. All multiplied values are then are then totaled up to give value for Technical Factor (TF).

**Table 1** Unadjusted Use Case Weights (UUCWs) based on number of operations

| Use case complexity | Operations (No.) | Weight |
|---|---|---|
| Simple | <3 | 5 |
| Average | 4–7 | 10 |
| Complex | >7 | 15 |

**Table 2** Unadjusted Actor Weight (UAW) based on actor type

| Actor type | Weight |
|---|---|
| Simple | 1 |
| Average | 2 |
| Complex | 3 |

**Table 3** Technical Complexity Factor (TCF) with weights

| ID | Technical Factor | Weight |
|---|---|---|
| I. | Distributed system | 2 |
| II. | Response | 1 |
| III. | End-user efficiency (online) | 1 |
| IV. | Complex internal processing | 1 |
| V. | Reusable code | 1 |
| VI. | Install ease | 0.5 |
| VII. | Use ease | 0.5 |
| VIII. | Portable | 2 |
| IX. | Easy to change | 1 |
| X. | Concurrent | 1 |
| XI. | Security features | 1 |
| XII. | Third parties access | 1 |
| XIII. | Requirement of user training facility | 1 |

**Table 4** Environmental Complexity Factor (ECF) with weights

| ID | Environmental Factor | Weight |
|---|---|---|
| I. | Familiarity with the Project | 1.5 |
| II. | Experience | 0.5 |
| III. | OO Programming Experience | 1 |
| IV. | Lead Analyst Ability | 0.5 |
| V. | Motivation | 1 |
| VI. | Stable Requirements | 2 |
| VII. | Part Time Staff | −1 |
| VIII. | Complex Programming Language | −1 |

4. Compute Environment Complexity Factor (EF): In this step, 8 Environmental Factors are considered and assigned weights in the range from 0 (no experience) and 5 (expert) as shown in Table 4. Each factor's given value from range is then multiplied by the corresponding weight. All multiplied values are then are then totaled up to give value for Environmental Factor (EF).
5. Compute Unadjusted Use Case Points (UUCP):

$$UUCP = UUCW + UAW \tag{3}$$

6. Compute Complexity Factor, where:

$$TCF = 0.6 + (0.01 * TF) \qquad (4)$$

$$ECF = 1.4 + (-0.03 * EF) \qquad (5)$$

7. Calculate the Use Case Point (UCP):

$$UCP = UUCP * TCF * ECF \qquad (6)$$

8. Calculate Effort Estimation

$$Effort\ Estimation = UCP * ER \qquad (7)$$

ER in staff-hours/UCP

The value of ER = 20 staff − hours/UCP as proposed by Karner [18].   (8)

## 3.1  A Case Study Using Existing Methodology

To use the existing methodology, Use Case Point, for estimating effort on mobile applications, five mobile applications are taken into consideration. The Technical and Environmental Factors for all five apps are studied. Actor Weights and Use Case Weights are estimated from Use Case Diagrams of these apps. The detailed description of one app is given, starting from Use Case Diagram and following UCP calculation steps. "Weather forecast" mobile app is to be developed. Steps followed for calculating UCP for this app are as follows:

1. Compute Unadjusted Use Case Weight (UUCW): Table 5 shows calculation of total UUCW.
2. Calculate Unadjusted Actor Weight (UAW): Table 6 shows calculation of total UAW.
3. Calculate Technical Factor (TF) = 49.5
4. Calculate Environment Factor (EF) = 15.66

**Table 5** Unadjusted Use Case Weights (UUCW) based on no. of operations for weather forecast mobile app

| Use case complexity | Operations (No.) | Weight | Weight * No. of use cases |
|---|---|---|---|
| Simple | <3 | 5 | 5 * 6 = 30 |
| Average | 4–7 | 10 | 10 * 2 = 20 |
| Complex | >7 | 15 | 15 * 0 = 0 |
| Total UUCW | | | 50 |

**Table 6** Unadjusted Actor Weight (UAW) based on actor type for weather forecast mobile app

| Actor type | Weight | Weight * No. of actor type |
|---|---|---|
| Simple | 1 | 1 * 0 = 0 |
| Average | 2 | 2 * 2 = 4 |
| Complex | 3 | 3 * 0 = 0 |
| Total UAW | | 4 |

**Table 7** Calculation of Use Case Point for five mobile apps

| [a]Mobile App | UUCW | UAW | UUCP | TCF | ECF | UCP |
|---|---|---|---|---|---|---|
| WeatherForecast | 50 | 4 | 54 | 1.095 | 0.93 | 54.99 |
| TrafficAnalyser | 90 | 4 | 94 | 1.05 | 0.89 | 87.843 |
| ChatMsgr | 75 | 5 | 80 | 0.98 | 0.92 | 72.128 |
| FileTrans | 80 | 7 | 87 | 1.05 | 0.92 | 84.042 |
| CamCorder | 74 | 6 | 80 | 1.04 | 0.96 | 79.872 |

[a]http://www.sourcecodester.com

**Table 8** Calculation of estimated effort for five mobile apps

| Mobile App | UCP | Estimated effort (Person-month) | Actual effort (Person-month) |
|---|---|---|---|
| WeatherForecast | 54.99 | 6.87 | 8.01 |
| TrafficAnalyser | 87.843 | 10.98 | 12.4 |
| ChatMsgr | 72.128 | 9.01 | 10.3 |
| FileTrans | 84.042 | 10.50 | 11.4 |
| CamCorder | 79.872 | 9.984 | 11.3 |

5. Calculate Unadjusted Use Case Points (UUCP) = 50 + 4 = 54
6. Calculate Complexity Factor, where:

$$TCF = 0.6 + (0.01 * 49.5) = 1.095$$
$$ECF = 1.4 + (-0.03 * 15.66) = 0.93$$

7. Calculate the Use Case Point (UCP) = 54 * 1.095 * 0.93 = 54.99
8. Effort Estimation = 54.99 * 20 = 1099.8 (Person-hours) = 6.87 (Person-months)

The calculation of UCP for other four mobile applications is shown in Table 7. The effort estimation using UCP and actual effort incurred for all five mobile applications is shown in Table 8.

## 3.2 Result Evaluation

Figure 2 shows comparative chart of actual effort and estimated effort using UCP. From this figure, it can be assessed that effort estimation is underestimated for all the mobile applications as compared to actual effort incurred. For evaluating the results, a common measure for accessing accuracy of estimate, magnitude of relative error (MRE) [19], is used. Table 9 shows calculation of MRE (%) for all five mobile applications.

$$MRE(\%) = (|Actual\,Effort - Estimated\,Effort|/Actual\,Effort) * 100 \qquad (9)$$

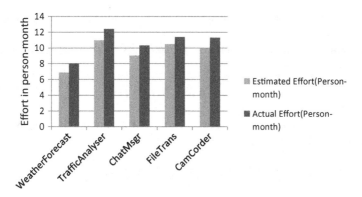

**Fig. 2** Comparative chart of actual effort and estimated effort using UCP

**Table 9** Calculation of estimated effort for five mobile apps

| Mobile App | Estimated effort (Person-month) | Actual effort (Person-month) | MRE (%) |
|---|---|---|---|
| WeatherForecast | 6.87 | 8.01 | 14.2 |
| TrafficAnalyser | 10.98 | 12.4 | 11.45 |
| ChatMsgr | 9.01 | 10.3 | 12.5 |
| FileTrans | 10.50 | 11.4 | 7.89 |
| CamCorder | 9.984 | 11.3 | 11.64 |

## 4  Proposed Methodology

In Sect. 3, the estimated efforts are underestimated as compared to actual efforts for mobile applications. Mobile applications have specific attributes which are different from traditional software. So the estimation models for traditional software need to be modified to apt them to mobile software by considering mobile-specific characteristics. De Souza et al. [20] addressed these characteristics and by adding them into existing Use Case Point calculation procedure, it is believed that more accurate and precise result for estimation of effort in mobile application domain can be achieved. Figure 3 shows a modified UCP or can be named M-UCP for effort estimation.

The following steps are used to estimate the size and effort in developing mobile applications in proposed methodology:

1. The first and second step is same as that of calculating UUCW and UAW in Use Case Point method.
2. Calculation of Technical Complexity Factor (TCF) considering 14 Technical Complexity Factors.

$$TCF = 0.65 + (0.01 * TDI) \qquad (10)$$

3. Calculate Environment Complexity Factor (ECF): This step is same as step 4 of UCP considering eight Environmental Factors.

$$ECF = 1.4 + (-0.03 * EF) \qquad (11)$$

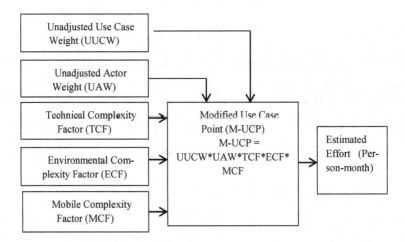

**Fig. 3**  Proposed M-UCP for effort estimation

4. Calculate Mobile Complexity Factor (MCF): In this step, 14 Mobile Factors proposed by De Souza et al. [20] are considered and assigned weights in the range from 0.90 (worst) and 1.10 (Excellent) situations. The degree of influence rating of each characteristic is summed up to give mobile total degree of influence (MTDI).

$$MTDI = \sum_{i=1}^{14} Fi \qquad (12)$$

where Fi is factor weighted from 0.90 to 1.10

$$MCF = C1 + (C2 * MTDI) \qquad (13)$$

where C1 and C2 will be constants based on results of interviews with mobile app developers.
5. To find the Mobile Adjusted Function Points (MAFPs), UFP is multiplied by Technical Complexity Factor (TCF), Environmental Complexity Factor (ECF), and Mobile Complexity Factor (MCF).

$$MAFP = UFP * TCF * ECF * MCF \qquad (14)$$

6. Calculate Effort

$$Effort = FP/Productivity \text{ factor in person month.} \qquad (15)$$

## 5   Conclusion and Future Scope

Early estimation of software project helps in planning software resources efficiently. In this paper, existing effort estimation metric Use Case Point is implemented on estimating size and effort for mobile software. The results show that the estimated efforts are underestimated as compared to actual efforts for mobile applications. As mobile applications have different attributes compared to traditional software that run on desktop or laptops, we cannot ignore this while computing estimations. So the existing Use Case Point (UCP) can be modified to include mobile-specific attributes. This paper proposes a new model incorporating mobile attributes along with other Technical and Environmental Factors for computing UCP.

In future, the proposed model will be implemented considering all the above factors on mobile applications.

**Ethics Approval and Consent to Participate** All procedures performed in studies involving human participants were in accordance with the ethical standards of the institutional and/or national research committee and with the 1964 Helsinki declaration and its later amendments or comparable ethical standards. Informed consent was obtained from all individual participants included in the study.

# References

1. https://newsroom.cisco.com/press-release-content?articleId=1741352 (As accessed on 25/2/ 2017)
2. http://www.androidauthority.com/google-play-store-apple-app-store-downloads-673499/ (As accessed on 25/2/2017)
3. Yinhuan Zheng, Yilong Zheng, Beizhan Wang, Liang Shi. Estimation of software projects effort based on function point. In: 4th International Conference on Computer Science and Education (2009).
4. P. Jodpimai, P. Sophatsathit and C. Lursinsap. Analysis of effort estimation based on software project models. In: 9th International Symposium on Communications and Information Technology, Icheon, pp. 715–720. https://doi.org/10.1109/iscit.2009.5341149 (2009).
5. M. Nasir. A Survey of Software Estimation Techniques and Project Planning Practices. In: Proceedings of the Seventh ACIS International Conference on Software Engineering, Artificial Intelligence, (SNPD'06), IEEE Computer Society (2006).
6. L. S. de Souza and G. S. de Aquino. The applicability of present estimation models to the context of mobile applications. In: 2014 9th International Conference on Evaluation of Novel Approaches to Software Engineering (ENASE), Lisbon, Portugal, pp. 1–6 (2014).
7. De Souza, L. S. & de Aquino Jr, G.S. Estimating the Effort of Mobile Application Development. In: Proceedings of Second International Conference on Computational Science and Engineering, 45–63. https://doi.org/10.5121/csit.2014.4405 (2014).
8. A. Nitze, A. Schmietendorf and R. Dumke (2014) An Analogy-Based Effort Estimation Approach for Mobile Application Development Projects. In: Joint Conference of the International Workshop on Software Measurement and the International Conference on Software Process and Product Measurement, Rotterdam, pp. 99–103. https://doi.org/10.1109/ iwsm.mensura (2014).
9. G. Schneider and J. P. Winters. Applying Use Cases, Second Edition. Addison Wesley (2001).
10. Robiolo, G., and Orosco, R. Employing use cases to early estimate effort with simpler metrics. In: Innovations in Systems and Software Engineering, Vol. 4 (1), 31–43 (2008).
11. Schneider, G. and winters, J. P. Applied use Cases, Second Edition, A Practical Guide. Addison-Wesley (2001).
12. Damodaran, M., and Washington, A. Estimation using use case points. Computer Science Program. Texas–Victoria: University of Houston. Sd (2002).
13. Carroll, E. R. Estimating software based on use case points. In: 20th annual ACM SIGPLAN conference on Object-oriented programming, systems, languages, and applications ACM. (pp. 257–265) (2005).
14. Ali Bou Nassif, Luiz Fernando Capretz, Danny Ho. Estimating Software Effort Based on Use Case Point Model Using Sugeno Fuzzy Inference System. In: International Conference on Tools with Artificial Intelligence, IEEE, pp. 393–398 (2011).
15. Ali Bou Nassif, Luiz Fernando Capretz, Danny Ho. Estimating Software Effort Using an ANN Model Based on Use Case Points. International Conference on Machine Learning and Applications, IEEE, pp. 42–46 (2012).
16. J. Smith. The Estimation of Effort Based on Use Cases. Rational Software white paper (1999).
17. Jonathan Lee, Hen-Tin Lee, Jong-YihKuo. Fuzzy Logic as a Basic for Use Case Point estimation. In: International Conference on Fuzzy Systems, IEEE, pp. 2702–2707 (2011).
18. Karner, G. Resource estimation for objectory projects. Objective Systems (1993).
19. L. C. Briand, K. E. Emam, D. Surmann, I. Wieczorek and K. D. Maxwell. An assessment and comparison of common software cost estimation modeling techniques. In: ICSE'99, vol. 0, pp. 313–322 (1999).
20. De Souza L.S., De Aquino G.S. Mobile Application Development: How to Estimate the Effort? In: Murgante B. et al. In: Computational Science and Its Applications – ICCSA 2014. ICCSA 2014. Lecture Notes in Computer Science, Vol. 8583. Springer, Cham (2014).

# Part II
# Intelligent Communications and Networking

# Authentication, KDC, and Key Pre-distribution Techniques-Based Model for Securing AODV Routing Protocol in MANET

Sachin Malhotra and Munesh C. Trivedi

**Abstract** Open wireless communication medium makes MANETs susceptible to various attacks. Security scheme implementation needs an adequate amount of memory and processing power of networking and communicating devices. But devices' uses in MANET not satisfying this requirement make implementation of security scheme more challenging. In MANET, absence of networking devices (communicating node works as router as well as host) makes this task more difficult. In this paper, authentication, key distribution center (KDC) along with key pre-distribution techniques-based model has been proposed to secure the AODV routing protocol against most dangerous and frequently happed attacks (black hole, gray hole, rushing attacks, message impersonation, spoofing, fabrication attacks, etc.). In this scheme, two levels of authentication using message authentication code are used. At first level, share key (key table distributed at the time of deployment) is used for authentication, and for second level, second key distributed by KDC at run time is used for generating the authentication code. Key pre-distribution (key table distributed at the time of deployment) is uses for speedup the our algorithm and reduces the overhead of each node to distribute the key during the communication while KDC based second key distribution is uses to enhances attacks prevention capability of our proposed scheme. Simulation results showing in this paper our proposed algorithm working efficiently well in the presence and absence of malicious nodes as compared to state of art AODV protocol and recently published well defined protocols. NS2.35 on Ubuntu 12.04 LTS with 4 GB RAM is used for simulation. PDR, AE2ED, and average TP parameters have been used for testing the performance of our proposed algorithm.

S. Malhotra · M. C. Trivedi (✉)
Department of Computer Science & Engineering, PAHER University,
Udaipur, India
e-mail: munesh.trivedi@gmail.com

S. Malhotra
e-mail: sachin_malhotra123@yahoo.com

M. C. Trivedi
Department of Computer Science & Engineering, ABES Engineering
College, Ghaziabad, India

© Springer Nature Singapore Pte Ltd. 2019
B. K. Panigrahi et al. (eds.), *Smart Innovations in Communication and Computational Sciences*, Advances in Intelligent Systems and Computing 669, https://doi.org/10.1007/978-981-10-8968-8_15

175

**Keywords** AODV routing protocol · MANETs · Attacks · Secure communication · KDC · Key pre-distribution · PDR · AE2ED Average TP · Message authentication code

# 1 Introduction

With increasing the number of applications of MANET [1, 2] (from home applications to defense applications), the demand for making the communication more secure is also increasing rapidly. Achieving this requirement in MANET is very difficult due to its nature as discussed in number of research papers [3–6]. In this work, we are basically focusing on explaining the working of proposed algorithm. Because today, the number of research work is available online in which introduction about MANET, working of routing protocols, security challenges, and possible MANET attacks are given in details, so no need to explain these things again and again [7, 8].

As far as classes security algorithms are concerned, they are divided into two classes: traditional security scheme based on symmetric key and modern security scheme based on asymmetric key. These two schemes have own advantages and limitations. The tradition symmetric key-based security schemes are easy to understand (uses simple mathematics) and simple and easy to implement (do not require much processing and storage space), while modern asymmetric key-based security schemes are difficult to understand and complex and complicated in implementation (need much more processing and memory power). As we know, the processing and storage capability of the MANET devices are limited, so we cannot go for asymmetric-based security scheme. In this paper, symmetric key-based authentication scheme is used for securing the AODV routing protocol against most dangerous and frequently happened attacks (black hole, gray hole, rushing attacks, message impersonation, spoofing, fabrication attacks, etc.).

In this paper, we have used two levels of authentication, at first level using SHA1 with first key (select key from key table deployed at each node) for hop-to-hop authentication [9, 10] and at second level using MD5 algorithm using second key (distributed by KDC between sender and receiver) for end-to-end authentication. First_Digest of size 160 bits is generated by the SHA1 algorithm, and Second_Digest of size 128 bits is generated by MD5 algorithm.

The rest of this paper is organized as follows: Sect. 2 introduces related work with relation to previous work and motivation to doing this research. Section 3 describes the proposed methodology. Section 4 explains simulation result and discussion, and finally conclusion is defined in Sect. 5.

## 2 Relation to Previous Works in Same Direction and Motivation

Research in the field of securing routing protocols in MANET is getting much more attention in recent years. The numbers of research works have been published in this field in last few years. In this section, the works which are similar and related to our proposed work are considered for showing the relation of our proposed model to existing works and for comparison purpose. Especially, we cover the symmetric key-based security mechanisms that have been used by Arya K. V. and Rajput S. S. [3] to secure AODV routing protocol using nested MAC. This model also used the key pre-distribution (key table distributed at the time of deployment) technique to overcome the limitation of model proposed by P. Sachan et. al. [11] that distribute the keys at run time (when communication connection establishing between the sender and receiver). Arya K. V. and Rajput S. S. [3] model significantly prevents the networks from many attacks (impersonation, modified routing information, black hole). When attacker is outsider it works efficiently but when attacker is insider i.e. our genuine node is compromised by the attacker work little bit ineffi-cient, this is the limitation of the method. Similar concept is used by Rajput S. S. [4] to protect ZRP routing protocol in MANETs against frequently occurred attacks. We are motivated to do the research in this field by these two papers. Detailed comprehensive survey about security issues and challenges are given in [12, 13].

## 3 Proposed Security Mechanism

The proposed model is using the traditional, easy to understand and simple and easy to implement security scheme, i.e., symmetric key-based authentication scheme. To boost up our model in terms of handling the attacks, two levels (hop-to-hop and end-to-end) of authentication with different algorithms and with different keys are used. In this paper, authentication, key distribution center (KDC) along with key pre-distribution techniques-based model has been proposed to secure the AODV routing protocol against most dangerous and frequently happened attacks (black hole, gray hole, rushing attacks, message impersonation, spoofing, fabrication attacks, etc.). In this scheme, two levels of authentication using message authen-tication code are used. At first level, share key (key table distributed at the time of deployment) is used for authentication, and for second level, second key distributed by KDC at run time is used for generating the authentication code. Key pre-distribution (key table distributed at the time of deployment) is uses for speedup the our algorithm and reduces the overhead of each node to distribute the key during the communication while KDC based second key distribution is uses to enhances attacks prevention capability of our proposed scheme. The working model of proposed model is shown in Fig. 1. As shown in figure, at sender side we are using two authentication algorithms: First, SHA1 with first key (select key from key

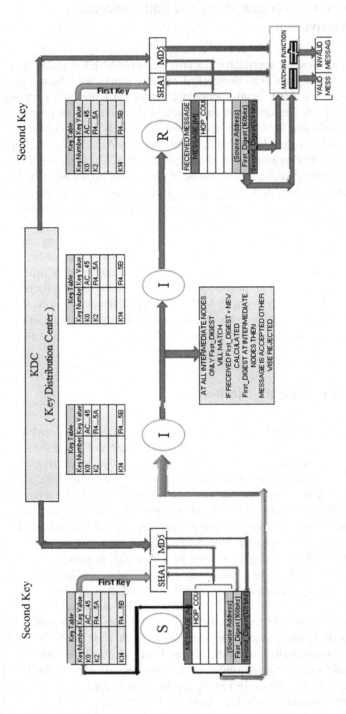

**Fig. 1** Working model of proposed scheme

table deployed at each node) is used for hop-to-hop authentication code generation [9, 10] and then MD5 algorithm using second key (distributed by KDC between sender and receiver) for end-to-end authentication code generation. First_Digest of size 160 bits is generated by the SHA1 algorithm, and Second_Digest of size 128 bits is generated by MD5 algorithm. At the destination side, same procedure is followed to check the validity and integrity of the message. It means only at sender and receiver side two codes (First_Digest & Second_Digest) are generated otherwise at intermediate side only first_digest is verified for the purpose of testing the integrity of the message. For speedup the algorithm 10 keys (K0–K9) in the form of key table stored at the time of deployment at each node. One key out of these keys is used for generating First_Digest of size 160 bits using SHA1 algorithm. For generating Second_Digest, of size 128 bits using SHA1 algorithm second key uses i.e. key generating by KDC at run time (when route is established).

The advantage of using two different keys (Key_1 from the shared symmetric key table and Key_2 from KDC) for making two levels of authentication is to speed up the algorithm and improve the security level as compared to Method [1], Method [2]. This mechanism helps us to protect our network from external as well as internal attacks. The concept of KDC in this model helps us to protect our network from black hole, gray hole, rushing attacks, message impersonation, spoofing, and fabrication attacks which are not possible in previous methods. For simulation purpose, same behavior of the attacker is used as discussed in Method [1] and Method [2]. As in Fig. 1, four nodes (S: sender, I: intermediate, R: receiver) are used to illustrate the working of our proposed scheme.

*At the sender side*: Sender S uses Key_1(from Key table) to generate First_Digest of size 160 bits using SHA 1 algorithm. Key_1 selects via same concept used in Method [1] and Method [2]. (Key_1 = Key_ number (Hop_count mode 10) send the request to KDC to generate key_2 for generating Second_Digest of size 128 bits using MD 1 algorithm then whole message (Message + First_Digest_1+ Seccond_Digest) send to the next node in the route.

*At the intermediate node*: On receiving the message, First_Digest is generated using SHA 1 algorithm and matches with First_Digest value coming in the input packet. If the value of new calculated First_Digest and received First_Digest is same, then message is treated as valid message, and then again First_Digest is created using next key in the key table and message along with calculated First_digest is forwarded to the next node in the route. If the value of new calculated First_Digest and received First_Digest is not same, then message is treated as invalid message and simply discard the message.

*At the receiver side*: On receiving the packet prom the previous node, new First_Digest and new second_Digest are created by using Key_1 and Key_2, respectively. Received message is valid only if both new digest matches with received digest otherwise received message is invalid and discard it. Step-by-step description of our model in the form of algorithm is giving in Algorithm 1.

## 4  Simulation and Result Analysis

NS2.35 [14] tool has been used to implement proposed model. Simulation parameters are given in Table 1.

---

**ALGORITHM 1:** PROPOSED SECURITY MECHANISM ALGORITHM

---

**Abbreviations:**

**KDC:** key distribution center. Make 5% of total nodes in the network as KDS. KDC nodes have more processing and memory power then remaining other nodes in the network.

**First_Digist:** 160 bits Message Authentication code generates by SHA 1 Algorithm

**Second_Digist:** 128 bits Message Authentication code generates by MD5 Algorithm

**H_Count:** Hop Count value in the route packet

**Key_1:** Shared key selected for hop to hop Authentication from Key table
    Key_1 = Key number (H_Count mode 10)   for ex. K0, K1, --- K9

**Key_2:** provided by KDC on the request of source node for end to end authentication
    It uses by MD5 Algorithm

**S:** Source Node, **R:** Destination Node, **I:** Intermediate node or nodes

**At Sender Node (S):**

**Step1:** Source node create message **M** for sending to **R**

**Step2:** Send request to **KDC** for **Key_2**, then **KDC** send **Key_2** to **S** and **R**

**Step3:** Select **Key_1 = K_0** (because **H_count** = 0 at sender) from the Key table

**Step4:** Receive **Key_2** form **KDC**

**Step5:** Calculate **Firest_Digest** = SHA1 (M, Key_1)

**Step6:** Calculate Second_**Digest** = MD5 (M, Key_2).

**Step7: Send (M+First_Digest+Second_Digest)**

**At Intermediate Nodes (I):**

**Step1:  Receive (M+First_Digest+Second_Digest)**

**Step2:** Select **Key_1** = Key_(H_Count mode 10) from the Key table

**Step3:** Calculate **First_Digest** = SHA1 (M, Key_1)

        IF (**New First_Digest**= = **Received First_Digets**)  **THEN:**
            H_Count = H_Count + 1;   Key_1 = Key_(H_Count mode 15)
            Calculate **First_Digest**= SHA1 (M, Key_1)
            **Send (M+First_Digest+Second_Digest)**
        **ELSE D**iscard **M** (M is invalid);

**At Receiver Node (R):**

**Step1:  Receive (M+First_Digest+Second_Digest)**

**Step2:** Select **Key_1** = Key_(H_Count mode 10) from the Key table

**Step3:** Received **Key_2** from **KDC**

**Step4:** Calculate **First_Digest = SHA1** (M, Key_1)

**Step5:** Calculate **Second_Digest = MD5** (M, Key_2).

    IF (New First_Digest= = Received First_Digets    **&&**
        New Seond_Digest = = Received Second_Digets)
    **THEN M** is valid and accepts
    **ELSE         Discard M (M is invalid);**

---

In the work, previous two models proposed by K. V. Arya and S. S. Rajput [3] (named as Method [1]) and another model proposed by S. S. Rajput et al. [4]

**Table 1** Simulation parameters

| Simulator | NS2(v-2.35) |
|---|---|
| Simulation time | 150 s |
| Performance parameters | Throughput (TP), PDR, Average End-To-End Delay (AE2ED) |
| Area size | 800 m × 600 m |
| Transmission range | 100 m |
| Number of nodes | 10–150 |
| Previous models | Method [1], Method [2] |
| Protocol | AODV |
| Transmission range | 250 m |
| Maximum speed | 0–20 m/s |
| Application traffic | CBR |
| Packet size | 512 bytes |
| Traffic rate | 4 packet/s |
| Node mobility model | Random waypoint model |
| Pause time | 10, 20, 60, 100 to 140 s |
| Mac model | 802.15.4 |

(named as Method [2]) are used to compare the performance of our proposed model. Method [2] is originally proposed for the ZRP routing protocol, but in this work we have changed this model to make compatible with the AODV routing protocol. Performances of all the models are measured in terms of AE2ED, TP, and PDR with the presence and absence of the malicious nodes.

Performance comparisons of all four models, i.e., our proposed model, Method [1], Method [2], and AODV in terms of all performance parameters (AE2ED, PDR, and Average TP) versus different pause time are shown in Figs. 2, 3, and 4, respectively. Results show that proposed model works better than Method [1] and

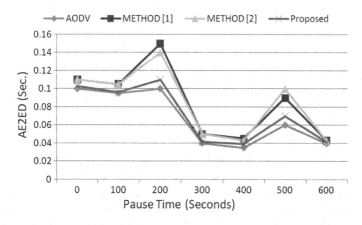

**Fig. 2** Pause_Time versus AE2E Delay

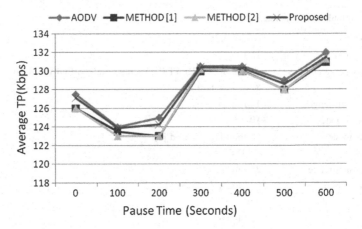

**Fig. 3** Pause_Time versus PDR

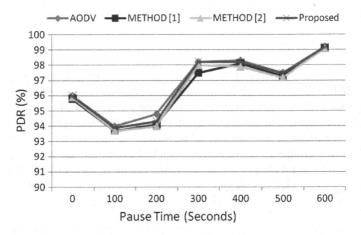

**Fig. 4** Pause_Time versus Average_TP

Method [2] and almost similar than original AODV in the absence of malicious nodes. This is because only First_Digest is generated at all intermediate nodes as compared to previous models (Method [1] and Method [2]) two digests using NMAC have generated at all intermediate nodes. It means proposed mechanism reduces the overhead by 50% as compared to previous models.

Simulation results in Figs. 5, 6, and 7 show the performance comparison of proposed model, Method [1] and Method [2] and AODV in terms of average AE2ED, PDR, Average TP with increasing number of malicious nodes (No_Mal_Nodes). For this simulation number of connections fixed to 10 and pause time fixed to 300 s. Results show that proposed model performing outstanding as comparing to the other three models in the presence of malicious nodes. Out of all malicious nodes in each simulation 50% nodes taking as inside attackers and

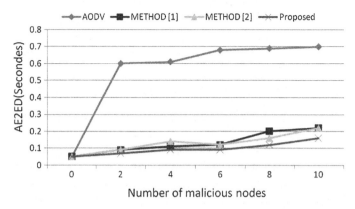

**Fig. 5** AE2E Delay versus No_Mal_Nodes

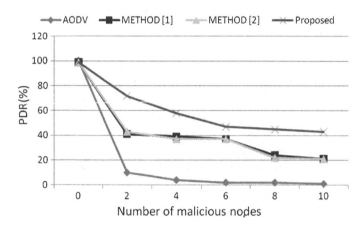

**Fig. 6** PDR versus No_Mal_Nodes

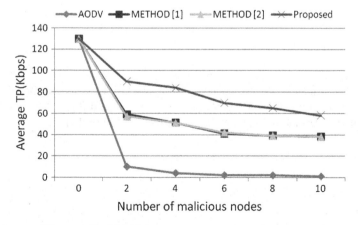

**Fig. 7** Average_TP versus No_Mal_Nodes

**Table 2** Pause time versus all performance parameters

| Parameters | Protocol | Pause time | | | | | | |
|---|---|---|---|---|---|---|---|---|
| | | 0 | 100 | 200 | 300 | 400 | 500 | 600 |
| AE2ED (s) | AODV | 0.1 | 0.095 | 0.1 | 0.04 | 0.035 | 0.06 | 0.04 |
| | METHOD [1] | 0.11 | 0.105 | 0.15 | 0.05 | 0.045 | 0.09 | 0.043 |
| | METHOD [2] | 0.11 | 0.105 | 0.14 | 0.05 | 0.043 | 0.1 | 0.043 |
| | Proposed | **0.103** | **0.096** | **0.11** | **0.041** | **0.039** | **0.07** | **0.041** |
| PDR (%) | AODV | 96 | 94 | 94.8 | 98.2 | 98.3 | 97.5 | 99.2 |
| | METHOD [1] | 95.8 | 93.7 | 94.1 | 97.5 | 98.1 | 97.3 | 99.1 |
| | METHOD [2] | 95.9 | 93.7 | 94 | 98 | 97.9 | 97.2 | 99.1 |
| | Proposed | **96** | **93.9** | **94.3** | **98.2** | **98.2** | **97.45** | **99.16** |
| Average TP (kbps) | AODV | 127.5 | 124 | 125 | 130.5 | 130.5 | 129 | 132 |
| | METHOD [1] | 126 | 123.5 | 123 | 130 | 130 | 128 | 131 |
| | METHOD [2] | 126 | 123 | 123 | 130.2 | 130 | 128 | 131.3 |
| | Proposed | **127.1** | **123.9** | **124.3** | **130.4** | **130.3** | **128.6** | **131.5** |

**Table 3** All performance parameters versus number of malicious nodes

| Parameters | Protocol | Number of malicious nodes | | | | | |
|---|---|---|---|---|---|---|---|
| | | 0 | 2 | 4 | 6 | 8 | 10 |
| AE2ED (s) | AODV | 0.05 | 0.6 | 0.61 | 0.68 | 0.69 | 0.7 |
| | METHOD [1] | 0.05 | 0.09 | 0.11 | 0.12 | 0.2 | 0.22 |
| | METHOD [2] | 0.05 | 0.09 | 0.14 | 0.12 | 0.16 | 0.22 |
| | Proposed | **0.05** | **0.07** | **0.09** | **0.091** | **0.12** | **0.16** |
| PDR (%) | AODV | 100 | 10 | 4 | 2 | 2 | 1 |
| | METHOD [1] | 99 | 41 | 39 | 37 | 24 | 21 |
| | METHOD [2] | 99 | 43 | 37 | 37 | 22 | 21 |
| | Proposed | **99** | **72** | **58** | **47** | **45** | **43** |
| Average TP (kbps) | AODV | 130 | 10 | 4 | 2 | 2 | 1 |
| | METHOD [1] | 130 | 59 | 51 | 41 | 39 | 38 |
| | METHOD [2] | 130 | 57 | 51 | 42 | 39 | 38 |
| | Proposed | **130** | **90** | **84** | **70** | **65** | **58** |

remaining 50% as outside attacker Simulation results also showing our model perform better than in the presence of inside attacker it means our model significantly overcome the drawback of Method [1] and Method [2].

Simulation results in tabular format are given in Tables 2 and 3.

# 5 Conclusion

In this paper, authentication, key distribution center (KDC) along with key pre-distribution techniques-based model has been proposed to secure the AODV routing protocol against most dangerous and frequently happened attacks (black hole, gray hole, rushing attacks, message impersonation, spoofing, fabrication attacks, etc.). In this scheme, two levels of authentication using message authentication code are used. At first level, shared Key_1 (key table distributed at the time of deployment) is used for authentication, and for second level, second key (Key_2) distributed by KDC at run time is used for generating the authentication code. Key pre-distribution (key table distributed at the time of deployment) is uses for speedup the our algorithm and reduces the overhead of each node to distribute the key during the communication while KDC based second key distribution is uses to enhances attacks prevention capability of our proposed scheme. Simulation results showing in this paper our proposed algorithm working efficiently well in the presence and absence of malicious nodes as compared to state of art AODV protocol and recently published well defined protocols i.e. Method [1] and Method [2].

# References

1. Anupam Kumar Sharma, Munesh C. Trivedi, "Performance Comparison of AODV, ZRP and AODVDR Routing Protocols in MANET", is accepted for publication in 2nd IEEE International Conference CICT 2016, 12–13 Feb 2016.
2. Munesh C. Trivedi, Anupam Kr. Sharma, "QoS Improvement in MANET using Particle Swarm Optimization Algorithm", is presented and published in Proceedings of the 2ND Springer's International Congress on Information and Communication Technology, pp 181–189 (ICICT-2015), Udaipur, Rajasthan, India.
3. K. V. Arya and S. S. Rajput, "Securing AODV routing protocol in MANET using NMAC with HBKS Technique," IEEE International Conference on SPIN, pp. 281–285, Feb 2014.
4. S. S. Rajput and M. C. Trivedi, "Securing ZRP routing protocol in MANET using Authentication Technique," IEEE International Conference on CICN, pp. 872–877, Nov. 2014.
5. D. Djenouri et al. "A survey of security issues in mobile ad hoc networks", IEEE Communications Surveys & Tutorials, Fourth Quarter 2005.
6. M. Charvalho, "Security in Mobile Ad hoc Networks" Published by the IEEE Computer Society, pp: 72–75, 2008.
7. Zheng Ming Shen and Johnson P. Thomas, "Security and QoS Self-Optimization in Mobile Ad Hoc Networks" IEEE TRANSACTIONS ON MOBILE COMPUTING, VOL. 7, NO. 9, SEPTEMBER 2008.
8. M. Guizani et al., "SECURITY IN WIRELESS MOBILE AD HOC AND SENSOR NETWORKS", IEEE Wireless Communications. October 2007.
9. S. S. Rajput, V. Kumar and S. K. Paul, "Comparative Analysis of Random Early Detection (RED) and Virtual Output Queue (VOQ) in Differential Service Networks" IEEE International Conference on SPIN, pp. 281–285, Feb 2014.
10. William Stalling, Cryptography and Network Security, 4th Ed. Pearson Education, India, 2006.

11. P. Sachan and P. M. Khilar, "Securing AODV routing protocol in MANET based on cryptographic authentication mechanism," International Journal of Network Security and Its Applications (IJNSA), vol. 3, no. 5, 2011.
12. D. Djenouri, L Khelladi, N Badache, "A survey of security issues in mobile ad hoc networks", IEEE communications surveys 7 (4), 2–28, 2005.
13. R. Sheikh and Mahakal Singh Chande and D. K. Mishra, "Security issues in MANET: A review", 2010 Seventh International Conference on Wireless and Optical Communications Networks - (WOCN), PP: 1–4, 2010.
14. E. H. T. Issariyakul, "Introduction to Network Simulator NS2," Springer Science and Business Media, NY, USA, 2009.
15. C. Perkins, E. Beldingroyer, and S. Das, "AODV RFC 3561," Internet Engineering Task Force (IETF), 2003. Available at http://datatracker.ietf.org/doc/rfc3561/.
16. S. S. Rajput, V. Kumar and K. Dubey, "Comparative Analysis of AODV and AODV-DOR routing protocol in MANET" International Journal of Computer Application, vol. 63, no. 22, pp. 19–24, Feb 2013.
17. N. Sharma, A. Gupta, S.S. Rajput and V. Yadav, "Congestion Control Technique in MANET: A Survey" 2nd IEEE International Conference on CICT, pp. 280–282, Feb. 2016.
18. B. A. Forouzan, Cryptography and Network Security, 2nd Ed., Tata McGraw-Hill Higher Education, India, 2008.

# Sensor Nodes Localization for 3D Wireless Sensor Networks Using Gauss–Newton Method

Amanpreet Kaur, Padam Kumar and Govind P. Gupta

**Abstract** Node localization is an essential difficulty in area of 3D Wireless Sensor Network (WSN). In the literature, there are different types of localization algorithms proposed for 3D WSN. But all the existing methods suffer from poor localization accuracy. This is due to the fact that all the existing methods use a less precise linear method called the least square method, for computing coordinates of nodes. This paper presents an improved localization which replaces the linear method with the highly accurate Gauss–Newton method to improve the estimation of 3D coordinates of the nodes and thus improve the localization accuracy. The performance of the proposed algorithm is evaluated by considering the effect of various parameters, and it is finally proved that the proposed algorithm is more effective than the existing algorithms through simulation results.

**Keywords** Sensor node localization · Gauss–Newton method
DV-Hop · RSSI · Ranging error

## 1 Introduction

In the 3D Wireless Sensor Networks (3D WSNs), there are many applications that require sensor reading along with the location information of the sensor nodes. For example, underwater event detection and fire surveillance system in forests, detection of the events has less significance if the placement position of node that detected the given event is not revealed [1]. Since the sensor nodes are arbitrarily sent in three-dimensional (3D) zone, it is not possible to manually implant sensor

A. Kaur (✉) · P. Kumar
Department of Computer Science & Engineering, Jaypee Institute
of Information Technology, Noida, India
e-mail: amanpreet.kaur1410@gmail.com

G. P. Gupta
Department of Information Technology, National Institute of Technology,
Raipur, Raipur, India

© Springer Nature Singapore Pte Ltd. 2019
B. K. Panigrahi et al. (eds.), *Smart Innovations in Communication
and Computational Sciences*, Advances in Intelligent Systems
and Computing 669, https://doi.org/10.1007/978-981-10-8968-8_16

node location information in every sensor node [2]. There is also a possibility that if mobile sensors nodes are used that regularly change their location, then manual configuration of node is impractical [2]. Sometimes, old nodes are replaced by new nodes in the WSN resulting in making manual configuration a big trouble [2]. Although global positioning system (GPS) [3] can be implanted in every sensor node in the network to decide their location information, putting GPS on every node is not practical being expensive and consuming high power. Due to aforesaid reasons, a low-cost and power-constrained sensor node localization algorithm is required for 3D WSN. The localization algorithms work in two phases. In the primary phase, the nodes compute their distances from anchor nodes (nodes that are aware of their location). Then, in second phase, the nodes compute their coordinates using distance information obtained in first phase. The localization algorithms can be grouped into two classifications based upon type of knowledge requirement to estimate distance between nodes: range-based and range-free [4, 5]. Range-based localization methods evaluate distances by exploiting physical properties of signals: signal strength [6], time information [7, 8], and angle information [9]. Examples of range-based are received signal strength indicator (RSSI) [6], time of arrival (TOA) [7], time difference of arrival (TDOA) [8], and angle of arrival (AOA) [9], etc. Range-based techniques usually require expensive hardware and are not preferred. Only, RSSI technique is preferred as it only requires radio and no hardware. The range-free methods use connectivity data to find distances between nodes. Examples of range-free localization algorithms are centroid algorithm [10], DV-hop [11], amorphous [12], MDS [13], and APIT [14]. DV-Hop algorithm is most popular among all range-free algorithms due to its high accuracy and simplicity and ability to find locations of all types of nodes that have at least one neighboring node. All these localization algorithms use least square method to compute location of the node which results in low localization accuracy. If somehow this least square method is replaced by nonlinear method, then accuracy can be improved further.

This paper proposes optimizations of both types of localization algorithms by applying nonlinear method called Gauss–Newton method to improve the localization accuracy. The main contribution of the paper is to propose an improvement of existing localization algorithms and then contrast the proposed algorithm with the existing algorithm to prove its effectiveness.

## 2 Localization Problem

A WSN comprises two sorts of nodes: regular and anchor. The regular refer to nodes that are not aware of their location, whereas anchor nodes have knowledge of their location. They are either placed manually or are equipped with GPS. In the localization problem, a small WSN network is examined in which $n$ sensor nodes are randomly deployed in a 3D space. There are few anchor sensor nodes, say $m$. The problem requires determining x, y, and z coordinates of $(n - m)$ regular nodes. To obtain position, the algorithm is divided into two phases.

i. **Distance/Angle Estimation**: This phase helps in finding the distances/angles of each regular node from each anchor node. The techniques used are RSSI, TOA, TDOA, AOA, or hop-counting technique.

ii. **Position Estimation**: This phase helps in estimating the position of regular node by using distance/angle information obtained from first phase. Usually, trilateration or multilateration technique is used.

The problem is to estimate x, y, z coordinates $\{x_i, y_i, z_i : i = 1, ..., (n - m)\}$ of regular nodes, given the coordinates of the anchor nodes, $\{x_i, y_i, z_i : i = 1, ..., m\}$ in WSN with high accuracy. The accuracy is specified in terms of ranging error. We propose to reduce ranging error by changing method used for position estimation phase. This method can be used for any type of range-based and range-free method. However, in this paper, we aim at RSS measurement and DV-Hop method as a result of its cost viability nature and furthermore for its accessibility for small and medium WSN applications.

## 3 Related Work

In this section, we have discussed the working of some most popular range-free-based localization algorithms. All the algorithms discussed in this section were proposed for 2D WSN. We have transformed them to be used for 3D WSN.

## 3.1 RSSI-Based Localization

This technique was first proposed by L. Girod et al. in [6], and it is the most common range-based technique. The concept used here is to estimate distance between transmitter and receiver by using received power strength and a specific path loss model. The path loss models used are free space propagation model, two-ray ground reflection model, log-distance path loss model, log-normal shadowing model, etc., [15]. The most common model used is log-normal shadow fading model which is described using Eq. (1).

$$PL(d) = PL(d_{ref}) - 10n \ \log_{10}\left(\frac{d}{d_{ref}}\right) + X_\sigma \qquad (1)$$

where PL(d) is signal power loss at the receiver side, d is the distance between sender and receiver nodes, $d_{ref}$ is the reference distance, usually taken as 1 m, $PL(d_{ref})$ is the power loss for the reference distance($d_{ref}$), n is the path loss exponent. The value of n varies from 2 to 6. $X_\sigma$ is power loss due to shadowing effect [15].

In RSSI-based localization, first the distances are computed using Eq. (1) by regular nodes. After obtaining distances, the nodes compute their location using least square method (LS). The RSSI method gives very accurate results in first phase, but this accuracy is sacrificed to some extent in the second phase while deploying least square method.

## 3.2  DV-Hop Algorithm

The DV-Hop algorithm is one of the prestigious range-free methods in WSN. It was presented by Dragos Niculescu et al. DV-Hop runs in two phases. In the principal phase, every node in the WSN ascertains the minimum number of hops to each anchor node. Then, the average distance per hop is calculated by anchor node using a simple equation and sent to every other node in the WSN. The nodes after receiving distance per hop find out their distances from each anchor by taking product of average distance per hop with the hop count. In the second phase, the location of a regular node is computed by applying least square method.

### 3.2.1  Phase 1: Distance Estimation from Each Anchor by Regular Node

This phase comprises of two sub-phases: Getting the minimum number of hops by each node from every anchor and estimating distance of every node from every anchor node. These sub-phases are described as follows:

**Sub-phase 1.1: Getting minimum number of hops by each node from every anchor**

Initially, all the anchors broadcast their location information along with their hop counts to whole network in the form of a packet. Thus, the packet contains $(x_i, y_i, z_i, h_i)$ information where $<x_i, y_i, z_i>$ is the coordinate information of anchor with id i and $h_i$ is the number of hops. The starting value of $h_i$ is 0. The node that receives this packet checks $h_i$ from its maintained table containing $(i, x_i, y_i, z_i, h_i)$ for every anchor i. If the received $h_i$ is less than the stored value, then it updates its $h_i$ value in its table. In the end, every node has minimum hops from all anchors.

**Sub-phase 1.2: Estimating distance of every node from every anchor node**

To obtain distance using minimum number of hops, the anchors first calculate average distance per hop using Eq. (2) and pass it to all nodes.

$$AvgHopDistance_i = \frac{\sum_{j=1 j \neq i}^{m} \sqrt{(x_i - x_j)^2 + (y_i - y_j)^2 + (z_i - z_j)^2}}{\sum_{j=1 j \neq i}^{m} h_j} \tag{2}$$

where i, j are ids of computing anchor and other anchors, and $(x_i, y_i, z_i)$ and $(x_j, y_j, z_j)$ are the coordinates of anchors i and j, respectively. Then, the receiving nodes estimate distance by multiplying hops with the average hop distance (refer Eq. (3)).

$$d_i = AvgHopDistance_j \times h_i \qquad (3)$$

### 3.2.2 Phase 2: Estimation of Position of Regular Node

The last phase is comprised of multilateration method which uses LS method to get the location of all regular nodes. This phase works as follows:

If $(x_r, y_r, z_r)$ is the location of regular node R and $(x_i, y_i, z_i)$ is the location of anchor $A_i$, then the problem can be mapped into following set of equations:

$$\left. \begin{aligned}
(x_r - x_1)^2 + (y_r - y_1)^2 + (z_r - z_1)^2 &= d_1^2 \\
(x_r - x_2)^2 + (y_r - y_2)^2 + (z_r - z_2)^2 &= d_2^2 \\
&\cdots \\
&\vdots \\
(x_r - x_m)^2 + (y_r - y_m)^2 + (z_r - z_m)^2 &= d_m^2
\end{aligned} \right\} \qquad (4)$$

Equation (5) can be generated from Eq. (4) by taking difference of all equations from last equation.

$$\left. \begin{aligned}
x_1^2 - x_m^2 + y_1^2 - y_m^2 + z_1^2 - z_m^2 - d_1^2 - d_m^2 &= 2 \times x_r \times (x_1 - x_m) + 2 \times y_r \times (y_1 - y_r) + \\
& \qquad 2 \times z_r \times (z_1 - z_r) \\
x_2^2 - x_m^2 + y_2^2 - y_m^2 + z_2^2 - z_m^2 - d_2^2 - d_m^2 &= 2 \times x_r \times (x_2 - x_m) + 2 \times y_r \times (y_2 - y_r) + \\
& \qquad 2 \times z_r \times (z_2 - z_r) \\
&\vdots \\
x_{m-1}^2 - x_m^2 + y_{m-1}^2 - y_m^2 + z_{m-1}^2 - z_m^2 - d_{m-1}^2 - d_m^2 &= 2 \times x_r \times (x_{m-1} - x_m) + 2 \times y_{m-1} \times (y_{m-1} - y_r) + \\
& \qquad 2 \times z_r \times (z_{m-1} - z_r)
\end{aligned} \right\} \qquad (5)$$

The matrix representation of Eq. (5) is as follows:

$$AX_r = B \qquad (6)$$

where

$$A = 2 \times \begin{bmatrix}
x_1 - x_m & y_1 - y_m & z_1 - z_m \\
x_2 - x_m & y_2 - y_m & z - z_m \\
\vdots & \vdots & \vdots \\
x_{m-1} - x_m & y_{m-1} - y_m & z_{m-1} - z_m
\end{bmatrix}, \quad X_r = \begin{bmatrix} x_r \\ y_r \\ z_r \end{bmatrix},$$

and

$$B = \begin{bmatrix} x_1^2 - x_m^2 + y_1^2 - y_m^2 + z_1^2 - z_m^2 - d_1^2 - d_m^2 \\ x_2^2 - x_m^2 + y_2^2 - y_m^2 + z_2^2 - z_m^2 - d_2^2 - d_m^2 \\ \vdots \\ x_{m-1}^2 - x_m^2 + y_{m-1}^2 - y_m^2 + z_{m-1}^2 - z_m^2 - d_{m-1}^2 - d_m^2 \end{bmatrix}$$

Equation (6) is converted to Eq. (7) as follows:

$$X_r = (A^T A)^{-1} A^T B \tag{7}$$

The regular node estimates its position by applying least square method on Eq. (7). Although DV-Hop algorithm has advantages that it is cheap, simple, and scalable, it gives low accurate results due to use of least square method (used for linear problems). In this paper, both the algorithms (RSSI and DV-Hop) that use least square method are called as least square localization algorithm (LSLA).

## 3.3   Hyperbolic DV-Hop Localization Algorithm

H. Chen et al. [16] proposed a hyperbolic DV-Hop localization algorithm (HLA) to increase accuracy. The working of algorithm is comprised of two phases. The first phase finds out the minimum hops of each node to each anchor in the similar way as in DV-Hop and then computes average of average hop distances.

### 3.3.1   Sub-phase 2 of Phase 1

Each anchor applies Eq. (2) to compute its average hop distance and broadcasts it to every other node. The regular nodes then compute average of these entire average hop distances using Eq. (8).

$$HopSize = \frac{\sum AvgHopDistance_i}{m} \tag{8}$$

Then, regular nodes compute its distance from each anchor i using Eq. (9).

$$d_i = h_i \times HopSize \tag{9}$$

### 3.3.2   Phase 2

In last phase, each regular node determines its location using hyperbolic [16] location algorithm instead of LS method. The accuracy of HLA is improved due to the use of nonlinear hyperbolic method.

## 3.4   Quadratic DV-Hop Algorithm

In [17], S. Tomic et al. proposed a variant of DV-Hop (QLA) that uses quadratic programming in the last phase of DV-Hop algorithm to reduce localization error.

### 3.4.1   Phase 2

The last phase is modified to improve accuracy. The least squares problem Eq. (7) is first considered and converted into quadratic problem (QP), which is then solved using quadratic programming method. Simulation results prove that the QLA gives better performance when compared with original DV-Hop algorithm.

## 4   Description of the Gauss–Newton Method

In [18], W. Navidi et al. prove that Gauss–Newton method is the best solution to be used for trilateration/multilateration among all statistical methods. The Gauss–Newton method is explained using Algorithm 1. For localization problem, the objective function is formulated as:

$$f(x, y, z) = \text{Min}\left(\sum_{i=1}^{n}\left((x_i - x_r)^2 + (y_i - y_r)^2 + (z_i - z_r)^2 - d_i^2\right)\right) \qquad (10)$$

The Jacobian matrix J is described using Eq. (11) as given below:

$$J = \begin{bmatrix} \frac{\partial f(x_1, y_1, z_1)}{\partial x} & \frac{\partial f(x_2, y_2, z_2)}{\partial x} & \frac{\partial f(x_3, y_3, z_3)}{\partial x} \\ \frac{\partial f(x_1, y_1, z_1)}{\partial y} & \frac{\partial f(x_2, y_2, z_2)}{\partial y} & \frac{\partial f(x_3, y_3, z_3)}{\partial y} \\ \frac{\partial f(x_1, y_1, z_1)}{\partial z} & \frac{\partial f(x_2, y_2, z_2)}{\partial z} & \frac{\partial f(x_3, y_3, z_3)}{\partial z} \end{bmatrix}$$

where $(x_i, y_i, z_i)$ are the positions of anchor i.

In this algorithm, $J^T$ is the transpose of matrix J.

---

Algorithm1: Gauss-Newton method

Step 1: $x = x_0$, $y = y_0$, $z = z_0$, $i = 0$, error=min_error; {Initialize position information , i used for iteration and error value}

Step 2: while($x - x_0$) || ($y - y_0$) || ($z - z_0$) >=error && iteration <=max_iteration {max_iteration denotes maximum number of iterations}

Step 3: $x = x_0 - (J^T J)^{-1} J^T f(x,y,z)$;   {Update x value}

Step 4: $y = y_0 - (J^T J)^{-1} J^T f(x,y,z)$;   {Update y value}

Step 5: $z = z_0 - (J^T J)^{-1} J^T f(x,y,z)$;   {Update z value}

Step 6: iteration=iteration +1;

Step 7: end

---

# 5   Proposed Localization Algorithm

In this section, we have described the working of the proposed improved localization algorithm for 3D WSN. The process of the proposed Gauss–Newton-based
localization algorithm (GNLA) is illustrated in Fig. 1. The proposed algorithm
works in two phases. The working of each phase is described as follows.

## 5.1   Phase 1: Distance Estimation from Sensor Node to Every Anchor Node

In this phase, each sensor node finds its distance from every anchor. The distance
estimation can be done in two ways. The first approach is by applying RSSI
technique. Another approach is by using DV-Hop method. Phase 1 has been
explained in Sect. 3. In this phase, all regular nodes also find their three nearest
anchors. This information of three nearest anchors is obtained by recording the first
three distinct packets received from anchors.

## 5.2   Phase 2: Position Estimation by Using Gauss–Newton Method

In the second phase, the regular nodes get their initial approximate value ($x_0$, $y_0$, $z_0$)
using centroid formula as depicted in Eq. (11):

$$x_0 = \frac{\sum_{i=0}^{3} x_i}{3}, \ y_0 = \frac{\sum_{i=0}^{3} y_i}{3}, \ z_0 = \frac{\sum_{i=0}^{3} z_i}{3} \tag{11}$$

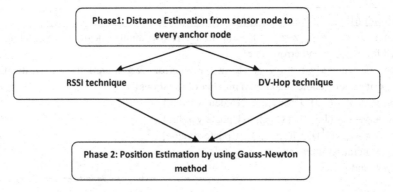

**Fig. 1** Proposed methodology

where $(x_i, y_i, z_i)$ represents the coordinates of first three near anchors from the given node.

Then, the regular node applies Gauss–Newton method instead of linear method (least square method) to get their position as described in Algorithm 1. The purpose behind utilizing Gauss–Newton method is that the localization problem constitutes nonlinear equations. Thus, if nonlinear problem is settled by applying linear solution, then it results in high error. This localization problem should be tackled by applying nonlinear method. In [18, 19], it is shown and proved that Gauss–Newton method gives more exactness if compared with other nonlinear methods. Thus, we have utilized this method for our proposed methodology.

# 6 Simulation Results

In this section, we use performance metric 'ranging error' which is used to compare proposed methodology (GNLA) with least square localization algorithm (LSLA), HLA, QLA. Then, we describe performance of GNLA algorithm by varying number of anchor nodes, communication radius of nodes, and total number of nodes in the network and border area. The algorithms have been simulated in MATLAB2013 [20]. Table 1 shows the simulation parameters used for given simulation. We have considered random topology for this environment.

**Table 1** Simulation parameters

| Simulation Parameters | Value |
| --- | --- |
| WSN area | Vary from $100 \times 100 \times 100 \, \text{m}^3$ to $300 \times 300 \times 300 \, \text{m}^3$ |
| Node count (n) | Vary from 200 to 500 |
| Anchor count (m) | Vary from 10 to 40% of total nodes |
| Total rounds | 50 |
| Communication radius (r) | Vary from 15 to 35 |
| Iterations | 50 |

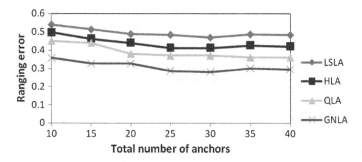

**Fig. 2** Comparison by changing anchor count

Figures 2, 3, 4, and 5 show the performance of proposed algorithm in terms of ranging error by varying various parameters. The simulation results prove that the GNLA algorithm performs better than other algorithms. It gives improvement of accuracy by about 20% over its closest rival.

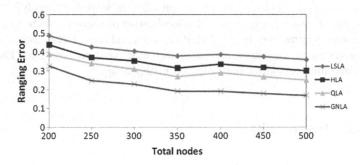

**Fig. 3** Comparison by changing node count

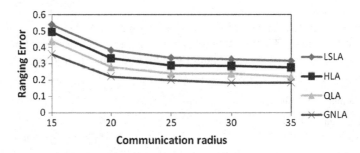

**Fig. 4** Comparison by changing communication radius

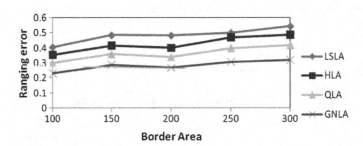

**Fig. 5** Comparison by changing border area

# 7 Conclusion

We have presented an improved localization algorithm for 3D WSN using Gauss–Newton method. In the proposed localization algorithm, we have used the good feature of the DV-Hop and centroid algorithm, and in the second phase for estimation of the location coordinate, we have used Gauss–Newton method. The algorithm is implemented and simulated in MATLAB2013, and simulation results prove that the proposed algorithm enhances the localization accuracy by 20% when compared with algorithm using least square method for calculating 3D position of the sensor nodes.

# References

1. Y. Sabri, and N. E. Kamoun. "Forest Fire Detection and Localization with Wireless Sensor Networks", in Networked Systems. Springer, Berlin, pp. 321–325, 2013.
2. V. K. Chaurasiya, N. Jain, and G. C. Nandi, "A novel distance estimation approach for 3D localization in wireless sensor network using multi dimensional scaling", Information Fusion 15, pp. 5–18, 2014.
3. N. Bulusu, J. Heidemann, and D. Estrin, "GPS-less low-cost outdoor localization for very small devices", IEEE personal communications, Vol. 7(5), pp. 28–34, 2000.
4. T. Kunz, and B. Tatham, "Localization in wireless sensor networks and anchor placement", Journal of Sensor and Actuator Networks, Vol. 1(1), pp. 36–58, 2012.
5. A. Kaur, P. Kumar, and G. P. Gupta. "A Weighted Centroid Localization Algorithm for Randomly deployed Wireless Sensor Networks", Journal of King Saud University-Computer and Information Sciences, 2017.
6. L. Girod, V. Bychobvskiy, J. Elson and D. Estrin, "Locating tiny sensors in time and space: a case study", in Proceedings of the 2002 IEEE International Conference on Computer Design: VLSI in Computers and Processors, Los Alamitos, pp. 214–219, 2002.
7. A. Harter, A. Hopper, P. Steggles. A. Ward, and P. Webster, "The anatomy of a context-aware application", Wireless Networks, Vol 8, no. 2, pp. 187–197, 2002.
8. X. Cheng, A. Thaeler, G. Xue, and D. Chen, "TPS: a time-based positioning scheme for outdoor wireless sensor networks," in Proceedings of the 23rd IEEE Annual Joint Conference of the IEEE Computer and Communications Societies (INFOCOM '04), pp. 2685–2696, Hong Kong, China, March 2004.
9. D. Niculescu, and B. Nath, "Ad hoc positioning system (APS) using AoA", In Twenty-Second Annual Joint Conference of the IEEE Computer and Communications. IEEE Societies, Vol. 3, pp. 1734–1743, 2003.
10. N. Bulusu, J. Heidemann, D. Estrin, "GPS-less low cost outdoor localization for very small devices", IEEE Personal Communications Magazine, Vol 7, no. 5, pp. 28–34, 2000.
11. D. Niculescu and B. Nath, "Ad-hoc positioning system.", In IEEE on Global Telecommunications Conference, Vol 5, pp. 2926–2931, 2001.
12. R. Nagpal, "Organizing a global coordinate system from local information on an amorphous computer", A.I. Memo1666, MIT A.I. Laboratory, 1999.
13. Y. Shang andW. Ruml, "Improved MDS-based localization," in Proceedings of the IEEE Conference on Computer Communications (INFOCOM'04), pp. 2640–2651, HongKong, March 2004.
14. T. He, C. D. Huang, B.M. Blum, J.A. Stankovic, and T. Abdelzaher, "Range-free localization schemes for large scale sensor networks," in Proceedings of the 9th the Annual International Conference on Mobile Computing and Networking, pp. 81–95, ACM, San Diego, Calif, USA, 2003.

15. T.S. Rappaprt, "Wireless Communications: Principles and practice (2nd edition)", Upper Saddle River, New Jersey: Prentice Hall PTR, 2002.
16. H. Chen, K. Sezaki, P. Deng and H.C. So, "An improved DV-hop localization algorithm for wireless sensor networks," In Proceedings of IEEE Conference on Industrial Electronics and Applications (ICIEA 2008), Singapore pp. 1557–1561, June 2008.
17. S. Tomic, and I. Mezei, "Improvements of DV-Hop localization algorithm for wireless sensor networks." Telecommunication Systems, pp. 1–14, 2016.
18. W. Navidi, W.S. Murphy, and W. Hereman, "Statistical methods in surveying by trilateration.", Computational statistics & data analysis, Vol 27(2), pp. 209–227, 1998.
19. K. Madsen, H.B. Nielsen, O. Tingleffm, "Methods for non-linear least squares problems", 2004.
20. S. Chapman, "J. MATLAB programming for engineers", Nelson Education, 2015.

# ERA-AODV for MANETs to Improve Reliability in Energy-Efficient Way

Baljinder Kaur, Shalini Batra and Vinay Arora

**Abstract** The nodes in MANETs communicate with each other through wireless medium and are constantly moving. To maintain the links in a continuously changing topology is a challenge, especially for longer duration of time, and in such scenarios, the reliability of the links becomes the prime issue. Further, the reliability must be maintained in an energy-efficient way. The paper presents energy-efficient version of RA-AODV named as ERA-AODV which measures reliability in terms of end-to-end delay, bandwidth, and mobility of the intermediate nodes. Since the prime factor which determines the consumption of energy of the node is the distance of communication between them, the proposed scheme takes into account adjustment of the transmission range of the mobile nodes according to residual energy, to provide improved energy efficiency. The parameters considered as performance measure which include packet delivery ratio, throughput and remaining energy demonstration improvement over the existing scheme.

**Keywords** MANET · RA-AODV · End-to-end delay · Bandwidth

## 1 Introduction

Mobile ad hoc networks (MANETs) are mix of mobile nodes lacking the presence of any master or control by any previous foundation. Such sort of networks utilizes multi-hop ways and wireless radio communication channel for majority of communication. In such networks, message exchange amid nodes is set up by multi-hop routing. Fresh nodes join or leave the network whenever desired. Due to the

B. Kaur (✉) · S. Batra · V. Arora
CSE Department, Thapar University, Patiala, India
e-mail: baljinder.neha@gmail.com

S. Batra
e-mail: sbatra@thapar.edu

V. Arora
e-mail: vinay.arora@thapar.edu

© Springer Nature Singapore Pte Ltd. 2019                                     199
B. K. Panigrahi et al. (eds.), *Smart Innovations in Communication and Computational Sciences*, Advances in Intelligent Systems and Computing 669, https://doi.org/10.1007/978-981-10-8968-8_17

dynamic nature, the arrangement of the nodes is frequently evolving and execution of network disintegrates quickly. In such scenarios, the development of a safe routing convention is a basic point of concern. A wireless ad hoc network is self-sorting, self-restraining, and self-adaptive which is a congregation of devices linked through wireless communication in addition to networking ability. Such devices can speak with other gadgets which are inside its radio range or to the one which are beyond its radio range.

There are two genera of networks namely wired and wireless networks. Wired networks are those networks wherein PC mechanical assemblies are associated individually with another with the use of a wire. This wire acts as a medium of communication for transmitting data from one fragment of the network to other portion of the network. However, a wireless network is a network wherein PC gadgets speak with other nodes without the use of any wire, i.e., the medium of data exchange amid the PC devices is wireless. When a PC device wishes to transfer data to other devices, the target devices are required to be in the radio range of each other. The transmission and gathering of information in wireless sensor networks is done by utilizing the EM signals. Mobile ad hoc networks come under the category of wireless networks where the devices functioning on battery attempt to consume the energy while performing the various tasks on devices or nodes. It is desired that the energy paths should be minimum since the devices in such kind of paths deplete their energy very quickly whereby the nodes cannot accomplish their desired jobs due to the limited energy availability. Since the nodes communicate wirelessly, the reliability of the data is of prime importance. This paper describes an algorithm to achieve reliability of the data in an energy-efficient way. Section 2 provides insight into the past researches that have been done focusing on the increasing network lifetime. Section 3 defines the existing problem, and finally, the paper shows the proposed work and results of the simulation in Sect. 4.

## 2  Existing Work

**Devaraj et al. (2016)** proposed ECDSR convention in an attempt to multiply the lifetime of the network where routing is required to be proficient in terms of energy consumption. Routing conventions like DSR, ESDSR, ECDSR, AODV, TORA, EEAODR, and EPAR are put forward for MANET. ECDSR resolution chooses nodes on the premise of minimum edge energy. Since ECDSR convention has overhearing and old routes issue, which results in parcel misfortune and over vitality utilization in this paper authors proposed the answer for address overhearing and old routes issue by suggesting change in ECDSR convention. MANET is utilized as a part of ongoing basic applications [1].

**Kuo et al. (2016)** investigated EE optimization as measured in bits per joule for MANETs in light of the cross-layer outline paradigm. They show this issue as a non-arched blended integer nonlinear programming (MINLP) definition by jointly considering routing, activity scheduling, and power control. Since the non-raised

MINLP issue is NP-hard by and large, it is exceedingly hard to all-inclusive upgrade this issue. They devise a modified branch and bound (BB) calculation to productively take care of such issue. The oddities of the proposed BB calculation include upper and lower bounding plans and branching standard that is designed by using the qualities of the nonconvex MINLP issue. Numerical outcomes demonstrate that the proposed BB calculation plot, individually, diminishes the optimality hole by 81.98% and increases the best achievable arrangement by 32.79%. Further, the outcomes not just give insight into the plan of EE amplification calculations for MANETs by employing cooperation's between various layers but also serve as execution benchmarks for distributed conventions produced for genuine applications [2].

**Aashkaar and Sharma (2015)** proposed an improvement in an AODV convention which is an upgrade in the current AODV convention. The convention computation which is achieved by energy-efficient ad hoc distance vector tradition (EE-AODV) has upgraded the RREQ and RREP taking consideration of system to save the essentialness in phones. In this paper, AODV convention is executed by using 30 nodes. The execution estimations used for evaluation are conveyance proportion, throughput, framework lifetime, and typical energy expended. The regeneration is done using NS2 [3].

**Choukri et al. (2014)** define a routing framework for a correspondence network established by a few ad hoc mobile nodes. This framework improves energy consumption by isolating the network into groups and then recognizing the most ideal path in terms of energy. It tries to identify the energy requirements mandatory for each accessible path and chooses the ideal passages. Every group is distinguished by a channel which is chosen consistent with its location and its remaining energy by means of a clustering calculation. The major goal of this paper is to upgrade the quantity of living nodes by conveying task to every suitable node in the network [4].

**Badal and Kushwah (2015)** used DSR protocol which otherwise has a little usage in practical applications in comparison to other routing protocols. This paper provides a modified version of the DSR protocol as a result of which it improves the energy efficiency and the lifetime of the nodes in MANET. Proposed modified DSR provides energy-efficient route establishment by amendment of the route discovery mechanism where performance has been using NS2 [5].

**Kumar and Dubey (2016)** considered the problem of limited energy associated with the nodes in MANETS. As soon as the energy is exhausted, the nodes are not able to communicate with each other. To provide complete usage of energy of the nodes and network connectivity, they proposed link stability control maximum (MAX) energy with multipath routing scheme. The algorithm always chooses the node with the highest energy among the neighbors and builds multiple paths between source and destination thus increasing the lifetime of the network. The energy is consumed more for the data delivery rather than route request [6].

**Palaniappan and Chellan (2015)** proposed a steady and energy-efficient routing procedure. In the suggested technique, nature of administration (QoS) monitoring operators is gathered to ascertain the connection unwavering quality measurements, for example link lapse time (LET), probabilistic link dependable time (PLRT), LPER and link got flag quality (LRSS). Furthermore, lingering battery control (RBP) is executed to preserve the energy proficiency in the network. To conclude, course determination likelihood (RSP) is figured in view of these assessed parameters using fuzzy logic [7].

# 3 Existing Problem

The nodes in mobile ad hoc networks are battery driven and are mobile in nature. They continue to move from one position to another, and their routing is governed mostly by the reactive routing protocols such as DSR and AODV. Since the reliability of data is very important, so all the packets sent by the source node must reach the destination node intact; otherwise, it results in loss of data packets. The packet loss in the network can be due to packet collision; it may result due to the presence of some malicious node in the network, or the packet loss may result due to the link breakage, etc. The existing routing protocols provide the shortest path from source to destination node in terms of hop count. They do not consider the energy-efficient routes neither they consider the quality of the links while deciding the route between source and destination node.

The mobile nature of nodes in ad hoc networks results in link breakage between the nodes. This is one of the factors that leads to packet loss and thus results in decrease of throughput which is defined as amount of data that is received at the destination node. In the study done by Tyagi et al. in [8], the authors have worked on the reliability of the data transfer from source to destination node. For this, the paths are first arranged in decreasing order of the hop count. Then, end-to-end delay and bandwidth for the shortest path is calculated. Again in the third step, the mobility of the intermediate nodes is considered for selection of a particular path to send data from source to destination node.

However, there can be few noticeable issues in the above-described approach. Firstly, the authors have not considered remaining energy of the intermediate nodes to select the path. Secondly, the authors have considered hierarchical approach; for example, first the shortest path is checked, then end-to-end delay and bandwidth, and then speed of the intermediate nodes. Then for the shortest path if the end-to-end delay is lesser than the defined limit and bandwidth is greater than the defined limit, the path is selected for data routing. However, it can be argued that a path which is not shortest path can have lesser end-to-end delay and more bandwidth.

# 4 Proposed Work

The proposed work aims at making network more energy efficient and at the same time achieving reliability. While the reliability of the network is measured in terms of end-to-end delay, bandwidth, and speed of the intermediate nodes, the energy efficiency of the node is usually determined by two parameters namely distance between sender and the receiver and the packet size.

For instance, consider a scenario where less energy nodes are communicating over a distance, say x meters and the high-energy nodes are also communicating over the same distance. The low-energy nodes will tend to die out soon. But if the low-energy nodes can adjust their transmission range in accordance with the remaining residual energy, they might be able to work for longer duration of time. So the proposed work aims to modify the way traditional broadcasting process takes place where all the nodes are forwarding the packets over the same distance neighborhood irrespective of their remaining energy levels. Thus, the nodes will be adjusting the transmission range of the nodes according to the residual energy. This inclusion in the existing RA-AODV is expected to being a positive change in terms of reduced energy consumption in the network in addition to its reliability.

When the source node has some data to forward to the destination, it will execute the route discovery phase in a way defined below:

```
{
Source S starts forwarding the RREQ packets to the neigh-
boring nodes.
    If (Destination node exists in the Routing Table)
    {
        Execute Route Reply phase
    }
    Else
    {
        Check the residual energy
        Adjust the Transmission Range accordingly and find
        neighbors // in case of lesser residual energy,
        find the neighbors in the smaller communication
        range

        Put the neighbors in the set - N
For each node in N {
    If (Neighbor satisfies LAR Condition) // Neighbor lies
in the quadrant towards the destination node
        {
        Forward the RREQ packet until destination is found
        }
    Else
        {
        Do not forward the RREQ packet
        }
END
    }
}
```

When destination has been found, compute the reliability parameters for each route found as mentioned in RA-AODV namely hop count of the paths, delay, and bandwidth. In order to achieve energy efficiency, residual energy of the intermediate nodes in the paths is also considered. If the intermediate nodes are not moving out of range of each other, then choose the path having highest residual energy and bandwidth, minimum hop count, and delay.

## 5  Simulation Results

Both the schemes were implemented in NS2, and the results were compared on the basis of packet delivery ratio, throughput, and energy remaining in the network. Table 1 below shows the simulation parameters used.

- Packet Delivery Ratio: It measures the percentage of packets that are successfully delivered in the network. Mathematically, it is defined as (Fig. 1)

**Table 1** Simulation parameters

| Simulation parameters | Value |
|---|---|
| Channel | Wireless |
| Propagation model | Two-ray ground |
| Mobility | Random waypoint |
| Routing protocols | AODV |
| Number of nodes | 50 |
| MAC | 802.11 |
| Antenna | Omnidirectional |
| Network area | 1100 m * 1100 m |
| Queue | Drop tail |

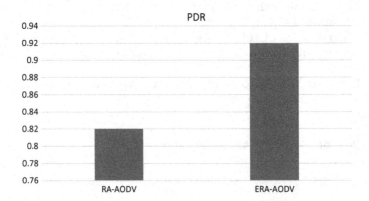

**Fig. 1** PDR comparison

$$PDR = \frac{Number\ of\ data\ packets\ received}{Number\ of\ datapackets\ sent} \tag{1}$$

- Throughput: It is defined as the amount of data received at the destination node per unit of time. It is normally measured in Kbps. Mathematically, it is computed as (Fig. 2):

$$Throughput = \frac{Number\ of\ datapackets\ received * Packetsize\ in\ bits}{1000} \tag{2}$$

- Remaining Energy: This represents amount of energy remaining in the network. More is the value of this parameter, better is the network lifetime (Fig. 3).

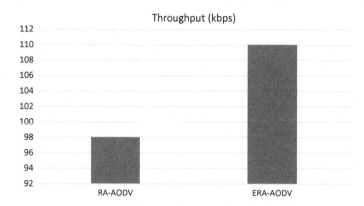

**Fig. 2** Throughput comparison

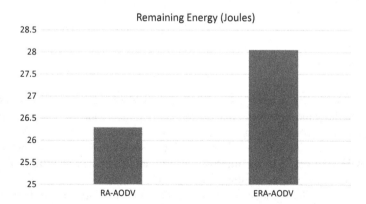

**Fig. 3** Remaining energy comparison

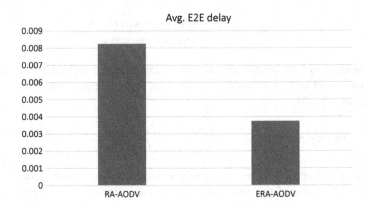

**Fig. 4** Avg. E2E delay comparison

**Table 2** Comparison of values of packet delivery ratio, throughput, remaining energy, and Avg. E2E delay

| | RA-AODV | Proposed scheme |
|---|---|---|
| Packet delivery ratio | 0.82 | 0.92 |
| Throughput (Kbps) | 98 | 110 |
| Remaining energy (J) | 26.2932 | 28.0526 |
| Avg. E2E delay (s) | 0.008243 | 0.003743 |

$$Remaining\ Energy = Initial\ Energy - Energy\ Consumed \qquad (3)$$

- Average End-to-End Delay: This represents time taken for the packets to reach from the source to destination node over the selected path (Fig. 4 and Table 2).

$$End\ to\ End\ Delay = Packet\ receiving\ time - Packet\ Sending\ time \qquad (4)$$

# 6 Conclusion

The paper presented an energy-efficient approach that improves the network lifetime and at the same time maintains the reliability of data transfer. The performance of the network is compared on the basis of remaining energy in the network, packet delivery ratio, and throughput. These parameters showed an improvement over the RA-AODV. The credit to the improved energy efficiency can be attributed to the concept of location-aided routing and dynamic transmission range used in the proposed scheme. If any node has lesser remaining energy, then forwarding packets over a shorter distance would balance the energy consumed in a better way. Since mobile ad hoc networks suffer from number of attacks, in future the security of the network can also be worked upon by considering different kinds of attacks.

# References

1. Hridya V Devaraj, Anju Chandran, Suvanam Sasidhar Babu: MANET protocols: Extended ECDSR protocol for solving stale route problem and overhearing. In: International Conference on Data Mining and Advanced Computing (SAPIENCE), pp. 268–272, 2016.
2. W. K. Kuo and S. H. Chu: Energy Efficiency Optimization for Mobile Ad Hoc Networks. In: IEEE Access, vol. 4, no., pp. 928–940, 2016.
3. Mohammed Aashkaar, Purushottam Sharma: Enhanced energy efficient AODV routing protocol for MANET. In: IEEE 2016.
4. Choukri, A. Habbani and M. El Koutbi: An energy efficient clustering algorithm for MANETs. In: International Conference on Multimedia Computing and Systems (ICMCS), Marrakech, 2014, pp. 819–824.
5. Deepti Badal and Rajendra Singh Kushwah: Nodes energy aware modified DSR protocol for energy efficiency in MANET, In: India Conference (INDICON), 2015 Annual IEEE, pp. 1–5, 2015.
6. Mukesh Kumar and Ghansyam Prasad Dubey: Energy efficient multipath routing with selection of maximum energy and minimum mobility in MANET, In: ICT in Business Industry & Government (ICTBIG), International Conference, pp. 1–6, 2016.
7. Senthilnathan Palaniappan, and Kalaiarasan Chellan: Energy-Efficient Stable Routing Using Qos Monitoring Agents in Manet. In: EURASIP Journal on Wireless Communications and Networking 2015.
8. Shobha Tyagi, Subhranil Som, Q. P. Rana: A Reliability Based Variant of AODV In MANETs: Proposal, Analysis And Comparison. In: 7th International Conference on Communication, Computing and Virtualization 2016.

# A Novel Multi-patch Triangular Antenna for Energy Harvesting

**Saikat Paul, Aakash Ravichandran, Mukul Varshney and Sujata Pandey**

**Abstract** Harvesting energy from the ambient supply of radio frequency has become a potential area for research in the past few years. This paper presents a three triangular patch antenna designed for multiband RF energy harvesting. Three triangles have been used connected to each other through the feedline. Antenna operates between 5 and 5.5 GHz. The antenna has been used to harvest energy in the range of 50–100 mV in the operational band.

**Keywords** Multi-patch · Microstrip · Energy harvesting

## 1  Introduction

In the near future it is expected that with developments in the field of IOT, billions of low-powered devices like sensors will be working all over the world 24 × 7. These devices need to be charged in an efficient, hassle-free way. To solve these problems, scientists have come up with the idea of wireless charging. To achieve this, microwave frequencies in the environment which otherwise goes unused are being harvested with the philosophy of accumulate and use. This system is meant to power these low-powered devices throughout its lifetime.

The system can be divided into two parts: Design a microstrip antenna with the help of HFSS in the GSM band, GPS band, ISM band or the UWB. The antenna is then fabricated and tested in an anechoic chamber for return loss, VSWR and radiation pattern.

S. Paul · A. Ravichandran · M. Varshney · S. Pandey (✉)
Amity Institute of Telecom Engineering & Management, Amity University, Noida,
Uttar Pradesh, India
e-mail: spandey@amity.edu

S. Paul
e-mail: saikatpaul2395@gmail.com

A. Ravichandran
e-mail: arcyo95@gmail.com

© Springer Nature Singapore Pte Ltd. 2019                                               209
B. K. Panigrahi et al. (eds.), *Smart Innovations in Communication
and Computational Sciences*, Advances in Intelligent Systems
and Computing 669, https://doi.org/10.1007/978-981-10-8968-8_18

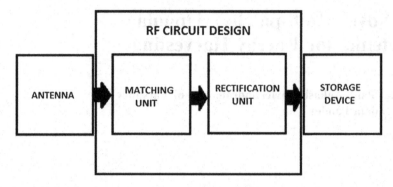

**Fig. 1** Block diagram of RF energy harnessing unit

Part 2 consists of impedance matching circuits and RF to DC rectifiers. The circuit is made to accommodate a wide band of frequencies and facilitate frequency tuning.

Kenneth Gudan, Shuai Shao and Jonathan J. Hull have harvested RF energy at 2.4 GHz band along with a system for storage with a sensitivity of −25dBm [1]. Here, they charge a battery from an antenna of −25dBm sensitivity. Their system works in the ISM band of 2.4 GHz. The circuit can charge a battery in 1 h by providing a power of 150 μJ.

AntwiNimo in his paper [2] of passive RF to DC power rectification and energy harnessing for μWatt sensors has studied the linearization of the rectifier circuit. The antennas have shown a return loss of -27dBm in the GSM bands. Lazare Noel in his paper [3] for portable charging with radio frequency energy harnessing has discussed the various building blocks of a harvesting system. The different components like Input Impedance Matching Circuit, RF-DC rectifier circuit and Operational Amplifiers. Other different methods of energy harvesting [4–7] and different antenna designs [8–10] are available in the literature.

Figure 1 shows the generalized block diagram of the RF harvesting system. To achieve maximum power transfer, additional passive matching networks are connected between the source and load. For increasing the efficiency of RF conversion and harvested DC power, we can also use some special RF source to feed the antenna.

In this paper, we have introduced a tri-triangular antenna and discussed different parametric studies of the same. It will also be tested for harvesting RF energy.

## 2 Antenna Design

The proposed design is as shown below, in which we have placed three triangular antennas connected to each other via feedline. Figure 2 shows the proposed antenna. The dimensions that are discussed ahead and after a lot of trials and trying

**Fig. 2** Proposed antenna

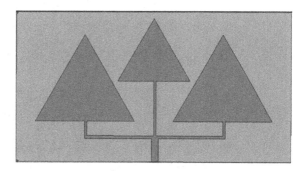

multiple combinations are used such that it provides maximum efficiency and values of various other parameters that has been discussed further ahead in the paper.

Antenna Dimensions

Substrate-65mm × 120 mm
Thickness-1.6 mm
Substrate material-FR4-epoxy
Ground-65 mm × 120 mm
Patch dimensions (Height × Base)
Triangle1-37.5 mm × 43.3 mm
Triangle2-27 mm × 31.1 mm
Triangle3-37.5 mm × 43.3 mm

There are three patches in total placed over the substrate with the above-mentioned dimensions together with a lumped port with 50 Ω impedance.

## 3 Design Results

Ansoft HFSS was used for performing the simulations given in this section. Important antenna parameters like return loss, VSWR and radiation pattern have been plotted. Figure 3 shows the S11 return loss in the designed antenna.

This graph gives us fair idea about the frequency range in which the antenna operates the best and also the frequency at which the antenna the best, also known as the resonant frequency. In accordance with this graph, we can see the bandwidth of this antenna being between 5 and 5.5 GHz and the resonant frequency being 5.2 GHz at which the return loss value comes out to be −26.59 dB.

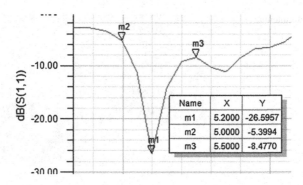

**Fig. 3** Frequency (GHz) versus $S_{11}$

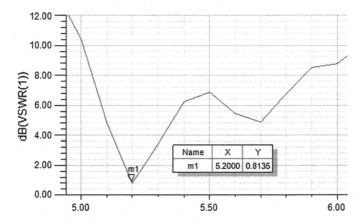

**Fig. 4** Frequency (GHz) versus VSWR

Voltage standing wave ratio (VSWR) is a method to measure the transmission line imperfections. Figure 4 shows the VSWR characteristics of the antenna.

A good antenna for commercial purposes should have a VSWR less than 2 which in this case is 0.813.

A radiation pattern of an antenna is the variation of power radiated by an antenna. For measuring the radiation pattern, we can either use far-field or near-field range. In the present case, we have used far-field range. In far-field range, the radiation pattern of the antenna becomes planar. Figure 5a, b shows the radiation pattern of the antenna.

The three-dimensional polar plot gives us an idea of the radiation pattern in 3D. The plot in Fig. 6 is in agreement with Fig. 5a. The distance in radial direction from the origin represents the strength of emitted radiation in a direction.

Fig. 7a shows the magnitude of the electric field variation on the patches of the antenna. Fig. 7b shows the magnitude of the magnetic field variation on the patches

**Fig. 5 a** Radiation pattern (theta). **b** Radiation pattern (phi)

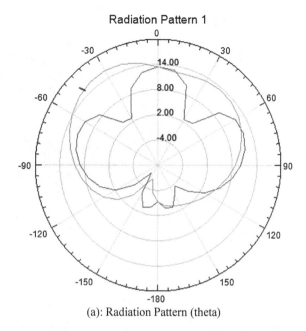

(a): Radiation Pattern (theta)

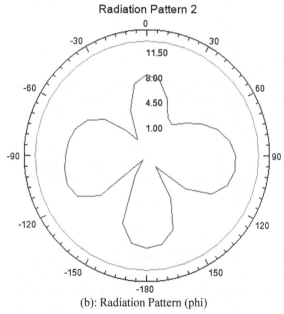

(b): Radiation Pattern (phi)

**Fig. 6** 3D polar plot

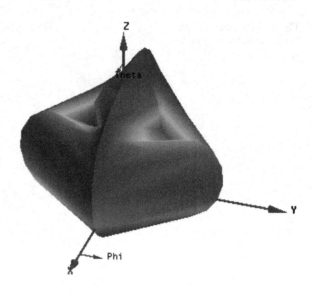

of the antenna. Fig. 7c shows the magnitude of the electric field variation on the patches of the antenna.

## 4 Antenna Fabrication Process

Fabrication of the antenna is done with MITS PCB machine. The machine is very helpful to manufacture antennas. Then, Design Pro connects the computer and MITS PCB machine. The fabrication is done according to the following block diagram (Fig. 8).

The antenna is designed by use of HFSS software. Then, this design is exported in .dxf file. The Design Pro Software has a converter which is used to edit the PCB pattern data. Then, CAM is made for controlling the prototyping device. Then at last the antenna gets fabricated.

## 5 Energy Harvesting Using Prototype Antenna

Experimental setup has been done using transmitting antenna, receiving antenna (designed antenna), tripod stand, transceiver device (0.04–4.4 GHz), RG 141 connectors and scale as shown in Fig. 9.

A horn-shaped antenna is used as transmitting antenna. The two antennas are connected to the antenna training device using RG141 wire power of +10dBm which is transmitted through the planner horn-shaped antenna as shown in Fig. 10.

**Fig. 7 a** Mag_E field plot.
**b** Mag_H field plot.
**c** Mag_Jsurface field plot

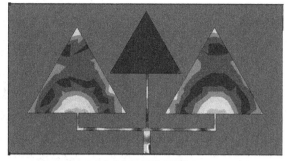

(a): Mag_E field Plot

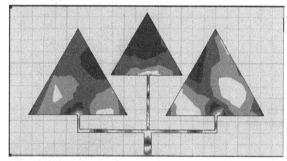

(b): Mag_H field Plot

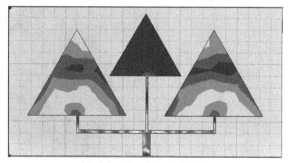

(c): Mag_Jsurface field Plot

**Fig. 8** Fabrication process
block diagram

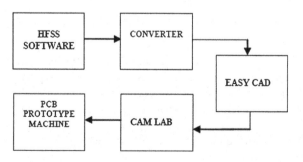

**Fig. 9** Setup for RF energy
harvesting

**Fig. 10** Planar horn-shaped
transmitting antenna

The target frequencies are 5100–5800 MHz. The antenna has been tuned, and the received power is measured at each frequency. The distance between the transmitting antenna and receiving antenna is initially kept 50 cm and readings are taken. A RF to DC converter has been used to convert the captured RF energy to DC voltage as shown in Fig. 11.

We have obtained the harvested energy in the range of 50–100 mV.

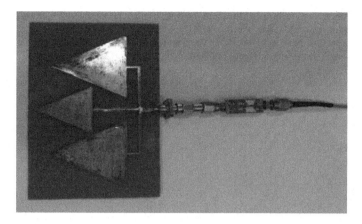

**Fig. 11** Energy harvesting using triangular patch antenna

## 6 Conclusion

This paper discusses the design simulation of a three triangular multi-patch antenna and its different parameters which works in a bandwidth of 5 GHz to 5.5 GHz with resonating frequency close to 5.2 GHz. The important antenna parameters like return loss, VSWR and radiation pattern of the antenna have been analyzed. Further, better characteristics can be obtained by forming an array and then measuring the antenna characteristics.

## References

1. K. Gudan, S. Shao, and J. J. Hull, "Ultra low power 2.4 GHz RF Energy Harvesting and storage system with −25dBm sensitivity", IEEE RFID Conference, San Diego, CA, 2015.
2. A. Nimo, T. Beckedhal, T. Ostertag and L. Reindl, "Analysis of passive RF-DC Power Rectification and Harvesting Wireless RF energy for Micro-wave sensors", AIMS Energy, Vol. 3(2) pp. 184–200, 2015.
3. Lazare Noel, "Portable Charger Using RF energy harvesting", ICEETS Conference, 2016.
4. D. Gunduz, K. Stamatiou, N. Michelusi, and M. Zorzi, "Designing intelligent energy harvesting communication systems," IEEE Communications Magazine, Vol. 52, no. 1, pp. 210–216, January 2014.
5. H. J. Visser and R. J. M. Vullers, "RF energy harvesting and transport for wireless sensor network applications: principles and requirements," Proceedings of the IEEE, Vol. 101, no. 6, pp. 1410–1423, 2013.
6. C. Mikeka and H. Arai, "Design issues in radio frequency energy harvesting system," Sustainable Energy Harvesting Technologies - Past, Present and Future, pp. 235–256, published by InTech, 2011.
7. A. Mavaddat, S. H. M. Armaki, and A.R. Erfanian, "Millimeter-Wave Energy Harvesting Using Microstrip Patch Antenna Array," IEEE Antennas And Wireless Propagation Letters, Vol. 14, pp. 515–518, 2015.

8. G.P. Ramesh and A. Rajan, "Microstrip Antenna Designs for RF Energy Harvesting," ICCSP conference, pp. 1653–1658, April 3–5, 2014, India.
9. S. Shrestha, S-K Noh, and D-Y Choi, "Comparative Study of Antenna Designs for RF Energy Harvesting," International Journal of Antennas and Propagation Volume 2013, Article ID 385260.
10. K. Yuvaraj and A. Ashwin Samuel, "A patch antenna to harvest ambient energy from multiband rf signals for low power devices," International Journal of Emerging Technology in Computer Science & Electronics, Vol. 13, Issue 1, 2015.

# An Approach to Improve Gain and Bandwidth in Bowtie Antenna Using Frequency Selective Surface

**Sarika, Rajesh Kumar, Malay Ranjan Tripathy and Daniel Ronnow**

**Abstract** A novel approach of improving gain and bandwidth of bowtie antenna using Frequency Selective Surface (FSS) is presented. A $42 \times 66$ mm$^2$ bowtie antenna is used. The 5 mm $\times$ 5 mm FSS unit cell consists of metallic square loop and grid wires on FR4 substrate with permittivity of 4.4 and loss tangent = 0.02. The effect of FSS layer on bowtie antenna is investigated in terms of parameters like gain and bandwidth. Simulation results show that gain and bandwidth of the bowtie antenna gets increased by a factor of two when the FSS layer is kept on it. Further, the effect of change of height of the substrate on gain and bandwidth is also studied. The proposed structure shows applications in X-band and Ku-band range. High-Frequency Structural Simulator (HFSS) software is used for the simulation.

**Keywords** Frequency selective surface · Bowtie · Bandwidth enhancement
HFSS

## 1 Introduction

The key in designing any antenna lies in the compactness of the design [1]. In recent years, focus has turned on in reducing the size and alongside maintaining the properties like directivity and bandwidth [2]. However, various antennas like micro-strip patch antennas suffer from drawback that they have narrow bandwidth [3]. The main onus lies in enhancing the bandwidth which can be done by taking care of various factors like substrate thickness, its dielectric constant, and height of

Sarika (✉) · R. Kumar · M. R. Tripathy
Department of Electronics and Communication Engineering, ASET,
Amity University, Noida, Uttar Pradesh, India
e-mail: sarikauppal@gmail.com

D. Ronnow
Department of Electronics, Mathematics and Natural Sciences,
University of Gavle, Gävle, Sweden

© Springer Nature Singapore Pte Ltd. 2019
B. K. Panigrahi et al. (eds.), *Smart Innovations in Communication
and Computational Sciences*, Advances in Intelligent Systems
and Computing 669, https://doi.org/10.1007/978-981-10-8968-8_19

the substrate. Frequency Selective Surfaces (FSS) are planar periodic assembly of structures which repeat themselves in certain manner [4, 5]. When a FSS structure is combined with an antenna as a superstrate, various properties do get improved. The height of superstrate layer over the antenna plays a vital role in the design. A unit cell is first fabricated and then periodic assembly of unit cells forms a FSS layer. The transmission and reflection properties depend on the size of the unit cell and the patch. The complementary design of the same unit cell is also possible, when the patch is replaced with an aperture [4–6]. Instead of using a patch, an antenna can also be utilized as a substrate. Wide variety of applications which exists for FSS prove that the antenna used should have a high gain [7, 8]. Generally, a FSS is designed to achieve high gain and bandwidth [9]. According to study in [10], a rectangular slot is etched in a ring slot as a design of FSS, which is kept over an antenna to increase the overall gain and bandwidth of the structure. The bandwidth of FSS with antenna is around 5.3–6 GHz (17.2%) with $S_{11}$ and $S_{22}$ less than $-10$ dB.

This paper proposes a bowtie antenna which uses a new shape of FSS as a superstrate kept over it. Several simple and effective methods for increasing gain and bandwidth are available for planar antennas; however, the novelty of this paper lies in the approach of using Frequency Selective Surface having a new shape. The paper is organized as follows. Section 1 presents the introduction about FSS. Section 2 gives the design of bowtie antenna, unit cell, and simulation results. Section 3 presents the layered structure of bowtie antenna with FSS and their results. Section 4 finally concludes the research work.

## 2   Bowtie Antenna and FSS Unit Cell Approach

The dimensions of bowtie antenna of [11] have been used in this paper. Figure 1a depicts the dimensions of the bowtie antenna. Figure 1b shows the results of simulated reflection coefficient of the bowtie antenna according to which four resonant frequencies are obtained as in [11]. The frequencies are 6.6 GHz, 10.3 GHz, 13.9 GHz, and 17.8 GHz, respectively. The respective bandwidths obtained are 0.9, 1.1, 1.3, and 1.1 GHz. The maximum return loss obtained is $-30.7$ dB. The maximum gain obtained is 1.1 at 13.9 GHz. In order to increase the bandwidth and gain of the bowtie antenna, an approach of using a Frequency Selective Surface is applied.

To analyze the characteristics of FSS, a unit cell is designed as in Fig. 2a. The unit cell is symmetric to both x-axis and y-axis with dimensions of 5 mm × 5 mm. The unit cell structure consists of metallic square loop and grid wire on the same side of FR4 substrate having permittivity of 4.4 and loss tangent of 0.02 with the thickness of the substrate as h = 0.8 mm. Figure 2b represents the simulated reflection coefficient results of the FSS unit cell. When lumped port as a feed is

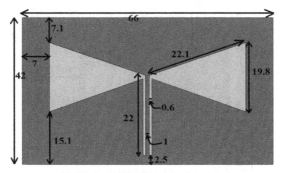

(a) Dimensions of Bowtie Antenna

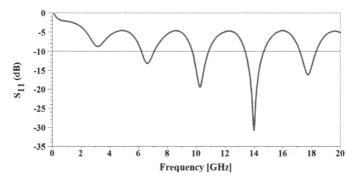

(b) Simulated Reflection coefficient of Bowtie Antenna

**Fig. 1** **a** Dimensions of bowtie antenna. **b** Simulated reflection coefficient of bowtie antenna

applied, the results show dual resonant frequencies. The structure first resonates at 8.4 GHz with return loss of −17.3 dB. The bandwidth at this frequency is 0.53 GHz. However, the second resonant frequency is at 13.4 GHz with a bandwidth of 2.7 GHz and return loss of −28.7 dB.

## 3 Bowtie Antenna with Frequency Selective Surface Array

An array of FSS is designed and is used as a superstrate for the bowtie antenna. The array consists of 6 unit cells in y-direction and 10 unit cells in x-direction. Figure 3a represents the array of FSS and the side view of bowtie antenna with FSS as a superstrate is shown in Fig. 3b, respectively. Simulation results for various heights of FSS array and bowtie antenna are studied as shown in Fig. 3c. When FSS is kept at 1 mm from the bowtie antenna, the structure resonates at tri-band frequencies at 6.5 GHz, 10.2 GHz, and 13.8 GHz yielding bandwidth of 2.0 GHz, 1.4 GHz, and 1.2 GHz, respectively. The gain obtained at each band is 0.6, 0.2, and 1.0,

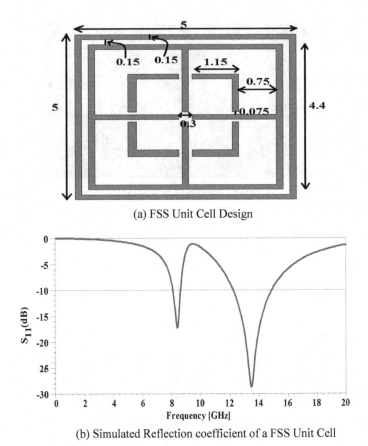

(a) FSS Unit Cell Design

(b) Simulated Reflection coefficient of a FSS Unit Cell

**Fig. 2** **a** FSS unit cell design. **b** Simulated reflection coefficient of a FSS unit cell

respectively. The variation in this height from 1 to 5 mm is further done for parametric study. Table 1 summarizes the results of different stages when the FSS layer is kept above the bowtie antenna at various heights.

Table 1 reveals that four resonant frequencies are obtained in each case except for 1 mm spacing. At 6.5 GHz, it can be seen that the bandwidth is highest (2 GHz) when the spacing is at 1 mm. Afterward, it decreases at 3 mm and 5 mm but increases at 4 mm. At 9.9 GHz, similar pattern can be observed for bandwidth. But with the increase in frequency at 13.8 GHz, increase in bandwidth starts at 3 mm instead of 4 mm as in previous case. The structure with 1 mm spacing does not resonate at 17.4 GHz as in case of rest spacing. In this case, the increase in bandwidth shifts to 2 mm instead of 3 mm.

Hence, it can be said that for lower frequency range, the distance between FSS and antenna can be of greater range; however, for achieving higher frequencies the distance between FSS and antenna should be kept small. The maximum return loss is achieved at 13.8 GHz frequency in all the cases with 3 mm spacing having value

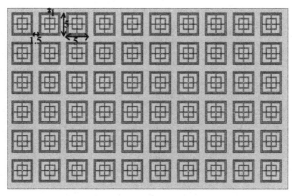

(a) Frequency Selective Surface Array

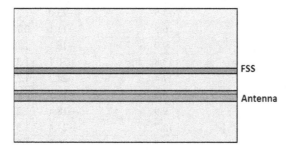

(b) Side View of Bowtie Antenna with FSS as a superstrate

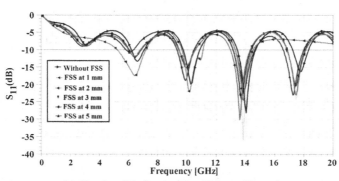

(c) Simulated Reflection Coefficient of Bowtie
Antenna with FSS at different heights

**Fig. 3** **a** Frequency Selective Surface array. **b** Side view of bowtie antenna with FSS as a superstrate. **c** Simulated reflection coefficient of bowtie antenna with FSS at different heights

**Table 1** Different parameters at different heights of FSS above the antenna

| FSS height above the antenna | Number of bands | Resonant frequency (GHz) | $S_{11}$ (dB) | Bandwidth (GHz) | Gain |
|---|---|---|---|---|---|
| Without FSS | 4 | 6.6 | −13.2 | 0.9 | 0.3 |
| | | 10.3 | −19.4 | 1.1 | 0.5 |
| | | 13.9 | −30.7 | 1.3 | 1.1 |
| | | 17.8 | −16.2 | 1.1 | 1.1 |
| 1 mm | 3 | 6.5 | −17.4 | 2.0 | 0.6 |
| | | 10.2 | −22.1 | 1.4 | 0.2 |
| | | 13.8 | −24.6 | 1.2 | 1.0 |
| 2 mm | 4 | 6.5 | −11.8 | 0.8 | 0.5 |
| | | 9.9 | −14.4 | 1.1 | 0.2 |
| | | 13.6 | −30.1 | 1.4 | 0.7 |
| | | 17.5 | −23.0 | 1.3 | 0.6 |
| 3 mm | 4 | 6.5 | −11.1 | 0.7 | 0.6 |
| | | 9.9 | −16.1 | 1.1 | 0.2 |
| | | 13.8 | −36.1 | 1.3 | 1.4 |
| | | 17.4 | −24.2 | 1.3 | 1.8 |
| 4 mm | 4 | 6.5 | −12.1 | 0.9 | 0.6 |
| | | 9.9 | −18.8 | 1.2 | 0.3 |
| | | 13.7 | −24.4 | 1.3 | 0.9 |
| | | 17.5 | −20.3 | 1.3 | 0.5 |
| 5 mm | 4 | 6.1 | −10.7 | 0.6 | 0.7 |
| | | 9.9 | −15.2 | 1.0 | 0.3 |
| | | 13.9 | −21.7 | 1.3 | 0.9 |
| | | 17.3 | −23.0 | 1.4 | 1.6 |

of −36.1 dB as compared to that of −30.7 dB as obtained from bowtie antenna without FSS. The improvement in gain can be observed predominantly at 6.6 GHz where it is increased by a factor of two; at 13.8 GHz, gain increases from 1.1 to 1.4, and at 17.4 GHz, a rise of 0.7 can be observed. However, the maximum gain obtained is 1.8 at 17.4 GHz with 3 mm spacing.

Figure 4a, b represents the E-plane and H-plane radiation pattern for structure without FSS and structure with 3 mm spacing at 13.8 GHz. It is observed that the structure fires predominantly in one direction with high bandwidth. Figure 5a, b represents the E-plane and H-plane radiation pattern for structure without FSS and structure with 1 mm spacing at 6.5 GHz.

**Fig. 4** **a** E-plane radiation
pattern at 13.8 GHz.
**b** H-plane radiation pattern at
13.8 GHz

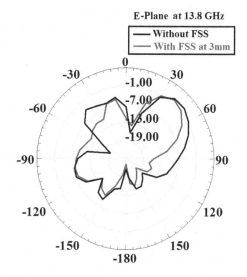

(a) E-Plane Radiation Pattern at 13.8 GHz

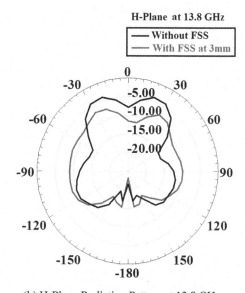

(b) H-Plane Radiation Pattern at 13.8 GHz

**Fig. 5 a** E-plane radiation
pattern at 6.6 GHz. **b** H-plane
radiation pattern at 6.6 GHz

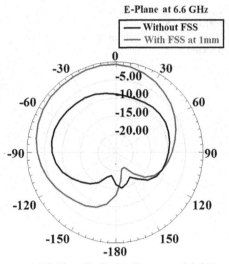

(a) E-Plane Radiation Pattern at 6.6 GHz

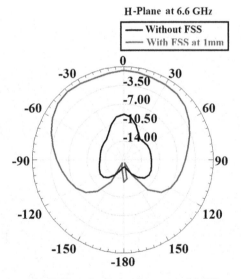

(b) H-Plane Radiation Pattern at 6.6 GHz

## 4 Conclusion

An approach of increasing gain and bandwidth of the bowtie antenna using Frequency Selective Surface has been presented. Integration of FSS with bowtie antenna has enhanced the bandwidth by a factor of two at 6.6 GHz, whereas the return loss is increased by −6 dB at 13.8 GHz, and gain has been increased by 63.6% at 17.4 GHz. This integrated structure can be used for wireless applications as well as in space applications as the obtained frequencies fall into X-band and Ku-band, respectively. The surface wave effect is reduced which further improves the performance of the antenna. The study of height variation suggests that improvement at higher frequencies can be attained by keeping smaller distances between FSS and the antenna.

## References

1. Munk, B.: Frequency Selective Surfaces: Theory and design, John Wiley & Sons, New York, (2000)
2. Wu T.K..: Frequency Selective Surfaces and Grid array: John Wiley & Sons Inc., New York, (2000)
3. Kiami, G.I., A.R., Wiley, Esselle K.P.: FSS absorber using resistive cross dipoles. In: IEEE Antennas and Propagation Society International Symposium, pp 4199–4204, (2006)
4. Munk B.A.: Finite Antenna Arrays and FSS: Hoboken, NJ, Wiley Interscience, pp 136–152, (2003)
5. Kat Z., Sadek S.: Wideband antenna for WLAN Applications. In: IEEE, pp. 197–199, Lebanon, (2006)
6. O. Manoochehri, S. Abbasiniazare, A. Torabi, K. Forooraghi.: A second order BPF using a miniaturized element Frequency Selective Surface. In: Progress in Electromagnetic Research C, vol. 31, 229–240, (2012)
7. Singh D, Kumar A, Meena S, Agarwala V.: Analysis of Frequency Selective Surface for Radar Absorbing Materials. In: Progress in Electromagnetic Research C, vol. 38, 297–314, (2012)
8. Liu H.T., Cheng H. F., Chu Z. Y., et al.: Absorbing properties of Frequency Selective Surface Absorbers with Cross-shaped resistive patches. In: Material Design, vol. 28, 2166–2171, (2007)
9. Cook B. S., Shamim A.: Utilising Wideband AMC structures for High Gain Ink-jet printed antennas on lossy paper substrate. In: IEEE Antennas and Wireless Propagation Letters, vol. 12, 76–79, (2013)
10. Hua z., Yang Yu., Li X., Ai Bo.: A wideband and High gain Dual-polarized Antenna Design by a Frequency selective Surface for a WLAN Applications. In: Progress in Electromagnetic Research C, vol. 54, 57–66, (2014)
11. Kumar R., Tripathy M.R., Ronnow D.: Multi-band Meta-material based Bowtie Antenna for Wireless applications. In: Progress in Electromagnetic Research, pp 2368–2371, (2015)

# Transmission of Electromagnetic Radiations from Mobile Phones: A Critical Analysis

Tanvir Singh, Rekha Devi, Amit Kumar, Santosh S. Jambagi
and Balwinder Singh

**Abstract** In recent years, the use of handheld wireless communication devices such as smart phones and tablets has been increasing enormously, thus ensuing surge in mobile data traffic which is expected to grow even more in the upcoming years. In mobile data traffic, the major drivers are audio and video processing which consume the major bandwidth hog, putting a great deal of demand on telecommunication networks. Telecom service providers are continuously meeting the demands for more bandwidth and faster broadband connections by stepping up site acquisitions to improve their infrastructure, but there is a huge public health concern looming from the threat of electromagnetic radiations (EMRs) from the mobile phones. It emits radio frequency (RF) energy, a form of non-ionizing electromagnetic radiation, which can be absorbed by tissues flanking to the handset and would be harmful to health. There is scope for study to evaluate prevailing radiation concern. In this paper, a critical analysis of electromagnetic radiations transmitted by the mobile phones is carried out in different scenarios such as incoming/outgoing calls (Global System for Mobile Communication (GSM) and Universal Mobile Telecommunication System (UMTS)), short message service (SMS), mobile data, Bluetooth, Wireless Fidelity (Wi-Fi), and Wi-Fi hotspot. It has been observed from

T. Singh (✉)
Spoken Tutorial – IIT Bombay, MHRD Government of India, Mumbai, India
e-mail: singhtanvir21@gmail.com

R. Devi
I.K.G. Punjab Technical University, Jalandhar, Punjab, India

A. Kumar
JNV Shimla, HP, MHRD Government of India, Shimla, India

A. Kumar
College of Information Science and Technology, Nanjing Forestry
University, Nanjing, China

S. S. Jambagi
Department of ECE, R.V. College of Engineering, Bengaluru, Karnataka, India

B. Singh
Centre for Development of Advanced Computing, Mohali, Punjab, India

© Springer Nature Singapore Pte Ltd. 2019
B. K. Panigrahi et al. (eds.), *Smart Innovations in Communication
and Computational Sciences*, Advances in Intelligent Systems
and Computing 669, https://doi.org/10.1007/978-981-10-8968-8_20

229

the results that GSM technology showed high radiation levels, whereas UMTS is more or less unchanged compared to idle standby level. Further, the possible precautionary measures to minimize the harmful effects of the EMR are proposed.

**Keywords** EMR · Bluetooth · GSM · Internet · Mobile radiation
Real-time applications · Mobile phones

# 1 Introduction

Currently, the mobile phone industry is the fastest growing industry owing to a huge subscriber base and tremendous growth in mobile phone technologies [1]. A whole new world of applications and content services are now available for every user. Due to continuously falling prices and expanding network services, mobile phones can be afforded by every individual. Recent studies conducted in this field have shown potential health hazards arising from prolonged use of mobile phones operating on RFs for voice and data purposes. These effects can be thermal or non-thermal. Thermal effects refer to heating of tissues due to radiation absorption by our body, whereas non-thermal effects are related to genes, cells, and DNA damage [2, 3]. The experimental findings suggest that mobile radiations may modulate with the activity or functionality of neural networks as the brain has larger exposure to mobile radiations as compare to rest of the body [4].

According to CISCO, global mobile data traffic astonishingly jumped up to 7.2 exabytes per month at the end of 2016, from 4.4 exabytes per month at the end of 2015 owing to 63% growth in global mobile data traffic in 2016 [5]. While these numbers are very exciting from the point of view of technology penetration but exhibits a horrifying sight of potential health dangers arising from radiations emitted by them. It has been observed that due to short distance and large inter-action, 30–70% of radiated power from the mobile phone is absorbed in the head, hand, or body [6]. Also, only thermal effects of radiations from mobile phones have been researched till now; study on non-thermal effects is still under consideration but not recognized yet [7]. It is pertinent to describe and measure the distribution of the EMRs and the energy levels absorbed by the human tissues. It is important to calculate the rates of RF energy absorption, called specific absorption rates (SARs), in order to evaluate impending health hazards and conformity with standards [8]. SAR is a dissymmetric quantity which is defined as the rate of RF power absorbed per unit mass by any part of the body [9]. SAR value is normally specified at the maximum transmission power and is generally determined for either 1 g or 10 g of simulated biological tissue in the shape of a cube. SAR is measured in W/kg, and the limit specified by Federal of Communication Commission (FCC) is 1.6 W/kg.

**Fig. 1** SAR distributions toward human head while using mobile phone

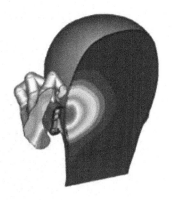

$$SAR = \frac{\sigma E^2}{\rho} \text{ W/kg}$$

Figure 1 shows the SAR distribution toward human head while using mobile phone. RF radiations have the ability to cause biological damage through heating effects as the human body is constituted of approximately 70% water, electrolytes, and ions. Further, weak electromagnetic fields (EMFs) such as RF radiations emitted from wireless mobile devices will interact with human body and affect the human body's own weak EMFs thus interfering with the body's immunity system [1]. This paper extends the above research and explores the mentioned threats in further new areas and usage scenarios. Also, some precautionary methods have been suggested to lessen the hazards from the unavoidable gadgets to some extent.

## 2 Mobile Wireless Communication: Present Scenario

The mobile phone boom continues its upward trend showing a largest ever increase in the adaptation of content and services provided. According to Statista, global smart phone shipments are expected to reach 1.47 billion units in 2016 [10]. A simple endorsement to the trend is of tablets which even out shelled PCs worldwide in 2015. Likewise, the global connection speed is also expected to grow to 3898 kbps in 2017 from 189 kbps in 2010 showing a CAGR of 54% [11, 12]. Figure 2 shows the global subscriptions and market shares by the end of December 2016. It is observed that all wireless technologies aimed at cohesive target of high performance and efficiency by following diverse evolutionary paths. In the 1980s, analog cellular technology was named as first generation (1G), and second generation (2G) started digital systems in 1990s with the short message service and low-speed data. After this, International Telecommunication Union (ITU) specified requirements for third-generation technology (3G) as a part of the IMT-2000 project, with much anticipated 2 Mbps in indoor environments at quite affordable

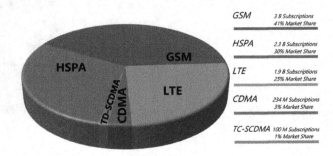

**Fig. 2** Global subscriptions and market shares. *Source* Ovum, WCIS, December, 2016

prices. Now, LTE and VoLTE have totally transformed the telecom market by the launch of 4G networks by various telecom giants and quenching the thrust of data-hungry consumer by 1Gbps speed for stationary device and now striving for versatile services through upcoming 5G networks [12].

## 3 Experimental Setup for Critical Analysis

There are numerous methods in order to compute the EMRs transmitting from mobile phone. Most commonly used analytical method is finite-difference time-domain (FDTD) which is especially used for brain with non-uniform character and body modeling in the time domain (each time point refers to the time node) as well as in the space domain (each volume unit is referred to as a cell). Apart from analytical techniques, there are various simulators available viz. Suite, Feko, Antenna Magus, HFSS, CST Studio, and electromagnetic meters [13].

In this paper, we have used CORNET ED-75 electrosmog meter for electromagnetic wave exposure detection and level measurement for both low and high frequency. It can operate in RF power meter mode and LF Gauss meter mode. Only RF meter mode is considered in this paper. Frequency range for RF meter is from 100 MHz to 6 GHz (useful up to 8 GHz) with dynamic range of 60 dB. LCD histogram and bar display is used to get the power density level. LED indications for various readings in the meter are as follows (Table 1).

To perform various experiments, we have used two smart phones. Smart phones are faster in processing and having better hardware and software configuration than normal cell phones and enable the users to access Internet and other online applications which were not available in conventional mobile phones. Wireless data usage is continuously rising with an unprecedented pace. Along with penetration

**Table 1** LED indications for various readings in the meter

| LED indications for various readings in the meter | | |
|---|---|---|
| −5 to −15 dBm | Red | Danger zone |
| −20 to −30 dBm | Yellow | Moderate zone |
| −35 to −40 dBm | Green | Safer zone |

rates of tablets and Apple iPad, smart phones such as the iPhone and BlackBerry are also increasing rapidly [2, 14]. Xiaomi Mi3 with SAR value 1.1 W/kg (MAX) and Motorola Moto G with SAR Value 1.35 W/kg (MAX) are used.

## 4 Experimental Analysis of Different Scenarios in Mobile Phones

Presently, mobile phones are operating on numerous technologies such as Global System for Mobile Communication (GSM), code-division multiple access (CDMA), Universal Mobile Telecommunication System (UMTS), Enhanced Data for Global Evolution (EDGE), Long-Term Evolution (LTE), General Packet Radio Service (GPRS), and Wideband Code Division Multiple Access (WCDMA). Further, technologies associated with GSM are EDGE, GPRS, UMTS, and WCDMA. GSM uses 900/1800 MHz frequency spectrum, whereas WCDMA uses 2100 MHz spectrum.

## 5 Results and Discussion

Results have been represented in graphical form (as shown in Figs. 3, 4, 5, 6, 7, 8, 9, and 10). With the help of electrosmog meter, RF/LF field strength power meter (100 MHz–6 GHz)/(50 Hz–10 kHz), radiation levels have been compared during different cases of mobile phone operation which includes incoming/outgoing calls (GSM and UMTS), SMS, mobile data, Bluetooth, Wi-Fi, and Wi-Fi hotspot. It has been analyzed that WCDMA (3G) is the most widely accepted and used carrier interface for UMTS and has many advantages as compared to GSM (2G) which are as follows:

**Fig. 3** Comparison between power transmitted (in dBm) by mobile phones on GSM and UMTS networks during incoming calls

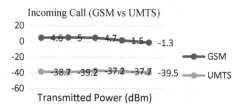

**Fig. 4** Comparison between power transmitted (in dBm) by mobile phones on GSM and UMTS networks during outgoing calls

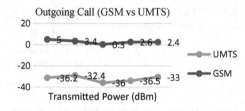

**Fig. 5** Comparison between power transmitted (in dBm) by mobile phones on EDGE and HSPA+ networks during usage of mobile data

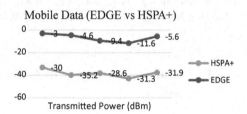

**Fig. 6** Comparison between power transmitted and signal strength (in dBm)

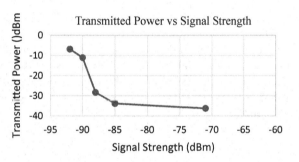

**Fig. 7** Comparison between power radiated (in dBm) by single and dual SIM mobile phones

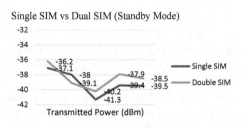

**Fig. 8** Comparison between power radiated (in dBm) by mobile phones during Wi-Fi and mobile data usage

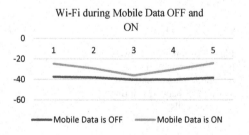

**Fig. 9** Power transmitted (in dBm) by mobile phone while using Bluetooth

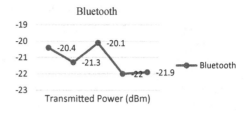

**Fig. 10** Power radiated (in dBm) by mobile phone while using a Wi-Fi hotspot

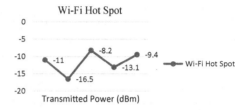

- Fast Power Control Mechanism: 3G has fast power control mechanism which adjusts the power in 0.6 ms, whereas in case of 2G, time period is 60 ms.
- Soft Handovers: 3G support soft and softer handover which means user equipment (UE) keeps always one radio link as frequency is same in WCDMA.
- Low Initial Power: 2G uses step-down technology in which mobile starts with maximum power for several seconds and step down its power to such extent so that call is established, whereas 3G uses step-up mechanism [15].

In this section, we compared the radiation levels of GSM and UMTS during incoming calls. The levels ranged between −1.3 and 4.6 dBm for GSM and −39.5 and −38.7 dBm for UMTS. The graph clearly shows high radiation levels for GSM, while the radiation for UMTS is more or less unchanged compared to the idle standby level.

In the above section, we compared the radiation levels of GSM and UMTS during outgoing calls. The levels ranged between −2.4 and 5.0 dBm for GSM and −33.0 and −36.2 dBm for UMTS. The outgoing call radiation level rises as soon as the call is placed. UMTS again shows little change in the radiation levels as compared to standby.

The above section compares the radiation levels of GSM and UMTS during a mobile data (GPRS for GSM and HSPA for WCDMA). The levels ranged between −5.6 and −3 dBm for GSM and −31.9 and −30 dBm for UMTS. Mobile data continuously uses the mobile network and so is visible in the graphs. UMTS is proving its safety here again.

The above section compares the power transmitted and signal strength (in dBm). The levels ranged between −36.3 and −6.9 dBm. As signal strength goes worse (i.e., at −92 dBm), power radiated goes across the safer limit (i.e., −6.9 dBm, which is not a safer zone).

In this section, we compared the radiation levels of a single SIM vs a double SIM configuration. The levels ranged between −38.5 and −36.2 dBm for GSM and

−39.5 and −37.1 dBm for UMTS. Multiple SIM Configuration in Dual SIM Dual Standby (DSDS) mode shows changes in comparison to the single SIM configuration.

Wi-Fi is wireless fidelity and is also known as wireless LAN or WLAN. It provides a very high speed and reliable wireless link which can be used as a substitute to wired LAN connection [16]. In this section, we analyzed the radiation levels of the mobile phone during a Wi-Fi session. The levels ranged between to dBm. Evidently, Wi-Fi has proved much safer than mobile data owing to high frequency of operation.

In this section, we analyzed the radiation levels of the mobile phone during a Bluetooth session. The levels ranged between −21.9 and −20.4 dBm. Bluetooth shows little to no deviation from standby levels.

In this section, we analyzed the radiation levels of the mobile phone during a Wi-Fi hotspot. The levels ranged between −9.4 and −11.0 dBm. In contrast to Wi-Fi, using Wi-Fi hotspot feature leads to higher radiation levels which can be more dangerous since the phone will be much closer to the body than a Wi-Fi router.

## 6  Precautionary Measures

Mobile phones have travelled a long distance from being a luxury to necessity. Thus, we can't avoid using these gadgets, but we can minimize their harmful effects by adopting the following precautionary measures.

One can limit phone calls to those which are absolutely necessary. One should never hold his/her phone close to his/her body. One should avoid calling in weak signal areas as it causes more radiation as per experimental readings described above. One should use mobile phones with lesser SAR because more SAR directly translates into more radiations as explained. Headphones or Bluetooth headsets should be used whenever possible.

Use of landline telephones should be preferred. One should prefer Internet calling (preferably via wired Internet or via Wi-Fi) to regular calls. Mobile phone should be avoided by people who have metallic objects in their head or body so as to avoid increase in radiations due to phenomena of reflection, amplification, resonance, passive re-emission, etc. Mobile devices should be kept away from children because of their lighter body weight, the radiation is more damaging. Mobile phone should not be kept beside bed because even in standby mode, they emit radiation.

## 7  Conclusion

The exceptional adoption of wireless mobile communication technologies by end-users and explosion of development of mobile applications is fueling the growth of global deployments in the telecom industry. The telecom operators around the world are rolling out the requisite services to help them meet the ever-growing demand of data-hungry end-user for more bandwidth, higher security, and faster connectivity ubiquitously. Mobile phone usage is altering our lifestyle as well as escalating the risks of many diseases which originated from the exposure to mobile phone radiations. There is an imperative need to address the issue, and in this paper, we have carried out number of experimental comparisons in different scenarios of mobile phone usage with the help of electrosmog meter, RF/LF field strength power meter (100 MHz–6 GHz)/(50 Hz–10 kHz) to compare the radiation levels. Further, the possible precautionary measures to minimize the destructive effects of the EMR are anticipated.

# References

1. Mat, D.; Kho, F.; Joseph, A.; Kipli, K.; Sahrani, S.; Lias, K.; Marzuki, A.S.W. (2010). The effect of headset and earphone on reducing electromagnetic radiation from mobile phone toward human head, 8th Asia-Pacific Symposium on Information and Telecommunication Technologies (APSITT), pp. 1,6.
2. Anoob, B.; Unni, C. (2014). Investigation on Specific Absorption Rate due to electromagnetic radiation. International Conference on Communication and Network Technologies (ICCNT), pp. 302,307.
3. Mat, D.; Tat, F.K.W.; Kipli, K.; Joseph, A.; Lias, K.; Marzuki, A.S.W. (2010). Visualization and Analytical Measurement of Electromagnetic Radiation from Handheld Mobile Phones. Second International Conference on Computer Engineering and Applications (ICCEA), pp. 246,250, https://doi.org/10.1109/iccea.2010.201.
4. Relova, J.L.; Pértega, S.; Vilar, J.A.; López-Martin, E.; Peleteiro, M.; Ares-Pena, F. (2010). Effects of Cell-Phone Radiation on the Electroencephalographic Spectra of Epileptic Patients [Telecommunications Health & Safety]. Antennas and Propagation Magazine, vol. 52, no. 6, pp. 173,179.
5. Cisco Whitepaper. Cisco Visual Networking Index: Global Mobile Data Traffic Forecast Update, 2014–2019. http://www.cisco.com/c/en/us/solutions/collateral/service-provider/visual-networking-index-vni/white_paper_c11-520862.pdf. Accessed 18 January 2015.
6. Gavan, J.; Haridim, M. (2007), "Mobile Radio Base Stations and Handsets Radiation Effects: Analysis, Simulations and Mitigation Techniques", *International Symposium on Electromagnetic Compatibility, EMC 2007*.
7. Li Yang; Lu guizhen (2008), "The discussion on the biological effect of mobile phone radiation", *International Conference on Microwave and Millimeter Wave Technology, ICMMT*.
8. Pinho, P.; Lopes, A.; Leite, J.; Casaleiro, Joao (2009). SAR determination and influence of the human head in the radiation of a mobile antenna for two different frequencies.
9. Sai-Wing Leung; Yinliang Diao; Kwok-Hung Chan; Yun-Ming Siu; Yongle Wu (2012). Specific Absorption Rate Evaluation for Passengers Using Wireless Communication Devices inside Vehicles with Different Handedness, Passenger Counts, and Seating Locations. IEEE Transactions on Biomedical Engineering, vol. 59, no. 10, pp. 2905,2912.
10. Global smartphone shipments forecast 2010–2021. Resource Document. The Statistics Portal. www.statista.com/statistics/263441/global-smartphone-shipments-forecast/. Accessed 10 July 2016.
11. Android Growth Drives Another Strong Quarter for the Worldwide Tablet Market, Resource Document. International Data Corporation.http://www.idc.com/getdoc.jsp?containerId=prUS24420613. Accessed 18 January 2015.

12. The Mobile Economy Report 2014. Resource Document. GSMA.www.gsmamobileeconomy. com/GSMA_ME_Report_2014_R2_WEB.pdf. Accessed 18 January 2015.
13. Kundu, A. (2013). Specific Absorption Rate evaluation in apple exposed to RF radiation from GSM mobile towers. Applied Electromagnetics Conference (AEMC), pp. 1,2.
14. Chang-xia Sun; Yong Liu; Fei Liu (2011). The research of 3G mobile phone radiation on the human head. 2nd International Conference on Artificial Intelligence, Management Science and Electronic Commerce (AIMSEC), pp. 4869,4872.
15. J. Baumann, F.M. Landstorfer, L. Geisbusch and R. Georg (2006). Evaluation of radiation exposure by UMTS mobile phones. Electronics Letter.
16. Resource Document. 4G Americas. http://www.4gamericas.org/index.cfm?fuseaction= page&sectionid=260.

# Performance Evaluation of Improved Localization Algorithm for Wireless Sensor Networks

**Santar Pal Singh and S. C. Sharma**

**Abstract** In the previous years, wireless sensor networks (WSNs) have attracted a lot of interest from the industrial and research community. In most of the WSN-based applications, the node's location information is of much significance. Numerous localization algorithms have been reported in the literature to precisely locate nodes in wireless sensor networks. These algorithms are broadly classified into range-based and range-free techniques. Range-based schemes exploit exact computation measure (distance or angle) among nodes in the network. So, range-based schemes need some additional hardware for such computation, while range-free methods do not need any specific hardware. Range-free schemes use connectivity information among nodes in network. So, range-free methods are treated as cost-effective alternatives to range-based methods. In this paper, we propose PSO-based algorithms for localization in WSNs. We perform the analysis of our proposed algorithms with other existing algorithms of its category on the basis of localization error and accuracy. Simulation results confirm that the proposed algorithm outperforms the other existing algorithms of its category.

**Keywords** Wireless sensor networks · Localization algorithms
Connectivity · DV-Hop · Performance analysis

## 1 Introduction

Recent advancement in MEMS and wireless communication technologies allows the micro-autonomous system comprised of small tiny devices known as sensors. These sensors can detect, compute, and communicate via suitable sensor technology that

S. P. Singh (✉) · S. C. Sharma
Electronics and Computer Discipline, DPT, IIT Roorkee,
Roorkee 247667, India
e-mail: spsingh78@gmail.com

S. C. Sharma
e-mail: scs60fpt@gmail.com

© Springer Nature Singapore Pte Ltd. 2019      239
B. K. Panigrahi et al. (eds.), *Smart Innovations in Communication and Computational Sciences*, Advances in Intelligent Systems and Computing 669, https://doi.org/10.1007/978-981-10-8968-8_21

gives birth to wireless sensor network [1, 2]. Deployment ease and low-cost sensors make wireless sensor network suitable for many applications like health care, transportation, smart building, and environmental monitoring. For such applications, the location of sensor is needed to detect the events in monitored area [3]. So, the location information is of much use in various applications and needed to build context-aware applications [4].

Most of the applications of sensor network used a model where some of the nodes are ware about their location (either manually placed or enabled with GPS) are recognized as anchors (beacon) nodes. The remaining nodes are normal nodes, deployed in random way due to the hostility of the region to be monitored. The nodes estimate their locations with the help of information they obtained from the anchors [5]. In this article, we presented an improved algorithm for node localization based on PSO scheme. This work makes three main contributions to the problem of localization in WSNs. First, we proposed a quick and user-friendly localization method with higher accuracy. Second, the proposed algorithm reduces the localization error and improves accuracy as compared to traditional DV-Hop method. Third, we investigated the effect of total no. of nodes, ratio of anchors and communication radius on the performance of DV-Hop, and our new algorithm.

The rest of the paper is planned as: Section 2 gives the review of the related literature. In Sect. 3, PSO algorithm is discussed briefly. Section 4 illustrates the proposed algorithm. In Sect. 5, simulation outcomes are described. Lastly, Sect. 6 concludes the work.

## 2 Related Work

In this part, we survey the literature related to our work. Lots of works reported in the literature proposed techniques for sensor network localization for various uses. Since every method was built up to accomplish a special objective, they vary usually in many parameters [3–5]. The localization schemes are usually classified based on certain measures like computational model, range measurements. Based on range measurements, node localization is mostly partitioned into range-based as well as range-free approaches. The range-based schemes need to measure real distance among nodes, while the range-free schemes need only the connectivity information among nodes [5, 6]. The ranging methods require distance (or angle) information among nearby nodes to find out the position. Some techniques to measure that are RSSI, ToA, TDoA, AoA based. The accuracy of range-based approaches is higher as compared to range-free schemes but needs extra hardware device. So, the overall cost becomes high in large-scale deployment. In range-free schemes, some restricted numbers of nodes are outfitted with GPS known as "anchors" or "beacons." Anchors broadcast their locality information into network, and unknown nodes estimate their positions on the basis of hop values from beacons [7, 8]. Distance estimation is vital in this scheme. In the range-free schemes, the popular algorithms are DV-Hop, APIT, Amorphous, and Centroid. Niculesue et al. propose the DV-Hop algorithm for localization that is alike to the conventional distance vector routing scheme [9]. In DV-Hop, node first count the minimum hop value from beacons and after that work out the distance among unknown node and anchor one with help of minimum hop value along with average hop distance. Lastly, node estimates its location with help of triangulation or maximum likelihood estimation (MLE) method. Additional device for localization is not required in this method. The range-free methods are less influenced by environmental factors, and additional ranging hardware is not required [10]. These characteristics make them suitable to sensor network localization.

## 3  PSO Algorithm

The particle swarm optimization is an optimization method, modeled after the societal behavior of a bird's flock [11]. In the PSO, a swarm corresponds to the amount of prospective solutions of problem, where every prospective solution is known as a particle. Main objective of this method is to determine the particle location that fallout within best estimation of the set fitness function. During the initialization stage, every particle is set initial parameters at random and is flown throughout the multidimensional search area. In every generation, every particle utilizes the information regarding its earlier person best position as well as globally best position for maximizing the possibility of moving on the way to a better solution space which will result in a better fitness together with updation of its candidate solutions as per the equations given as:

$$V_{id}(t) = \omega \times V_{id}(t-1) + C_1 \times \varnothing_1 \times (P_{id} - X_{id}(t-1)) + C_2 \times \varnothing_2 \times (P_{gd} - X_{id}(t-1))$$

$$(1)$$

$$X_{id}(t) = X_{id}(t-1) + V_{id}(t) \tag{2}$$

where $V$ and $X$ represent velocity and position of particle, $t$ denotes time, $C_1$, $C_2$ represent the learning coefficients, and $\omega$ is the inertia weight. $\phi_1$, $\phi_2$ are the random numbers belong to $(0, 1]$. $P_{id}$ is the particle individual best position, and $P_{gd}$ is the global best position. The flowchart of PSO algorithm is revealed in Fig. 1.

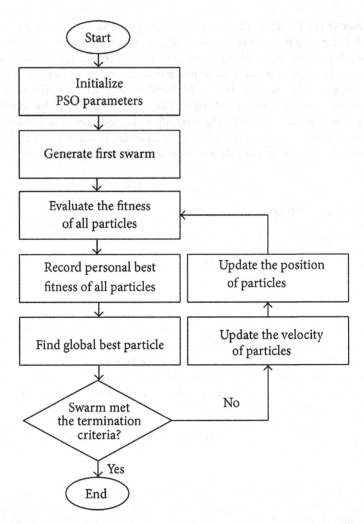

**Fig. 1** Flowchart of PSO [18]

## 4 Proposed Algorithm

Here, in this part of work, we presented an improved hop-based localization algorithm that comprised of four steps. Though various improved DV-Hop localization algorithms are reported in literature [12–17]. The step 1 and step 2 of the proposed method are same as of traditional DV-Hop algorithm. In step 3, unknown node positions are determined with help of 2D hyperbolic algorithm. The 2D hyperbolic location algorithm improves the estimated location precision. In step 4, we correct the estimated positions with help of PSO. So the step 3 of the new modified algorithm is same as the improved DV-Hop method [13]. However, in the step 4 of proposed algorithm, we use PSO for correction the position estimation.

*Step1*: Determine the minimal hop value between unknown and each anchor node.

In this step, every anchor broadcasts a message all over the network comprised the location of anchor with value of hop count, i.e., initially one. Every node that obtains the message records the each node's minimal hop count although ignoring the bigger one from same anchor, and then, hop count value is incremented by one and is passed on to neighbor nodes.

*Step2*: Determine the actual distance among unknown node and anchor node.
In this step, every anchor estimates the average hop distances by the following equation:

$$HopSize_i = \frac{\sum\limits_{i \neq j} \sqrt{(x_i - x_j)^2 + (y_i - y_j)^2}}{\sum\limits_{i \neq j} h_{ij}} \tag{3}$$

where $(x_i, y_i)$ denotes the coordinate or position of anchor $i$ and $(x_j, y_j)$ denotes the coordinate or position of anchor $j$. $h_{ij}$ is the hop value between anchors $i$ and $j$.

Beacon node disseminates the average hop value. Unknown node evidences the first collected average hops distance and passes on it to neighbors. Then, unknown node determines the distance to every beacon in accordance with hop counts.

*Step3*: Determine the position of unknown nodes.
An unknown node makes use of trilateration or MLE method to determine the coordinate of these nodes.

The distance from all beacon nodes to unknown node P $(x, y)$ is given by the formula:

$$(x_1 - x)^2 + (y_1 - y)^2 = d_1^2$$
$$\vdots \tag{4}$$
$$(x_n - x)^2 + (y_n - y)^2 = d_n^2$$

Meantime formula (4) can be stated as:

$$x_1^2 - x_n^2 + 2(x_1 - x_n)x + y_1^2 - y_n^2 - 2(y_1 - y_n)y = d_1^2 - d_n^2$$
$$\vdots \tag{5}$$
$$x_{n-1}^2 - x_n^2 + 2(x_{n-1} - x_n)x + y_{n-1}^2 - y_n^2 - 2(y_{n-1} - y_n)y = d_{n-1}^2 - d_n^2$$

Formula (5) steps the right equation for

$$AX = B \tag{6}$$

where

$$A = \begin{bmatrix} 2(x_1 - x_n) & 2(y_1 - y_n) \\ & \cdot \\ 2(x_{n-1} - x_n) & 2(y_{n-1} - y_n) \end{bmatrix}$$

$$B = \begin{bmatrix} x_1^2 - x_n^2 + y_1^2 - y_n^2 + d_1^2 - d_n^2 \\ \cdot \\ x_{n-1}^2 - x_n^2 + y_{n-1}^2 - y_n^2 + d_{n-1}^2 - d_n^2 \end{bmatrix}$$

$$X = \begin{bmatrix} x \\ y \end{bmatrix}$$

The geographic coordinate of unknown node P can be find out with help of the following formula:

$$P = (A^T A)^{-1} A^T B \tag{7}$$

*Step 4*: We will apply PSO to find the correct position of unknown nodes.

Let $(x, y)$ be the geographical position of an unknown node; the distance $d_i$ can be obtained as described in step 2 of improved DV-HOP method.

Thus, the positioning error may be described as:

$$f_i(x, y) = \left\| (x - x_j)^2 + (y - y_j)^2 - d_i^2 \right\| \tag{8}$$

$$fitness(x, y) = \sum_{j=1}^{n} \left( \frac{1}{Hop\ value_j} \right)^2 f_j(x, y) \tag{9}$$

Particle's updations are done with the help of (1) and (2), and (9). Equation (9) is the fitness function to estimate the fitness of particles. The total no. of iterations is set accordingly. After these iterations, the optimal solution is considered as the final estimated position of an unknown node. The flow sheet interpretation of proposed algorithm is revealed in Fig. 2.

**Fig. 2** Flowchart of
proposed algorithm

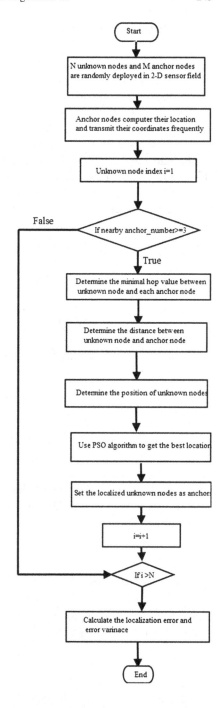

# 5   Results and Analysis

For the performance analysis of the proposed algorithm, we use Matlab2015a as a simulator for implementation of the scenario of networks and determination of final results. Parameter setting for simulation is as follows: initially, few beacon nodes are fixed in 100 × 100 m$^2$ area; then rest anchors and unknown nodes are distributed randomly in that area. We analyze the impact of total no of nodes, impact of anchor nodes, and impact of communication range on the localization results.

The parameters in PSO algorithms are $c_1 = c_2 = 2.0$, $\omega = 0.9$, no. of particles = 20, $V_{max} = 10$, and we use 30 iterations. The stability and accuracy of localization are analyzed by localization error and error variance, and these performance parameters are calculated by the following formula:

$$Error_i = \sqrt{\left(x_i^{eval} - x_i^{real}\right)^2 + \left(y_i^{eval} - y_i^{real}\right)^2} \tag{10}$$

Average localization error is considered as localization error and computed like:

$$Localization\,Error(LE) = \frac{\sum_{i=1}^{n} Error_i}{n \times R} \tag{11}$$

The accuracy of localization is computed as:

$$Localization\,Accuracy = \frac{\sum_{i=1}^{n} \sqrt{\left(x_i^{eval} - x_i^{real}\right)^2 + \left(y_i^{eval} - y_i^{real}\right)^2}}{n \times R^2} \tag{12}$$

where $n$ represents the value of unknown nodes. $(x_i^{real}, y_i^{real})$, $(x_i^{eval}, y_i^{eval})$ are the real and evaluated positions of unknown node $i$, respectively. The sensor node's radio range is represented by $R$.

We deploy 100 nodes out of these; few are beacon nodes, and others are unknown ones, randomly in 2D area of 100 × 100 m$^2$. The node's distribution is shown in Fig. 3.

## 5.1   Impact of No. of Anchor Nodes on Localization Results

The influence of beacons on the localization parameters like error and accuracy is given away in Fig. 4 and Fig. 5, respectively.

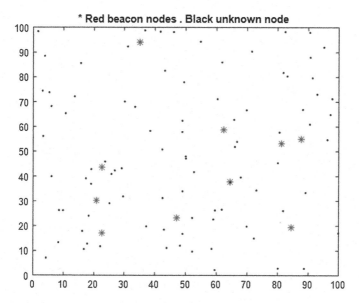

**Fig. 3** Nodes' distribution

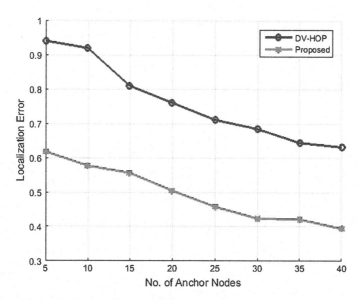

**Fig. 4** Localization error with changing no. of anchor nodes

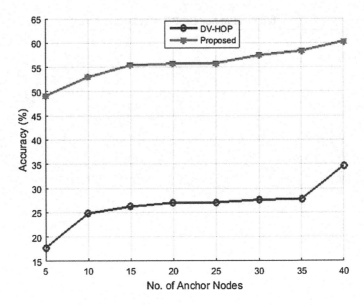

**Fig. 5** Localization accuracy with changing no. of anchor nodes

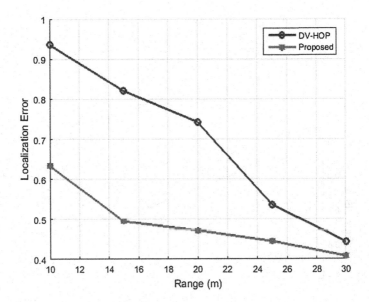

**Fig. 6** Localization error with changing radio range

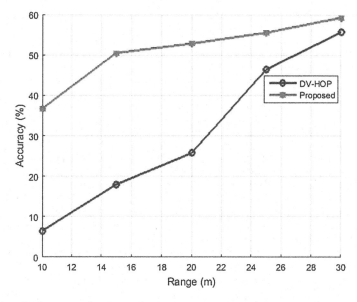

**Fig. 7** Localization accuracy with changing radio range

## 5.2  *Impact of Radio Range on Localization Results*

The effect of nodes' radio range on the localization parameters like error and accuracy is given away in Fig. 6 and Fig. 7, respectively.

## 6  Conclusion

For references, follow the given guidelines. We presented a new PSO-based algorithm, which improves the conventional DV-Hop algorithm considerably. The analysis of simulation result section states that our modified algorithm betters the localization error and accuracy as compared to basic hop-based localization algorithm. As revealed in the simulation section of this paper, it can be stated that proposed method is efficient and has excellent application forefront. We also investigated the impact of anchor nodes and radio range on the localization error as well as on variance. In proposed algorithm, we correct the position estimates with help of PSO. It is clear that modified algorithm boosts the precision and stability of localization method. But due to the use of PSO, there is a little increase in computation time.

**Acknowledgements**  This work is carried out with the help of the human resource development ministry (MHRD, Govt. of India) under research grant scheme for doctoral studies.

# References

1. Akyildiz, I.F., Su, W., Sankarasubramaniam, Y, Cyirci, E.: Wireless sensor networks: a survey. Comput. Netw. 38(4), 393–422 (2002).
2. Yick, J., Biswanath, M., Ghosal, D.: Wireless sensor network survey. *Comput. Netw.* 52(12), 2292–2330 (2008).
3. Mao, G., Fidan, B., Anderson, B. D.: Wireless sensor networks localization techniques. Comput. Netw. 51(10), 2529–2553 (2007).
4. Samira, A.: A Review of Localization Techniques for wireless sensor networks. J. Bas. Appl. Sci. Res. 2(8), 795–7801 (2012).
5. Han, G. Xu, H., Duong, T. Q., Jiang, J., Hara, T.: Localization algorithms of wireless sensor networks: a survey. Telecommun. Syst. 52(4), 2419–2436 (2013).
6. Zhang, Y., Wu, W., Chen, Y.: A Range Based Localization Algorithm for Wireless Sensor Network. J. Commun.Netw. 7(5), 429–437 (2005).
7. He, T., Huang, C., Blum, B., Stankovic, J., Abdelzaher, T.: Range-free localization schemes in large scale sensor networks. In: Proceeding of Ninth Annual International Conference on Mobile Computing and Networking, San Diego, CA, USA (2003).
8. Singh, S. P. Sharma, S. C.: Range Free Localization Techniques in Wireless Sensor Networks: A Review. Procedia Comput. Sci. 57, 7–16 (2015).
9. Niculescu, D., Nath, B.: DV Based Positioning in Ad Hoc Networks. Telecommun. Syst. 22 (14), 267–280 (2003).
10. Singh, S. P. Sharma, S. C.: Critical Analysis of Distributed Localization Algorithms in Wireless Sensor Networks. Int. J. Wirel. Micro. Tech. 6(4), 72–83 (2016).
11. Kulkarni, R. V., Venagyamoorthy, G. K.: Particle Swarm Optimization in Wireless Sensor Network: A Brief Survey. IEEE T. Syst. Man. Cy. C. 41(2), 62–267 (2011).
12. Tomic, S. Mezei, T.: Improved DV-Hop Localization Algorithm for Wireless Sensor Networks. In: Proceedings of IEEE 10th Jubilee International Symposium on Intelligent Systems and informatics (SISY 2012), Serbia (2012).
13. Chen, H., Sezaki, K. Deng, P., So, H. C.: An Improved DV-Hop Localization Algorithm for Wireless Sensor Networks. In: Proceedings of 3rd IEEE Conference on Industrial Electronics and Applications, USA (2008).
14. Chen, X., Zhang, B.: An Improved DV-Hop Node Localization Algorithm in Wireless Sensor Networks. Int. J. Distrib. Sens. N. (2012). https://doi.org/10.1155/2012/213980.
15. Hu, Y., X. Li, X.: An Improvement of DV-Hop Localization Algorithm for Wireless Sensor Networks. Telecommun. Syst. 53, 13–18 (2013).
16. Peong, B. Li, L.: An improved localization algorithm based on genetic algorithm in wireless sensor networks. Cognitive Neuro. 9(2), 249–256 (2015).
17. Li, D., Xian, B. W.: An improved PSO algorithm for distributed localization in wireless sensor networks. Int. J. Distrib. Sens. N. (2015). https://doi.org/10.1155/2015/970272.
18. https://www.researchgate.net/figure/261721236_fig1_Standard-flowchart-of-PSO.

# Dynamic Service Versioning: Network-Based Approach (DSV-N)

Prashank Jauhari, Sachin Kumar, Chiranjeev Mal, Preeti Marwaha
and Anu Preveen

**Abstract** The version of a service describes the service functionality, guiding client on the details for accessing the service. Service versioning requires optimal strategies to appropriately manage versions of the service which result from changes during service life cycle. However, there is no standard for handling service versions which leads to the difficulties in tracing changes, measuring their impact as well as managing multiple services concurrently without any backward compatibility issues. Sometime these services need to be modified as per the client's prerequisites which result in a new version of the existing service, and changes done in the services may or may not be backward compatible. If changes done in the services are not backward compatible, then it can create compatibility issues in the client side code which is using these service features. The problem aggregates even more when a product requires customization for its clients with minor differences in each version. This results in deploying multiple versions of the service for each one of them. This work describes DSV-N (Dynamic Service Versioning-Network-Based Approach) to handle issues related to change management of the service. DSV-N is also capable of handling both backward and incompatible changes. This paper also extends the functionality of dynamic service dispatching by using it with service versioning so that the versions do not need to reside in the memory permanently except the system bus which will execute the appropriate version at run time. Another advantage of using DSV-N is that multiple

P. Jauhari (✉) · S. Kumar · C. Mal · P. Marwaha · A. Preveen
Department of Computer Science, Acharya Narendra Dev College, University of Delhi,
New Delhi, India
e-mail: prashank.jauhari@gmail.com

S. Kumar
e-mail: sachin010496@gmail.com

C. Mal
e-mail: chiranjeevmal1995@gmail.com

P. Marwaha
e-mail: preetimarwaha@andc.du.ac.in

A. Preveen
e-mail: anupreveen@andc.du.ac.in

© Springer Nature Singapore Pte Ltd. 2019
B. K. Panigrahi et al. (eds.), *Smart Innovations in Communication
and Computational Sciences*, Advances in Intelligent Systems
and Computing 669, https://doi.org/10.1007/978-981-10-8968-8_22

service versions are required to be bound with the system bus merely (which is the address of all the service versions for the client). DSV-N automates the process of replicating identical modules in different versions of the service with the help of component file in which version id(s) is(are) prefixed to each module name. DSV-N also provides backward compatibility because the appropriate version that needs to be executed will be resolved at run time. Another advantage DSV-N provides is that the component file needs to be parsed only if the service modules of the component file are modified or only for the first request.

**Keywords** Web service · Serving versioning · Networks

# 1 Introduction

Client end uses the support of multiple internet services for generating the appropriate response. Since the user is dependent on the service, a change in the service will cause compatibility problem if the changes are not backward compatible. This can be resolved by deploying multiple versions of the same service at different socket pairs (IP address and port number). Changing the user's code will be an overhead because it will result in changing a large number of lines of code and redeploying the application again. Multiple approaches have been devised to solve the problem of service versioning. A naive approach to handling service versioning is whenever a new version of a service is ready to be deployed, it would be deployed on a different socket pair. Executing multiple versions of the same service concurrently provides no backward compatibility. On the other hand, this approach is not very cost-effective as deploying multiple versions of the same service at different socket pair is not feasible. Moreover, this approach does not support code reusability.

BROWN and ELLIS (2004) [1] described the best practices for dealing with service versioning. It presents a set of solutions to version services, such as maintaining service descriptions for compatible versions and using different namespaces for every compatible one, which is identified by a date or timestamp in accordance with W3C (World Wide Web Consortium) schema naming. For incompatible versions, it is suggested the implementation of intermediate routers in order to redirect client requests to older implemented versions. This approach provides a solution for multiple socket pair as multiple versions do not need to be deployed on a different socket pair, but code reusability is still not supported because a separate code would be written for each version of the service. Moreover, cost of implementing intermediate router redirection is an overhead.

ENDREI et al. (2006) [2] described the best practices for dealing with service versioning. The first step on introducing a new version is to access its compatibility with regard to the current one. The result can lead to the deprecation of a version or the need to simultaneously maintain different versions of the same service.

This approach provides the solution for backward compatibility, but due to which, it can break the client application when the changes are not backward compatible.

LEITNER et al. (2008) [3] suggested a version graph model, which stores and maintain the relationship between different service versions. This approach allows service customers to choose among the existing versions that can bind their applications to. This approach provides the choice for client to choose the version to which their application binds to, but the cost of maintaining the relationship between service versions is an overhead.

BANATI et al. (2012) [4] suggested a change in standard WSDL (Web Service Definition Language) structure and introduced WSDL-T (WSDL-Temporal) which solves the issue of code reusability. Whenever some change occurs in service version, they change the WSDL file and add a timestamp along with change. This can make the WSDL-T file bulky if change happens frequently. They have not provided any way to deal with this issue.

## 2  DSV-N Approach

In DSV-N approach, the appropriate version of the service is dispatched dynamically when a client requests for the service. Main component of this approach includes component file, system bus, cache, and parser.

### 2.1  Component File

The component file comprises of different modules present in different versions of the service(s). This component is responsible for code reusability as the whole source code is present in a single file, so no redundant code is present among multiple versions of the service. In component file, version id(s) is prefixed to each module name, which is a part of component file. If we identify version 1 of the service with "S1" and version 2 with "S2" identifiers, then the component file for a service will be of the format shown in Fig. 1. Now this single file has the source code for both versions of service. This structure will be helpful in reusability of the modules.

### 2.2  System Bus

This component listens to client's request(s) for the service and generates the appropriate response by parsing the component file as per the requested version of the service. System bus binds the appropriate executable with the descriptor of the connected client by injecting the descriptor into the executable file so that communication between client and requested service version can be performed. After binding, system bus creates a child process for handling the request and continues listening for incoming requests.

## 2.3 Cache

Parsing the component file for each request and generating the executable file will be an overhead. Thus, the support of cache is provided that will store the executable file. The component file is parsed only when service modules of the component file are modified.

## 2.4 Parser

The parser is a module used inside the system bus. This component is responsible for parsing the component file and generating the source of the service. It has two arguments as input, path of the component file and version number. Time requirement for parsing is linear by using KMP (Knuth–Morris–Pratt) algorithm [5]. The output of parser is described in Fig. 2 if inputs to the parser are path of component file (as shown in Fig. 1) and S1 (id for version 1).

Parser act as a window for handling change management in the component file. Parser first generates the source code for service version, and then changes will be made in the generated source code which will be reflected in the component file appropriately. As parser is generating the source code linearly [5], so it will handle the component file, if it gets bulky.

```
#S1|S2# void f1(){
....
}
#S1 void f2(){
....
}
#S2 void f2(int a ,int b){
....
}
```

**Fig. 1** Component file for a service

**Fig. 2** Version 1 of the
service

```
void f1 () {

...

}


void f2 {

...

}
```

## 2.5  Algorithm

The general algorithms for implementing DSV-N (to handle service versioning) and parser creation is as follows:

*Input*: IP address, port number for system bus and requested version numbers for each connected client.

Output: Final output is compatible version produced from the response returned by the loaded executable file, i.e., response returned by the loaded executable file.

Inputs to the system bus are socket pair over which we want to bind multiple versions of the service. System bus creates a socket and starts listening on the passed socket pair. Whenever a request for the connection arrives, system bus accepts the connection and reads the connection descriptor of the connected client and stores it in a variable named "connfd". It then reads the requested version and stores the values in a variable named "Vno". After that, system bus creates a child process to handle the request further and starts listening to other requests. Child process gets the connection descriptor from the parent and duplicates it to three other descriptors 3, 4, and 5 for reading, writing, and error, respectively. These duplicated descriptors are for input/output operation between the connected client and loaded executable file.

**SYSTEM-BUS(IP,PORT)**

- create socket from passed socket pair
- while(true)

    - System bus listening for connection request
    - ConnectionId = accept()
    - Read requested version number into Vno variable
    - Create child process to handle connection
    - Parent closes connection and starts listening again
    - Child duplicate connection id into 3, 4, and 5 descriptors for reading, writing and error communication.
    - Search the executable file for required Vno and store the path of executable in variable Path

- if(Path==null)
- Parse component file as per version number, generate executable file, store the generated executable file in the cache and store the path to executable file in path variable
- Load the executable file into memory and generate response
- After request handling gracefully terminate the child process.

The child process selects the appropriate executable file and loads it into memory (if successfully founds) and starts handling the request, otherwise it parses the component file as per the requested version number. After handling the request, child process terminates gracefully. In this way, "Dynamic Service Dispatch" is performed.

Parser uses the Knuth–Morris–Pratt algorithm [5] for parsing the component file and generating the appropriate source code. Running time for this algorithm is O(m +n) where m is the length of pattern, i.e., the length of service version string, and n is the length of component file, i.e., the total number of character present in the component file. Running time can be further reduced to O(m) by using Trie data structure which makes our parser fast enough to parse even a bulky file (Fig. 3).

## 2.6 Case Study

This case study is taken to explain the benefits of DSV-N approach for handling multiple versions of Internet services. The example describes student's record management service which is required for student's analysis system of a university (a fictional university). This university needs a way to analyze student's performance, and for that, university hires a developer to develop a system for student's analysis. As per the university guidelines, the criteria for analysis is based on student's past academic results which are stored in university database. As direct access to the university database is restricted, so for retrieving the student records, university has student record management service. This service returns student marks in the past semesters as a response. The response returned is used by the student's analysis system for generating student performance. But after some time university changed the criteria for analyzing student's performance, they added student's accomplishments, research paper published, and competition won along with their marks obtained in last semesters. Hence, student's record service needs to be modified for student's analysis system. But record service is being used by some other applications as well. So to fulfill this requirement, another version of student

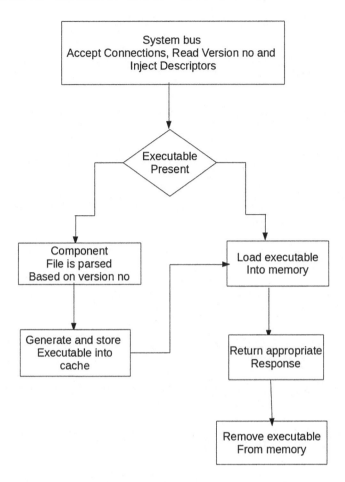

**Fig. 3** Flow diagram of DSV-N

record service needs to be deployed. As per the current approach, both services need to be deployed over different socket pair and need to be present in memory for preparing the response. Student analysis system will read an XML file containing the output. The code for both versions is of the following format.

```
V1|V2 void read_marks(file, student_id){////function def-
inition}
V1|V2 void send_file(File file, connection id){
//function definition}
V1 void main(){
// code above
listen (192.168.45.3,6000);
while(true){
con=accept();
read(student_id,con);
read_marks(file,student_id);
send_file(file,con_id);
close(con);}}
```
[Code for Student_record_service_1]

```
void read_marks(file,student_id){//function definition}
void send_file(file,connection id){//function definition}
void read_resarchPaper(file,student_id){//function defi-
nition}
V2 void read_competetionWon(file,student_id){//function
definition}
V2 void main(){// code above
listen (192.168.45.30,6001);
while(true){
con=accept();
read(student_id,con);
read_marks(file,student_id);
read_researchPaper(file,student_id);
read_competetion(file,student_id);
send_file(file,con);
close(con_id);}}
```
[Code for Student record service 2]

Service version 1 listens at IP address 192.168.45.3 (a fictional address) and port number 6000 (a fictional port), and service version 2 listens at IP address 192.168.45.3 and port number 6001. The read_marks(), read_research_Paper(), read competition() will access the university database and retrieve student marks, research paper published, and competition won which are uniquely identified by student's id within university database. Returned value is stored in the XML file referenced by file variable, and we return this XML file as the response and close the accepted connection. In service version 1, generated response only has marks information, while in service version 2, response contains marks, research paper published, and competition won information. For connection to appropriate service version 1, client uses the following code. Client passes the IP address and port number in the connect call and reads the appropriate response. In connect call to version 1, IP is 192.168.45.3 and port is 6000, and in connect call to version 2, IP is 192.168.45.3 and port is 6001.

```
connect(IP,port);
read(response_file);
[Connection with student_record_service]
```

Same changes in the student record service can easily be handled by DSV-N approach. The code of system bus, which will be responsible for accepting the client's connection and selecting the appropriate version to run, is as under.

```
void system_bus(){
listen (192.168.45.3,6000);
while(true){
con=accept();
child=createChild();
if(child){
read(version_no);
dup(con,3);
dup(con,4);
dup(con,5);
if(find(vno)){
file=parse(componentfile,vno);
file=compilef(file);
store(file);
}
else{
file=search(vno);}
load(file);
}else{
 close(con);}}}
```

[Code for system bus]

All the version of student's record management service runs over IP address 192.168.45.3 (a fictional address) and port number 6000 (a fictional port). All the connection requests for a user for all the versions are handled by the system bus. System bus has the connection descriptor which needs to be injected into the loaded executable file for performing input/output with the connected client. The value of connection descriptor will vary from one client to another. So we need a general descriptor for all clients, and for that purpose, we duplicated the connected descriptor into 3, 4, and 5 for reading, writing, and error handling, respectively. The choice of an integer value for the generalized descriptor is dependent on implementation choice. You can choose any group of available descriptors. After loading the appropriate file, system bus closes the connection. Below code explains how loaded file uses generalized descriptor:

```
V1|V2 void  read_marks(file, student_id){//code defini-
tion}
V1|V2  void send_file(File file, connection id//function
definition}}
V1 void main(){
read(student_id,3);
read_marks(file,student_id);
if(!error)
   send_file(file,3);
else
  write(5,"error try again...");
close(con)}}
V2 void read_resarchPaper(file,student_id){ //function
definition}}
V2 void read_competetionWon(file,student_id){//function
definition}
V2 void main(){
read(student_id,3);
read_marks(file,student_id);
read_researchPaper(file,student_id);
read_competetion(file,student_id);
 if(!error)
   send_file(file,3);
else
  write(5,"error try again...");
  close(con)}
  close(con);}}
```
[Structure of Component File]


V1 and V2 are the identifiers for version number prefixed with modules. They help in parsing the component file for generating the source code for the corresponding version. Notice the main method for corresponding versions, they are using generalized descriptors for performing input/output with the connected users. Code for student record service version 2 when requested is given below. Executable corresponding to this version is obtained either by parsing the component file and compiling the generated source code or by using the cache.

```
void read_marks(file, student_id){ //function definition}
void send_file(file, connection id){ //function defini-
tion}
void read_resarchPaper(file,student_id){ //function defi-
nition}
void read_competetionWon(file,student_id){ //function
definition}
void main(){// code  before read function call
read(student_id,3);
read_marks(file,student_id);
read_researchPaper(file,student_id);
read_competetion(file,student_id);
send_file(file,3);
close(3);}}
```

[Student record service 2 using DSV-N]

The above code clearly explains that after handling the request, loaded exe-cutable closes the connection and process gets terminated, which causes the dynamic dispatching. Student analysis system connects with the appropriate version of student record service as under.

```
connect(192.168.45.3,6000);
write(version_no);
write(student_id);
read(response_file);
```
[Connection code in student analysis system]

It is clear from the above code that client end uses only single socket pair for connecting with the service versions. The version number passed by the client end to the system bus will decide the appropriate record service version.

## 3   Results and Discussion

DSV-N provides the support of following functionalities, and in comparison, we discuss the results. Table 1 describes the results we get when we use DSV-N approach.

**Table 1** Algorithm for parsing component file

| S.no. | Function | Algorithm | Time complexity |
|-------|----------|-----------|-----------------|
| 1.    | Parse    | KMP       | $O(m+n)$        |

## 3.1   Socket Pair

It means the address of the service, i.e., IP address and port number. **Naive** and **WSDL-T** [4] have multiple addresses for multiple versions of the service, while DSV-N approach will run all the versions of the service over a single IP address and port number.

## 3.2   Code Reusability

It allows usage of unchanged modules in different versions of the service. **BROWN and ELLIS (2004)** [1] do not support code reusability. DSV-N supports code reusability by using component file along with parser module. The source code of different version is created dynamically by the parser. So instead of writing similar modules of source code by the developer, they will be generated by the parser.

## 3.3   Dynamic Dispatch

It allows the service to be present in memory only when it is requested. **BROWN and ELLIS (2004)** [1], **ENDREI et al. (2006)** [2], **LEITNER et al. (2008)** [3] do not support dynamic dispatching. DSV-N provides the support of dynamic dispatching by making the system bus to load the requested version only when it is needed.

## 3.4   Backward Compatibility

It means handling all the version's requests. The **Naive** approach does not provide the support for backward compatibility, while all other current approaches do. DSV-N supports backward compatibility as the version with which the client will bind will get selected dynamically by the system bus.

## 3.5   Handling Changes

It means providing the support for addition or removal of modules in a service version. **BROWN and ELLIS (2004)** [1], **ENDREI et al. (2006)** [2], **LEITNER et al. (2008)** [3] do not provide any specification for handling changes. In DSV-N, any change needs to be performed in the service version will be performed by the

**Table 2** Comparison result between Naive and DSV-N

| Factors | Naïve | DSV-N |
| --- | --- | --- |
| Socket pair | Multiple | Single |
| Backward compatibility | Yes | Yes |
| Dynamic dispatch | No | Yes |
| Code reusability | No | Yes |
| Handling changes | Each changes result separate version | Yes |

*Note* Readers might think that compiling the generated source code for corresponding versions will be an overhead. For solving this problem, we have provided the support of cache, and due to which, single version will be compiled only once and gets reused from the cache for future requests, so we can skip the compilation as much as possible. All service versions can be stored by the developer in the cache during the time of service creation

parser. Parser first generates the source code of the service; then, changes will be made in the generated source code which will be reflected in the component file appropriately. As parser is generating the source code linearly [5], so it will handle the component file, if it gets bulky (Table 2).

## 4 Conclusion

Presented approach (DSV-N) solves various problems that persist in the existing approaches. Code reusability is one of the key issues that is not present in the current approaches, and this paper tries to solve this issue by introducing component file along with parser module. Source code for different versions is created dynamically by parser. So instead of writing similar components of source code, they will be generated by the parser. Presented paper also covers the idea of running multiple service versions onto a single socket pair, so there is no need for deploying different versions over different socket pair. In normal client–server scenario, server should always be in the running state for generating the response for corresponding request and hence may consume extra memory (as different version of service executable file will be in memory for handling requests), so to overcome this difficulty, this paper discusses dynamic generation of the response by running the executable file on demand basis; i.e., different version of the service will be dispatched on demand basis. It is the responsibility of system bus to choose appropriate executable file for client and hence add backward compatibility among the service versions.

## References

1. BROWN, K.; ELLIS, M. *"Best Practices for Web services Versioning,"* http://www.ibm.com/developerworks/webservices/library/ws-version/. Access in: May 2012.

2. ENDREI, M. et al. *"Moving forward with web services backward compatibility,"* http://www.ibm.com/developerworks/java/library/ws-soa-backcomp/. Access in: May 2012.
3. LEITNER, P. et al. *End-to-End Versioning Support for Web Services*, In: *IEEE INTERNATIONAL CONFERENCE ON SERVICE COMPUTING–VOLUME 1*, 2008.,Washington, DC, USA. Proceedings... IEEE Computer Society, 2008. pp. 59–66. (SCC'08).
4. Preeti Marwaha, Punam Bedi and Hema Banati,*"WSDL-temporal: An approach for change management in web services,"* copyright 2012 IEEE.
5. https://en.wikipedia.org/wiki/Knuth-Morris-Pratt_algorithm
6. *"Service versioning and compatibility at feature Level,"* Yamashita Marcelo Correa, university: Universidade Federal do Rio Grande do Sul. Instituto de Informática. Programa de Pós-Graduação em Computação, http://hdl.handle.net/10183/78867.
7. W. Richard Stevens, Bill Fenner, Andrew M. Rudoff. 2003. *UNIX Network Programming, Volume 1, 3rd ed:* Addison Wesley.
8. *Xientd* (xtended network service), open source project.

# Smart Anti-theft System for Vehicles Using Mobile Phone

**Deepali Virmani, Anshika Agarwal and Devrishi Mahajan**

**Abstract** A vehicle is one of the expensive assets that a person can own. Its security has always been a major concern. Just like any other expensive asset, it brings along a secondary cost of the risk of theft. So, in this paper, a system is proposed to provide greater safety in order to reduce the probability of vehicle theft. This paper introduces a smart anti-theft system for vehicles using mobile phones. It deals with controlling the vehicle engine via mobile phone through Short Messaging System (SMS). The intention behind the system design is to provide communication between the owner and the vehicle so as to give him a remote access to control his/her vehicle's engine function. The proposed system is smart enough to detect the trusted owners. Every time the system senses an ignition, it makes an attempt to authenticate the owner. In case authentication fails, the system immediately turns off the engine and prevents the vehicle from being stolen.

**Keywords** Arduino · Engine control · Global system for mobile communication (GSM) · Security · Short Messaging System (SMS) Vehicle anti-theft system

D. Virmani · A. Agarwal (✉)
Department of Computer Science, Bhagwan Parshuram Institute of Technology,
New Delhi, Delhi, India
e-mail: anshika0103.ipu@gmail.com

D. Virmani
e-mail: deepalivirmani@gmail.com

D. Mahajan
Amity Institute of Space Science and Technology, Amity University,
Noida, UP, India
e-mail: libra.devrishi@gmail.com

B. K. Panigrahi et al. (eds.), *Smart Innovations in Communication and Computational Sciences*, Advances in Intelligent Systems and Computing 669, https://doi.org/10.1007/978-981-10-8968-8_23

# 1  Introduction

A report in Times of India [1] read that a vehicle is being stolen every 13 min in Delhi, India [2]. States that in 2014, the average interval between car thefts was 23.6 min which was reduced to 13 min in 2015. Of all the IPC crimes reported in Delhi, motor vehicle theft makes a fifth. It seems to be getting organized better by the day. High-end vehicles with built-in security systems are not much safe either before the professionalism of these skilled car boosters. Most security features in cars today like central locking, steering handle lock, anti-theft alarm system are very efficient, but increasing crime rates are an indicator of the inefficiency of such technology. Therefore, this paper proposes an automobile security system that provides the owner of the vehicle a remote access to immobilize the engine in case of a security threat. The vehicle can only be accessed once user authentication is successful. In the case of any unauthorized access to the vehicle, the system asks the user for either authentication or command to stop it, or if the user is unable to respond to the alerts in due time, it itself stops the vehicle. The system is so designed that if the intruder tries to restart the vehicle after it being shut down in the first attempt, the system will automatically shut down the engine further based on the vibrations sensed by the vibration sensor. Because of this functionality, the time delay caused by immobilizing the engine provides enough halt where law enforcement authorities can be alerted and the necessary vehicle recovery actions can be taken.

The following could be possible scenarios of vehicle access.

- Scenario 1 (When the user accesses his vehicle)

In the situation where the owner himself or someone in the knowledge of the owner accesses the vehicle, the moment an alert SMS, i.e. Short Messaging System, is sent to the owner, they may reply with the password message to turn off the microcontroller and not affect the working of the engine.

- Scenario 2 (When the intruder accesses the vehicle)

In the case where an intruder tries to steal the vehicle, the owner may once turn off the engine deliberately by responding to the alert SMS with any message. This turns off the engine. If the user has not replied to the alert message in a stipulated time period, an alert call is made to user's cell phone. Failure to respond to the alerts in due time triggers a security threat and the system shuts the engine off automatically.

After the engine getting shut down once, if the intruder tries to restart the engine, the microcontroller would automatically turn it off again, smartly detecting an unauthorized access. Hence, the vehicle does not get very far from the parking place.

## 2   Literature Review

There are a variety of security systems designed and proposed by many authors and researchers. These proposed ideas range from fairly simple to sophisticated designs having their own drawbacks. B Webb in [3] proposed the wheel and steering lock on unauthorized use of the vehicle. This is easily visible and can be easily overpowered.

In [4] and in [5], the authors have made use of smartphone and Android applications, respectively, to alert the owner in case of theft. This will not work in case of feature phones or phones with iOS and Windows operating system (OS).

Authors L. Wan, T. Chen in [6] have used GSM TC 35 module which is comparably expensive to GSM Sim 900a module used in the system proposed in this system. This makes the system cost-effective. Many other systems previously implemented and researched help keep track of the theft but do not prevent the theft. The authors in [7] have proposed the system which receives authentication from sensors on the door. But this is not possible in case of two- or three-wheeler vehicles.

In [8], authors have implemented theft control using the ARM7 microcontroller which is very expensive (as found in [9]) to implement and hence may not be a viable option for many consumers. The system in [10] is a sensor-based network. The sensors detect an unauthorized access as an unusual movement and alert the base station. As the sensors cannot communicate with the base station directly, it would fail to provide protection to the vehicle when in solitude.

In [11], a more sophisticated model is proposed by introducing the biometric technology. Fingerprints used in this system can be copied off from other surfaces of the vehicle, and facial recognition will not detect an intruder if a photograph is put in front of the camera. Thus, these situations render a sophisticated and expensive theft control mechanism useless before a skilled thief.

Keeping in mind all the drawbacks in these research papers, an effective anti-theft system for a vehicle has been proposed in this paper. An alert is sent to the owner in case of an access to the vehicle, and theft control is implemented via engine control irrespective of the type of phone and OS. The system grants the owner, remote control over engine function. If the access gained to the vehicle is unauthorized, the user can command the system to shut the engine off (or the system can do it automatically if the user fails to respond in due time).

The system is so designed that even if after shutting down the engine once, the thief tries to restart the vehicle, it will automatically shut down the engine based on the vibrations received from the vibration sensor. This implementation helps obtain enough time delay for the owner and regulatory authorities to be alerted and take action.

# 3   Features of the Proposed Anti-theft System

In [3–11], many systems have been proposed. These have their limitations. Some of these limitations are listed below:

1. The system in [3] is visible to everybody and can be damaged by the thief.
2. In [4, 5], the system is restricted to Android platform in mobile phones which would make it impossible to authenticate the user from another type of mobiles.
3. The system in [7] works on door locks and hence is not implementable on two-wheelers.
4. In both [6, 9], the hardware used is neither cost-effective nor cost-efficient.

Thus, based on the requirements of today's world, this paper presents an anti-theft system which overcomes these limitations to provide the following features to its users:

1. It is not visible to the third person and hence cannot be damaged or tampered with.
2. Its usage is not restricted to mobiles with Android platform.
3. It can be implemented on all type of vehicles (with engine).
4. It provides a remote access of engine control to the user.
5. It is very cost-effective and cost-efficient.
6. Since it senses the ignition, the car does not get to go far from the parking space and can be easily located.
7. It takes care of the worst-case scenario, i.e. when the user is unable to respond to theft alerts timely and shuts off the engine automatically.
8. Since the functioning time of the system is short in comparison with those working for as long as the vehicle is running, it does not heat up enough to pose any working issue.

# 4   Existing Systems

## 4.1   Mini Spy by Flashmen for Motorcycle Providing Real-Time Tracking

This system available at [12] is very much similar to the proposed system but is cost-ineffective as it is applicable to only bikes but not the cars or trucks.

## 4.2   Anti-theft Wheel Lock, Clamp Lock for Tires

Available at [13], it is an effective anti-theft lock but being heavy, it poses the problem of locking and unlocking every time the vehicle is accessed.

### 4.3 TP-1000 by AutoCop

This product released by [14] is an efficient one as it can raise almost all types of alarms, but being a tracking device it can only track a vehicle and notify users and not prevent theft.

### 4.4 TK103B and GPS103B as a Real-Time Door Lock Sensor for Car with Alarm

This GSM/GPRS-based product available at [15] provides security via door lock; i.e. it can be implemented only on the vehicles with doors. It does not consider two-wheelers or three-wheelers which do not have any doors.

Many other similar products are available which have their own respective disadvantages and constraints over the level of anti-theft security they provide.

## 5 Design and Flow Analysis of the Proposed System

### 5.1 Proposed Design

Figure 1 gives the overall project outline in terms of hardware connections. Microcontroller Arduino Uno R3 is the heart of the proposed system. With this

**Fig. 1** Hardware design of the proposed system

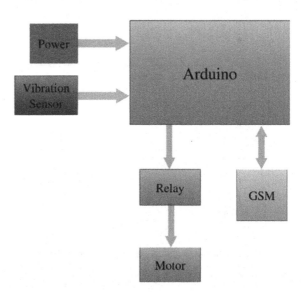

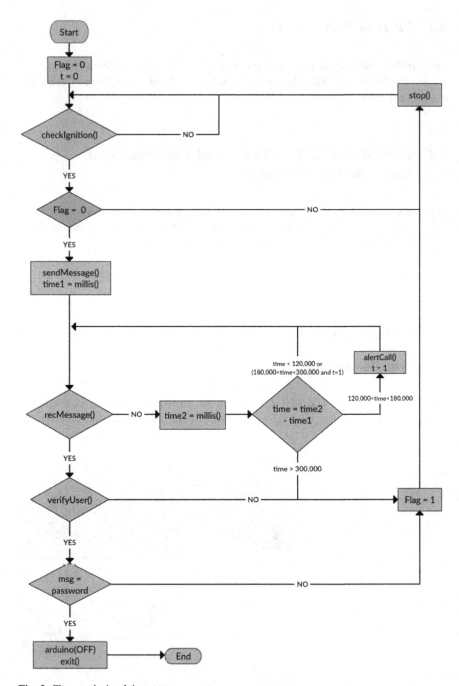

**Fig. 2** Flow analysis of the system

open-source microcontroller, the components that attached are the motor, the vibration sensor and the GSM. As the battery turns "ON" it activates the Arduino, hence activating the other peripherals.

Arduino receives a power of +5 V from the battery source and turns "ON" global system for mobile communication (GSM), enabling the remote control for the user. The GSM is responsible for notifying the user in case of theft and also accepts commands as a response from the user in order to control the engine function.

Thus, the system takes inputs from vibration sensor and GSM, the processing is done by Arduino and output is given by relay and GSM.

## 5.2  Flow Analysis

Figure 2 gives a flow analysis describing step-by-step working of the proposed system.

The system starts whenever the battery is turned "ON". After the battery turns on the Arduino, ignition of the vehicle is checked via vibration sensor. Engine's first ignition is sensed (if any), and the user is alerted via a notification message. The system then waits for a command from the user. The moment a response from the user is received, user's number is verified. After successful user verification, the message is mapped with a preloaded password. If password match is a success, then access to the vehicle is authorized and the system shuts down.

But if the user authentication fails or the message received is not the password, the system shuts the engine OFF.

Also, if the owner fails to reply in given time, he is alerted with a call, and the system waits for user's response. Any further delay by the user triggers a security threat, and the system shuts the engine OFF automatically.

Now, if the intruder tries to ignite the engine again after the first shut down, the smart system senses it and turns off the engine again without having to repeatedly notify the owner or wait for further authentication.

Hence, theft is controlled as shutting down of engine recurrently provides enough delay to alert the authorities and stop the vehicle from getting stolen.

In this proposed system, the developed prototype uses a motor, in place of which a realized and commercially implemented system would be the engine. That is, the use of a motor in the prototype is analogous to the engine in a real-time scenario. The motor is thus used to exhibit the functioning of the proposed system.

# 6 Algorithm

The system follows the following algorithm:

1. Checking the ignition

```
checkIgnition ()
{
  if (piezo = HIGH)
    if (flag = 0)
    sendMessage()
    return true
  else return false
}
```

2. Sending a message

```
sendMessage ()
{
  GSM ("AT + CMGS=\" +917845523478\"")
  GSM.msg ("Superman is Active!! Theft Alert! Engine
is ON")
}
```

3. Receiving a message

```
recMessage (msg)
{
  if (Message.available())
    if (verifyUser() = true)
      if(msg="#superman!")
        return 1
    else return 2
  else return 0
}
```

4. Verification of the owner

```
verifyUser ()
{
    for(int x = 4;x < 17;x++)
      if(RcvdMsg[x] == Owner[x-4])
              continue;
      else return false;
      return true;
}
```

5. Making an alert call

```
alertCall ()
  GSM ("ATD7845523478")
```

6. Shutting down the engine

```
stop ()
  pinMode (10, LOW)
```

7. Vehicle anti-theft system

```
main ()
{
  flag=0, msg='', t=0
  abc: ignition = checkIgnition ()
  if (ignition = true)
    time1 = millis ();
  else goto abc
  xyz: var = recMessage(msg)
  if(var=0)
    time2 = millis ()
    time=time2-time1;
    if (time< 120,000 || (120,000 < time < 300,000&&
  t=1))
      goto xyz;
    else if (120,000<=time<=180,000&& t=0)
      alertCall ()
      t=1;
      goto xyz;
    else flag = 1
  else if (var=1)
    arduino(OFF)
    exit(0)
  else flag = 1
  if (flag = 1)
    stop()
}
```

# 7  Results

The result in Fig. 3 was obtained when the system notified the user via message and call and the user replied with an "S", i.e. any message but the password in order to stop the motor.

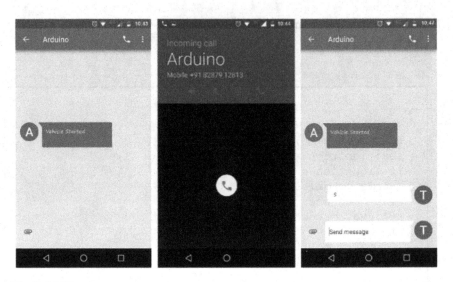

**Fig. 3** Mobile screens

Figure 4 shows the graph representing the vibrations at 9600 baud rate. The x-axis represents the time (ms), and the y-axis represents the amplitude of the vibrations.

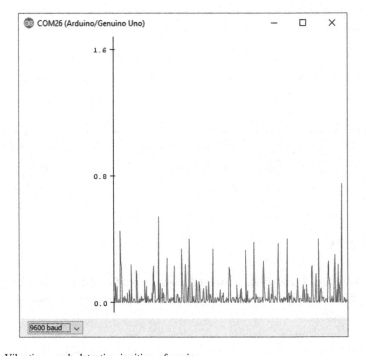

**Fig. 4** Vibration graph detecting ignition of engine

# 8 Conclusion

The purpose of developing this system is mainly to introduce an inaccessible and advanced security system that senses vehicle movement, alerts the owner, accepts owner's commands and also responds automatically to a security threat, compared to existing systems in the industry which merely either track the vehicle or just send information to the user but not take any action to prevent the theft.

It is comparably cost-effective. The system is easy to use and learn due to its simple message and call alert feature. Users require no special training to use this proposed system.

Figure 3 shows the effectiveness of the system. Thus, the system proposed in this paper proves to be an effective solution over the existing anti-theft systems with a user-friendly environment, ease of use and more security.

# 9 Future Work

In future, when the prototype system would be commercially implemented as a working system in modern day vehicles, it may pose some working problems due to heating concerns based on where the module is placed. Thus, proper thermal insulation must be done to protect Arduino from external heat. A security concern would also arise when an illegal access to the owner's mobile phone provides access to the password required to run the vehicle, rendering the vehicle unable to judge owner's identity smartly. This scenario can be tackled by further improving the system using IoT, i.e. Internet of Things. Further, the integration of the system to the vehicle's electrical and fuel supply system would be such that firstly, it is not easy to tamper with even by skilled car boosters or roadside mechanics. Secondly, it shuts down the engine cutting the fuel supply without fail.

# References

1. A vehicle is stolen every 13 mins in Delhi; rate up 44% since last yr. In: The Times of India. https://goo.gl/RbL2yN, last accessed 2016/10/05.
2. Gone in 13 minutes: Vehicle thefts in Delhi break all records. In: The Times of India. https://goo.gl/taw2Do, last accessed 2016/10/07.
3. Webb, B.: Steering Column Locks and Motor Vehicle Theft: Evaluations from Three Countries. In: Situational crime prevention: Successful case studies (1997).
4. Lee, S.J., Tewolde, G., Kwon, J.: Design and Implementation of Vehicle Tracking System Using GPS/GSM/GPRS Technology and Smartphone Application. In: IEEE Word Forum on Internet of Things (2014).
5. Laguador, J.M., Chung, M.M., Dagon, F.J.D, Guevarra, J.A.M., Pureza, R.J., Sanchez, J.D., Iglesia, D.K.I.: Anti Car Theft System Using Android Phone. International Journal of Multidisciplinary Sciences and Engineering 4(5), 12-14 (2013).

6. Wan, L., Chen, T.: Automobile Anti-Theft System Design based on GSM. In: International Conference on Advanced Computer Control (2008).
7. Boskany, N.W., Abdullah, R. M.: Intelligent Anti-Theft Car Security System based on Arduino and GSM network. International Journal of Multidisciplinary and Current Research 4, 538–541 (2016).
8. Dutta, I., Gogoi, D., Gayan, B., Rabha, J., Katta, K.: A Review on Advanced Vehicle Security System with Theft Control and Accident Notification. International Journal of Current Engineering and Scientific Research 1(3), 30–36 (2014).
9. Robomart ARM 7 Development Board, https://goo.gl/LEm7qb, last accessed 2017/01/30.
10. Song, H., Zhu, S., Cao, G.: SVATS: A Sensor network based Vehicle. In: INFOCOM: The 27th Conference on Computer Communications (2008).
11. Powale, P. K., Zade, G. N.: Real time Car Antitheft System with Accident Detection using AVR Microcontroller; A Review. International Journal of Advance Research in Computer Science and Management Studies Research Paper 2(1), 509–512 (2014).
12. Amazon Electronics, https://goo.gl/s01RTt, Last accessed 2017/03/13.
13. Snapdeal ARM Wheel Lock Heavy Duty Anti-Theft Tyre Wheel Clamp Lock, https://goo.gl/8UlPMJ, last accessed 2017/02/15.
14. Snapdeal Autocop Model TP-1000, https://goo.gl/5vTEUy, last accessed 2017/02/26.
15. Amazon Electronics, https://goo.gl/t7GeRI, last accessed 2017/03/12.

# Energy Efficient Data Aggregation Using Multiple Mobile Agents in Wireless Sensor Network

Mehak and Padmavati Khandnor

**Abstract** Gathering data effectively has always been of primary importance in wireless sensor network. Mobile agent paradigm has made it possible to collect and aggregate data in a manner which is appropriate for real-time applications. In static wireless sensor network, the sensor nodes forward the data to the sink node through intermediate sensor nodes while mobile agents have an advantage as they reduce the passing of results between the intermediate nodes so consumption of network bandwidth also reduces. But mobile agent paradigm brings along its own challenges, and one of the major issues is to achieve energy efficiency. This paper proposes a data aggregation approach using multiple mobile agents which takes into account aggregation ratio, network lifespan, and energy efficiency. The network is divided into four quadrants, and a mobile agent is dispatched for each quadrant to collect data from the quadrant assigned to it. Simulation results show that the proposed approach consumes optimal amount of energy; hence, the network lifespan is elongated.

**Keywords** Base station · Data aggregation · Information gain
Mobile agent · Network lifetime · Wireless sensor network

## 1 Introduction

Wireless sensor networks (WSNs) are formed from hundreds or thousands of wireless sensors with limited resources that are used in a wide range of fields. Past years have seen the coming up of WSNs as a new paradigm for collection of information, where a considerable number of sensors are deployed over a field to be

Mehak (✉) · P. Khandnor
PEC University of Technology, Chandigarh, India
e-mail: mehak789@yahoo.co.in

© Springer Nature Singapore Pte Ltd. 2019
B. K. Panigrahi et al. (eds.), *Smart Innovations in Communication and Computational Sciences*, Advances in Intelligent Systems and Computing 669, https://doi.org/10.1007/978-981-10-8968-8_24

observed and the information of interest is identified by understanding the real-time events from the physical surroundings [1]. The data-gathering approach acquires sensed information from the sensor nodes and forward to sink node.

Systems which can perform complicated work and analysis, basically dense and distributed systems, are substituting the conventional client–server-based designs at a very fast rate. A distributed sensor network (DSN) is a group of thousands of sensor nodes which may be heterogeneous or homogenous. The nodes may be deployed spatially, logically, or geographically in a region in which required information is present and are connected through wireless links [2]. The sensor nodes gather data from the environment without any interruption, process it, and then forward it through the network. Sensing nodes are autonomous entities which are self-configurable, self-aware and collect data and transmit it by themselves. Sensor nodes are deployed in unpredictable and reckless surroundings, so human involvement is not much needed.

WSNs may suffer from some drawbacks such as consumption of large amount of bandwidth, high traffic in the network, low network lifetime, and a single point of failure. Such problems usually occur in client–server-based architecture where client node sends a query to the server for execution and server executes this query with the help of existing data. To overcome the above-mentioned difficulties, mobile agent architecture [3, 4] was presented for WSNs. In such type of architecture, mobile agent moves between the sensor nodes in the network to perform some task intelligently especially data gathering [5]. This model has many advantages like reduced network load, robustness, fault tolerant and also reduces the network latency. A mobile agent selects a path to migrate among the sensor nodes, performs some processing at each visited node, and then transfers the processed information to the sink node rather than transferring the entire duplicate data. This process is referred to as data aggregation. The basic idea of data aggregation protocols is to collect and aggregate the sensed data with minimal loss of energy [6]. This approach shifts the focus from finding the shortest routes to finding routes which allow data to be grouped as a whole. The mobile agent fuses the information of each sensor node to reach a certain level of accuracy [7]. To increase the efficiency of the protocol, multiple mobile agents may be used instead of a single mobile agent.

In this paper, an energy efficient multiple mobile agent-based data aggregation technique (EEDA MM) has been proposed. The network is divided into four quadrants, and a mobile agent is dedicated to one quadrant for the collection of data. The mobile agents start their journey from the base station (BS) and return back here after collecting data. The energy consumed by the sensor nodes is balanced to increase the overall network lifespan. This approach is flexible with reference to deciding the effect of different factors involved in the cost function.

The rest of the paper is organized as follows: The protocols related to the proposed approach have been mentioned in Sect. 2. Section 3 gives the system model used for implementing the proposed approach. The proposed approach has been discussed in detail in Sect. 4. Simulation results are presented in Sect. 5. Finally, the paper is concluded in Sect. 6.

## 2 Related Work

One of the major research challenges in mobile agent-based WSNs is to discover an optimal path for mobile agent migration to collect the data [8]. Many protocols have been proposed to select the path of mobile agent in an efficient manner. In [9], authors have presented an itinerary planning approach for mobile agent-based WSN. It takes into account information gain in data aggregation and accuracy. Another approach which considers mobile agent in a multihop environment has been presented in [10]. The technique makes use of directed diffusion (DD) to dispatch mobile agent to various sensor nodes in the network. The gradient in DD gives an approximate idea to effectively migrate the mobile agent among the sensors present in the target region. The mobile agent prototype together with the DD anatomy is referred to as mobile agent-based directed diffusion (MADD). In [11], authors have proposed a multi-agent system-based WSN for crisis management. It does not have any infrastructure for the sink node. This method performs tracking and data aggregation in parallel. The proposed protocol is more efficient due to multi-agent system (MAS), so it helps the sensor nodes in the management of their batteries in a better way.

In [12], authors compute all the possible paths presented by data aggregation trees using genetic algorithm (GA). GA selects the optimal tree which has the ability to balance the energy and the data load among the sensor nodes. This technique suits well for a homogeneous WSN which has some spatial correlation to observe the surrounding area. The sensor nodes monitor a specific region to collect data and then eliminate the duplicate data so that the data is aggregating in accordance with the data aggregation spanning tree. After this, they transmit data packets to the suitable neighbor sensor node to forward the packets toward BS using a least costly route in terms of energy. BS is a powerful station where all the sensed data by the sensor nodes is collected and processed to achieve specified goals.

Two approaches for itinerary planning have been proposed in [13]. One is itinerary energy minimum for first-source-selection (IEMF) which selects the node with least energy cost as the next source node and applies local closest first (LCF) to choose the remaining source nodes. Itinerary energy minimum algorithm (IEMA) further enhances the energy efficiency of IEMF. Authors have proposed a multi-agent hybrid approach which combines the advantages of decision and value fusion by aggregating data at the source in a cluster-based WSN [14].

# 3 System Model

## 3.1 Energy Model

The energy model used for the proposed approach is same as given in [15]. The energy required for communication at a sensor node can be calculated using Eq. (1) [15].

$$e(Sr, St) = mr \cdot Sr + mt \cdot St \tag{1}$$

where $Sr$ is the size of packet received by a sensor node and $St$ is the size of packet transmitted by sensor node. $mr$ and $mt$ are the energy required to transmit and receive one bit of data.

Hop distance between two sensor nodes, the energy required for traversal of mobile agent from one sensor node to another, and the total energy required by mobile agent to cover the entire path can be computed using the model as given in [15].

## 3.2 Assumptions

The following assumptions have been made for the implementation of proposed protocol:

- The sensor nodes are static after deployment.
- Mobile agents are moving within the network.
- The deployment of sensor nodes is dense.
- Energy level at the starting of every sensor is same and limited.
- 2D plane has been assumed for simplicity.

## 3.3 Network Model

The network consists of '$n$' sensor nodes and four mobile agents. The sensor nodes are all stationary while the mobile agents have been provided controlled mobility in the network. The mobile agents start their journey from BS and return back there after collection of data from sensor nodes. Each sensor node has limited resources such as energy. All sensor nodes keep a track of their location in the form of (P,Q) coordinates and also aware of its residual energy. The sensor node being visited by mobile agent is denoted by $v_s$, and the nominee for the next sensor node to be visited by mobile agent is denoted as $v_d$. Every sensor node has a transmission range *TR*, within which it can communicate directly with its neighbors. There is a predefined value of threshold varying from application to application. Also, a value of minimal energy has been defined after exhaustion of which the sensor node dies.

## 4  Proposed Approach

The network is divided into four quadrants based on Algorithm 1. One mobile agent is dispatched for each quadrant from the BS. Each mobile agent finds the nearest sensor node which has minimum cost according to the cost function as shown in Eq. 2 [15].

Mobile agent migrates to the selected sensor node and aggregates the data there. Similarly, it finds the next node to be visited.

$$Cost_{sd} = a\left(1 - \frac{I_d(u,v)}{I_m}\right) + b(N_{visited} + 1)\left(\frac{e_d}{e_m}\right) + c\frac{E_{sd}}{E_m} + d(dist_d) \qquad (2)$$

---

Algorithm 1:
(x,y):Center location
(a,b): Sensor node location

---

C ← (x,y);                                    //C is the center
For any sensor node k in the network with location(a,b)
Let new coordinates (P,Q) ← (a-x, b-y);      // finding (P,Q) corresponding
                                                to the center C

**If** (P > 0 && Q > 0) then
          Sensor node with location(a,b) lies in the 1st quadrant.
**end if**
**If** (P < 0 && Q > 0) then
          Sensor node with location(a,b) lies in the 2nd quadrant.
**end if**
    **If** (P < 0 && Q < 0) then
          Sensor node with location(a,b) lies in the 3rd quadrant.
**end if**
    **If** (P > 0 && Q < 0) then
          Sensor node with location(a,b) lies in the 4th quadrant.
**end if**

---

The constraints for Eq. (2) are:

$$0 \le a, b, c, d \le 1, N_{visited} \ge 0$$

where

$I_d(u, v)$     represents the information gain of sensor node $v_d$ for migration of mobile agent.

$I_m$        is the maximum information gain of the sensor node.

$N_{visited}$    is the number of times the sensor node $v_d$ has been visited by mobile agent.

$E_{sd}$       is the energy consumed for movement of mobile agent from $v_s$ to $v_d$.

$E_m$      is the maximum energy needed for mobile agent migration from $v_s$ to $v_d$.

$e_d$        represents the residual energy of the sensor node $v_d$.

$e_m$      is maximum initial energy of the sensor nodes assigned at the starting.

$dist_d$     is the distance of the mobile agent from the sensor node $v_d$.

$a$, $b$, $c$, and $d$ are constants assigned to the parameters information gain, remaining energy, migration energy, and distance of the sensor node from mobile agent, respectively.

The mobile agent in each quadrant firstly checks all the one-hop neighbors of the BS in its specific quadrant and selects the sensor node which results in least cost. Then the difference of information gain for this sensor node and the current value of information gain are computed. If this value is more than the specified threshold value, the mobile agent migrates to that sensor node and performs the data aggregation; else it migrates to the closest sensor node which is two hops away from the present sensor node for which data has been aggregated. Here the information gain is checked again. If the difference is still less than the threshold value, mobile agent migrates to closest sensor node four hops farther from the present node. This process continues in all the $2^n$ hops sensor nodes until the difference satisfies the threshold value. Upon reaching the desired sensor node, mobile agent starts aggregating data again and searches for its next destination in one-hop neighbors. At last all the mobile agents (one from each quadrant) move to the BS after they have visited all the desired sensor nodes and further transmit this information to the BS.

There are two approaches in which a round can be completed. In the first approach, the number of sensor nodes to be visited by the mobile agent in each quadrant may be fixed, whereas in the second approach, the amount of data to be collected by the mobile agent may be fixed. In this work, second approach has been used. These both parameters may vary for different applications. Also, the threshold value can be kept different according to the requirement of the application.

# 5 Simulation Results

The performance of the proposed protocol has been analyzed and compared with the existing energy efficient data aggregation (EEDA) approach presented in [15]. The effect of weight factor '*a*' used in information gain has been evaluated on the performance of the proposed protocol. The simulation has been performed based on the parameters illustrated in Table 1 using NS2 simulator.

## 5.1 Aggregation Ratio

Aggregation ratio is the accuracy of the data aggregated by mobile agent [15]. The accuracy level approaches 1 if the mobile agents visit all sensor nodes in the network. Figure 1 clearly shows that information gain depends directly on the weight factor '*a*,' i.e., if the value of '*a*' is increased, the information gain also increases and the accuracy of data aggregation increases. Since the proposed protocol (EEDA-MA) uses multiple mobile agents, the number of sensor nodes visited by mobile agents in same time interval increases; hence, the aggregation ratio also increases.

**Table 1** Simulation parameters

| Parameter | Value |
|-----------|-------|
| Network area | 1000 m * 1000 m |
| Number of nodes | 100–400 |
| MAC | 802_11 |
| Simulation time | 600 s |

**Fig. 1** Aggregation ratio varying with weight factor '*a*'

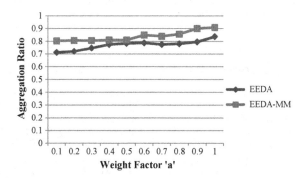

**Fig. 2** Remaining Energy of
sensor nodes

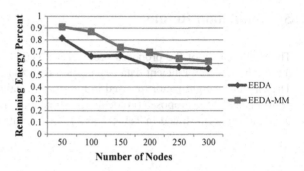

## 5.2  Remaining Energy

The residual energy of the network at the end has been plotted against the number of sensor nodes in Fig. 2. It varies according to the number of sensor nodes. Since the proposed protocol (EEDA-MA) partitions the network into sub-networks, reducing the number of times a sensor node transmits the data to mobile agent. Hence, energy lost during the transmission also decreases. The proposed approach improves energy efficiency by 16.11%.

## 6  Conclusion

In this paper, a technique for efficiently gathering data by using multiple mobile agents has been presented. The scheme has been modified to attain a good information gain and keep the energy consumption minimal so that the precision of collected data is improved. Results show that the proposed technique achieves a good performance in terms of network lifetime and aggregation ratio. Further work can be done to enhance the protocol by making the process of data transfer more secure and reliable.

## References

1. M. Zhao & Y. Yang. (2012). Bounded relay hop mobile data gathering in wireless sensor networks. IEEE Transactions on Computers, vol. 2, pp. 265–277.
2. Takaishi, D., Nishiyama, H., Kato, N. and Miura, R. (2014). Toward energy efficient big data gathering in densely distributed sensor networks. IEEE Transactions on Emerging Topics in Computing, 2(3), pp. 388–397.
3. Qi, H., Xu, Y. and Wang, X. (2003). Mobile-agent-based collaborative signal and information processing in sensor networks. Proceedings of the IEEE, 91(8), pp. 1172–1183.
4. Chen, M., Kwon, T., Yuan, Y. and Leung, V.C. (2006). Mobile agent based wireless sensor networks. Journal of computers, 1(1), pp. 14–21.

5. Tunca, C., Isik, S., Donmez, M.Y. and Ersoy, C. (2014). Distributed mobile sink routing for wireless sensor networks: A survey. IEEE communications surveys & tutorials, 16(2), pp. 877–897.
6. Patil, N.S. and Patil, P.R. (2010). Data aggregation in wireless sensor network. In IEEE international conference on computational intelligence and computing research (Vol. 6).
7. Biswas, P.K., Qi, H. and Xu, Y. (2008). Mobile-agent-based collaborative sensor fusion. Information fusion, 9(3), pp. 399–411.
8. Chen, M. and Gonzalez, S. (2007). Applications and design issues for mobile agents in wireless sensor networks. IEEE Wireless Communications, 14(6).
9. Lohani, D. and Varma, S. (2015). Dynamic mobile agent itinerary planning for collaborative processing in wireless sensor networks. In India Conference (INDICON), 2015 Annual IEEE (pp. 1–5). IEEE.
10. M. Chen, T. Kwon, Y. Yuan, Y. Choi, & V. Leung. (2007). Mobile agent-based directed diffusion in wireless sensor networks. EURASIP Journal on Applied Signal Processing, vol. 1, pp. 219–219.
11. Sardouk,A.,Mansouri,M.,Merghem-Boulahia,L.,Gaiti,D.and Rahim- Amoud, R. (2010). Multi-agent system based wireless sensor network for crisis management. In Global Telecommunications Conference (GLOBECOM 2010), IEEE (pp. 1–6).
12. Abdul-Salaam, G., Abdullah, A.H. and Anisi, M.H. (2017). Energy-Efficient Data Reporting for Navigation in Position-Free Hybrid Wireless Sensor Networks. IEEE Sensors Journal, 17 (7), pp. 2289–2297.
13. Wang, J., Zhang, Y., Cheng, Z. and Zhu, X. (2016). EMIP: energy-efficient itinerary planning for multiple mobile agents in wireless sensor network. Telecommunication Systems, 62(1), pp. 93–100.
14. Sethi, P., Chauhan, N. and Juneja, D. (2013). A multi-agent hybrid protocol for data fusion and data aggregation in non-deterministic wireless sensor networks. In Information Systems and Computer Networks (ISCON), 2013 International Conference on (pp. 211–214). IEEE.
15. Lohani, D. and Varma, S. (2016). Energy Efficient Data Aggregation in Mobile Agent Based Wireless Sensor Network. Wireless Personal Communications, 89(4), pp. 1165–1176.

# Bubble Rap Incentive Scheme for Prevention of Node Selfishness in Delay-Tolerant Networks

Sweta Jain and Ankit Verma

**Abstract** Delay-tolerant network (DTN), as its name suggests, is a network architecture that can tolerate delays in message delivery. In such an environment where there is a high rate of error and very large delay, routing becomes a challenging task due to the extreme conditions. DTNs incorporate store carry-and-forward routing technique to transfer data from source to destination using a new protocol called Bundle Protocol. While implementing this technique, it is assumed that each node cooperates in data forwarding and forwards the message irrespective of its sender and destination. But sometimes nodes may deny data forwarding and accepting messages of other nodes in order to save its own resources such as bandwidth and buffer. Such behaviour is termed as selfish behaviour whereby node may drop packets of other nodes or prevent them from forwarding further in order to save its own resources. Detecting selfish behaviour of nodes in DTN environments is a critical task. In order to overcome the selfish behaviour, nodes may be motivated by providing incentives in form of credits or virtual money. This paper presents a technique to detect selfishness of a node on the basis of its message forwarding and dropping behaviour, and a novel credit-based scheme to motivate nodes to cooperate in message forwarding. The proposed scheme is applied over Bubble rap routing algorithm which is a popular social-based routing algorithm.

**Keywords** Node selfishness · Incentive · Delay-tolerant networks

S. Jain (✉) · A. Verma
Computer Science and Engineering, Maulana Azad National Institute of Technology,
Bhopal, India
e-mail: shweta_j82@yahoo.co.in

A. Verma
e-mail: verma.ankit@ymail.com

© Springer Nature Singapore Pte Ltd. 2019        289
B. K. Panigrahi et al. (eds.), *Smart Innovations in Communication and Computational Sciences*, Advances in Intelligent Systems and Computing 669, https://doi.org/10.1007/978-981-10-8968-8_25

# 1   Introduction

Delay-tolerant network (DTN) deals with the heterogeneous environment where connectivity is not seamless. It provides an efficient way to communicate across the globe such as with satellite or other space-located entities. It deals with the varying nature environment where frequent changes in connectivity take place. Traditional network provides seamless connectivity, but when we move to space, there are various factors which affect the connectivity such as curvature of Earth and different alignments of orbits; thus, in order to provide transparent services, delay-tolerant network can be used. DTN architecture was introduced by Kevin Fall (RFC 4838 [1]). It is based on store and forward message switching. It is one of the oldest methods of message sending. In order to implement this mechanism, a new protocol called bundle layer protocol is used. As its name suggests, it bundles the regional-specific lower layers in order to provide communication with various regions. Here, bundles are stored and forwarded; i.e., they are the messages being sent. It adds one more layer to Internet layers which is bundle layer, and it is placed between Application and Transport layers. In DTN, a node can exhibit three different roles. It can be a host, a router or a gateway. All these entities have different functionalities; for example, a host can only send or receive a bundle but cannot forward it, whereas a router forwards a bundle into a single DTN but cannot forward it to another DTN. A gateway forwards a bundle to different DTN as it provides conversion of protocols of lower layers.

Earlier it was assumed that each node uses opportunistic data [2–6] propagation in which each node is assumed to be ready to forward the packet of others. But this assumption is easily violated in DTN, as a node can start behaving selfishly, whereby it may refuse to forward the message of other nodes. In case of cooperative nature, a node may forward the message of others. Hence, two extreme approaches can be experienced in a DTN. In order to overcome the selfish behaviour of nodes, an incentive scheme is required.

The best way to overcome the problems arising due to selfish behaviour of nodes is to encourage them to cooperate in network functioning. These schemes can be categorized as reputation-based schemes and credit-based schemes. In reputation-based approach, a node needs to keep an eye on the traffic of their neighbour nodes and have a record of reputation of each other. Whenever an uncooperative node is detected, that node can be excluded from the network [7–10]. On the other hand, credit-based schemes use virtual money to deal with the selfishness of nodes [11–13]. Reputation-based scheme is good for traditional ad hoc network, but when it comes to DTN, it does not perform so well due to poor connectivity and change in neighbourhood of nodes. In traditional ad hoc network, we have an end-to-end path between nodes (sender and receiver) that is determined before forwarding the data, but in DTN, there is no predetermined path exist; hence, reputation-based scheme does not fit well here. There is also an issue in DTN-related to the number of copies being forwarded. Reputation-based scheme is designed only for single copy routing strategies, while in DTNs, multiple copies of messages are forwarded in order to increase the reliability [2]. In this paper, a node selfishness detection scheme is proposed and a credit-based scheme is proposed in which node gets the credit, i.e. virtual money, for forwarding the messages. The credit rewarded is based on the message size and message TTL. The selfish node is rewarded extra credit based on its remaining buffer size.

## 2  Related Work

In this section, a discussion on cooperation stimulation in wireless networks and DTN data transmission protocols is presented. In wireless network, the concept of cooperation is used in various applications, especially in the field of decentralized wireless network such as ad hoc network. In decentralized network, nodes need to cooperate with each other so as to make communication possible, but nodes try to refrain from cooperation. Incentive schemes are used in such situation in order to promote the cooperation between nodes. Incentive schemes comprise of three categories, i.e. reputation based, credit based and barter based.

In reputation-based scheme [8], the reputation of each node is used as an incentive to promote its cooperation with other nodes in the network. Each node is assigned a value of reputation which shows its enthusiasm towards cooperation. This value quantifies the degree of reputation of a node. Reputation of a node is a dynamic entity which depends on the behaviour of node, if node is cooperating well with its neighbour, its reputation will increase. If node is exhibiting some sort of selfish behaviour, then the value of reputation would decrease. Routing is done on the basis of reputation of the node. In credit-based scheme [15–17], credit (virtual money) is used as an incentive to promote node cooperation in the network. As soon as a node forwards a packet of other nodes, it receives virtual money in the form of credit. This earned credit can be used by nodes in order to provide incentive to other nodes by offering them some credit. On the basis of credit-based concept, several such schemes have been proposed [14, 15]. Barter-based incentive scheme uses the concept of pairwise exchange [16–18]. In this approach, a node forwards the packets to its neighbour as much as its neighbour forwards for it.

As far as DTN is concerned, none of the incentive scheme discussed can be applied in its original form. However, there are several issues present in DTN which makes it difficult to apply any incentive scheme in their native form. In DTN, there is no end-to-end connectivity and network partitioning occurs frequently. Hence, it becomes difficult for a node to observe the behaviour of its neighbours and manage their reputation. Thus, reputation-based scheme cannot be used here directly. In the same way as the number of intermediate node is not predetermined, it becomes very difficult to set initial credit in credit-based scheme. Barter-based scheme also fails to be used in DTN as an incentive scheme in its native form due to lack of reliable third-party monitoring.

Nevertheless, several mechanisms have been proposed in the recent years for DTNs keeping in mind the characteristics of these networks. The Give-to-Get (G2G) Epidemic Forwarding and G2G Delegation Forwarding presented in [19] are the first works in DTNs for message forwarding that work under the assumption that all the nodes in the network are selfish. G2G protocols while being almost good in terms of success rate and delay and have less cost. The authors also show that both the protocols are Nash equilibria. The work in [20] studies the impact of selfish behaviour on the performance of DTNs and has proposed a tit-for-tat strategy to provide incentives to nodes. A modified version of TFT mechanism that incorporates generosity and contrition has been applied to overcome the problems of bootstrapping and exploitation. MobiCent, a credit-based incentive system for DTN presented in [21], uses a Multiplicative Decreasing Reward (MDR) algorithm to calculate payment and supports two types of client, namely clients that want to minimize cost or minimize delay. In SSAR [22], the nodes forward messages to relay nodes based on some probabilities such as meeting probability, delivery probability and buffer overflow probability. Here, meeting probability is the probability that TTL value of message will expire before intermediate nodes meet its destination node. If a message has less meeting probability than its delivery probability, then forwarding of that message will increase the network traffic. Buffer overflow is that a node's buffer got filled and a new message of higher priority came; then, the low priority messages will be dropped in order to accommodate those newly arrived high priority messages. Probability of this dropping is referred as buffer overflow probability. It also incorporates concept of node willingness which denotes a node's willingness to forward the message to a particular node which depends on the strength of its social tie with that node.

Due to scarcity of resources such as buffer and bandwidth, a node in a real social network might not always be willing to provide free service to other nodes; hence, it becomes essential to incentivize and motivate the nodes to cooperate and ensure proper network functioning.

The concept of social ties as prevalent in human community may also be used in DTNs to improve routing procedure. It is observed that humans behave differently with people in different relationships. Similarly, nodes in DTNs also have different social ties with different nodes in the network based on their meeting frequency, meeting duration, common interests, etc. The concepts of social-based networks such as community, centrality and friendship may be utilized in DTNs to improve the routing performance. Hui et al. proposed a social-based forwarding algorithm in [14] which combines the concept of two social metrics, namely community and centrality. Each node in the network is assumed to have two centrality values: global centrality and local centrality. The global centrality depicts the popularity of the node across the network, and the local centrality reflects the popularity of the node within its own community. Every node is assumed to belong to a community. The Bubble Rap forwarding includes mainly two phases: a bubble-up phase based on global centrality and a bubble-up phase based on local centrality. Suppose a node A has a message for node D which belongs to the community $C_D$. When node A has an encounter with node B, it first checks whether node B is of the same community $C_D$. If yes, then node A checks whether the local centrality of node B is greater than its own local centrality. If node B has greater local centrality, then the message is forwarded to node B, otherwise node A keeps the message. The idea is that if message is within a community member of the destination then message has to be relayed only if another community member with a greater popularity within the community is encountered. If node A is not in the same community and node B is from $C_D$, then the message is forwarded to node B. The Bubble Rap assumes that the members of the same community encounter each other more frequently and hence they can be made good relays. If node A is not in the same community and node B is also not from the same community, then node A checks the popularity of node B in the network. If the global centrality of node B is greater than that of node A, then the message is forwarded to node B, otherwise node A itself keeps the message. Bubble Rap combined the concept of community with centrality for making forwarding decisions. The bubble-up operations allow fast delivery of the message towards its destination.

# 3 Proposed Scheme

To prevent the network from the problem of selfishness, it is required to identify the selfish node and then stimulate them to provide their resources. For the detection of selfish behaviour, the node's forwarding and dropping activity is observed, i.e. how many messages are forwarded and dropped by each node. Generally, a node has messages which are generated by itself and the messages which it received from other nodes. In case of social-based routing, messages can be categorized in three ways: messages generated by a node itself, messages belonging to local community and lastly messages belonging to global community. In the current work, Bubble

Rap is used as the underlying social-based routing protocol and community formation is similar to the one used in Bubble Rap algorithm.

The basic idea for detecting the selfish node is derived from the concept that in DTN when a node meets another node it should forward the packet to enhance the probability of message delivery. Selfish node will not forward the other node's packet. A normal node drops the packet because either the buffer space is full or the TTL of message has been expired. On the other hand, a node behaves selfishly in order to save its resources, i.e. buffer, energy, bandwidth. A selfish node does not cooperate with other nodes in network, but it takes services from other nodes. To detect the type of selfishness of a node (i.e. individual or social), separate records are kept, that is how many messages of local community and global community are either dropped or forwarded by a node.

In this scheme, a node holds two pairs of observation parameters, i.e. LF and LD. These parameters are used to keep record of the number of messages of local community forwarded and dropped by it respectively. Similarly, GF and GD are the parameters used to record the number of messages of global community forwarded and dropped by a node respectively. Whenever a node forwards a message belonging to its local community, the value of LF is increased by one else GF will be increased by 1 if message belongs to the global community. Similarly, LD or GD will be increased by 1 if message being dropped belongs to its local community or global community. Using the values of (LF, GF) and (LD, GD), GV and LV are calculated for each node as shown in Eqs. 1 and 2, respectively:

$$G_V = \frac{G_F}{G_F + G_D} \tag{1}$$

$$L_V = \frac{L_F}{L_F + L_D} \tag{2}$$

The rating of $G_V$ and $L_V$ reflects the degree of trustworthiness of a node. $G_V$ and $L_V$ are node's individual trustworthiness values in global community and local community, respectively. A trustworthy threshold $G_T$ and $L_T$ is taken as the criteria to classify the node as selfish or cooperative.

A node should forward enough messages in order to get the higher values of GV and LV than the values of GT and LT, respectively, for it to be considered as normal node as shown in Eq. 2(a) and 2(b), respectively:

$$\text{If } G_V > G_T \text{ and } L_V > L_T \text{ Then Normal Node} \tag{3}$$

$$\text{Else If } G_V < G_T \text{ and } L_V < L_T \text{ Then Selfish Node} \tag{4}$$

If a node does not forward the messages, then the value of $L_F$ and $G_F$ will be low and the values of GV and LV which is directly proportional to them will also decrease. When these values become lesser than the value of GT and LT, then a node is considered as a selfish node.

Apart from detection of selfish node, nodes are also stimulated to cooperate with other nodes in the network.

## A. Proposed Credit-Based Incentive Scheme

Selfish nodes do not forward the other node's messages in order to save its own resources. There are some strategies such as barter-based strategy, credit-based strategies and reputation-based strategy for the prevention of selfish node behaviour. DTNs have intermittent connectivity; hence, barter-based strategy is not suitable for such type of networks. In reputation-based strategy, reputation of a node is based on the reporting of the neighbour node. As the neighbourhood of a node changes dynamically in DTN environment, continuous monitoring of a node's behaviour is still not possible.

In the proposed work, incentive mechanism is being used to motivate the nodes for forwarding messages of other nodes. For forwarding a message, a node will get a credit C and this credit will be given in a normal condition (when node is acting as normal node). C is not a constant value. Credit C is proportional to the ratio of message size and message TTL. Here, message size is used to indicate how much bandwidth is devoted to forward that message and message TTL is used to denote how important the message is in the network:

$$credit\ C \propto \textbf{message size}/\textbf{message TTL} \qquad (5)$$

A node that is acting selfishly will be given extra credit $e$ in order to motivate it so as to provide its resources and services to other nodes. To motivate a selfish node, a total credit "$C + e$" will be given, where "$e$" is extra credit that is given to selfish node for motivation. As a selfish node provides buffer space for the purpose of other nodes despite its own resource crisis, correspondingly e will be inversely proportional to its free buffer size; i.e., if a node has very less amount of free buffer space and it is still ready to accept and forward messages of other nodes, its extra credit $e$ will be more:

$$Credit\ e \propto \frac{1}{Free\ Buffer\ Size} \qquad (6)$$

The total credit gain (virtual money) of a node is stored in the temper proof hardware, and that node can encash that virtual money whenever it meets to the TA. TA is trusted third party which enchases the node's credit gain into the money or points which are to be used by that node for some work.

## B. Algorithm for detection of selfish node and incentive scheme

```
Step1. When forwarding a message:
If (local community message)
Then
L_F = L_F + 1
Else
   G_F = G_F + 1
Step2. When dropping a message
If (local community message)
Then
            L_D = L_D + 1
Else
         G_D = G_D + 1
Step3. Update G_v and L_v as:
```

$$G_V = \frac{G_F}{G_F + G_D} \text{ and } L_V = \frac{L_F}{L_F + L_D}$$

```
Step4. A Node is classified as
         Normal Node    if    G_v>G_T and L_v > L_T
         Selfish Node if    G_v> G_T and L_v > L_T
Where G_T and L_T are trustworthy thresholds.
Step5.
If (normal node)    //After detection of selfish node, in-
                      centive is given to motivate it //to
                      provide its resources for other nodes.
```

$$\text{Credit\_C} = \text{Credit\_C}_{old} + C$$

```
If (selfish node)
```

$$\text{Credit\_C} = \text{Credit\_C}_{old} + C + e$$

In the proposed incentive scheme, a node gets a credit for forwarding the messages, but it becomes selfish to save its resources and, when the higher credit is offered to selfish node, it sees that it gets some extra credit and then it gets ready to forward the message.

## C. System Model

A DTN environment consisting of various mobile wireless devices is considered. End-to-end connectivity is not available, and sender to destination path cannot be determined before the transmission. The transmission is based on store carry-and-forward method. Size of message may vary. Overhearing concept is not possible in DTN. When two nodes are communicating with each other, they would not be having information about other nodes in the network. Bubble Rap routing algorithm is used as the underlying routing algorithm. In bubble rap algorithm, forwarding is done on the basis of community and centrality; when two nodes belonging to different communities encounter each other, the node who is forwarding the message would check whether the destination node of outgoing message and encountered node belong to same community. If so, then the message would be forwarded and removed from forwarder's buffer else it would check the global centrality of the encountered node, and if global centrality is high, then it forwards the message. If encountered nodes belong to the same community, then forwarding node checks the local centrality of the node; if local centrality is high, then it forwards the message.

As discussed previously that DTN has no end-to-end connectivity, that's why the details of node and forwarded messages cannot be transferred from source to destination. So it is assumed that each node has a temper proof hardware in which all the details of credit earned by it and parameters used in algorithm will be stored. There will be a TA (trusted authority) where a node can encash its total credit (virtual money). A node will forward the message in order to earn more credit (virtual money). There is no malicious node present in the network which misbehaves intentionally. Every node will be ready to accept the credit.

# 4 Simulation and Results

Simulation environment has been set up using the Opportunistic Networking Environment (the ONE) simulator which is designed for evaluating DTN routing and application protocols. Three performance parameters, namely delivery probability, overhead ratio and average latency, have been evaluated by applying the selfish node detection and incentive scheme and without using incentive scheme. The Bubble Rap algorithm has been used as the reference algorithm to analyse the effectiveness of proposed scheme which is based on the concept of community and centrality. For extensive result analysis, different strengths of selfishness (up to 50%), i.e. existence of number of selfish nodes in the network, have been considered and simulated here.

Bubble Rap algorithm has been modified in ONE simulator to implement the proposed incentive scheme. We have computed and compared the results of simple Bubble Rap, where nodes in the network exhibit selfish behaviour but no strategy has been used to tackle them which has been denoted as simple bubble rap, while in other case the selfish nodes are motivated to forward the packet using an incentive-based scheme denoted as bubble rap incentive. The performance of both these protocols has been studied and compared in two different scenarios. In the first scenario, Cambridge traces have been used, while in the second scenario, a heterogeneous environment consisting of different types of node movement has been generated. In Cambridge scenario, 36 nodes are present and each node generates 35 messages to remaining other nodes in the network. Cambridge scenario

**Table 1** Cambridge scenario settings

| Scenario parameters | Values |
|---|---|
| Simulation time | 1036800 s |
| No. of nodes | 36 |
| Buffer size | 2 M |
| Transmit range | 10 |
| Transmit speed | 250 k |
| Message size | 200–600 KB |

**Table 2** Heterogeneous scenario settings

| Scenario parameters | Values |
|---|---|
| Simulation time | 700 k s |
| Number of nodes | 150 nodes |
| Number of node groups | 17 (8 WDM groups, 8 BBM groups, 1 SPMM) |
| Transmission speed | 250 KBps |
| Transmission range | 10 m |
| Buffer size | 2 m |
| Message size | 200–600 KB |
| WDM settings | Number of nodes = 16 per group<br>Wait time = 0 s<br>Speed = 0.8–14 m/s |
| SPMB settings | Number of nodes = 6<br>Wait time = 100–300 s<br>Speed = 8–14 m/s |
| BBM settings | Number of nodes = 2 per group<br>Wait time = 10–30 s<br>Speed = 7–10 m/s |

settings' details are shown in Table 1. Heterogeneous scenarios having 17 different types of group, group movement and number of nodes in group are different. Working day movement (WDM) [32], bus-based movement (BBM) and shortest path map-based movement (SPMM) models are used in heterogeneous scenario. Heterogeneous scenario settings' details are shown in Table 2.

From the simulation results shown in Figs. 1 and 2, it can be observed that delivery probability decreases with the increase in the number of selfish nodes in the network in both the scenarios, i.e. Cambridge and heterogeneous scenarios, as these nodes sometimes drop messages and prevent them from forwarding in order to save their own resources, namely buffer and bandwidth. However, the network performance is observed to improve when the incentive scheme is applied over Bubble Rap algorithm. As per the credit-based incentive scheme, the more they are incentivized to forward the messages to gain incentives and as the number of forwarded messages increases and the message replication increases and hence delivery probability improves.

**Fig. 1** Comparison of delivery probability for Bubble Rap routing algorithm with incentive scheme and without incentive scheme for Cambridge scenario

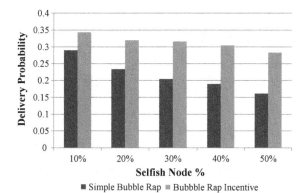

**Fig. 2** Comparison of delivery probability for Bubble Rap routing algorithm with incentive scheme and without incentive scheme for Heterogeneous scenario

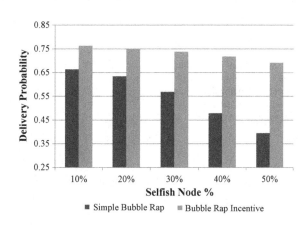

**Fig. 3** Comparison of Overhead Ratio for Bubble Rap routing algorithm with incentive scheme and without incentive scheme for Cambridge scenario

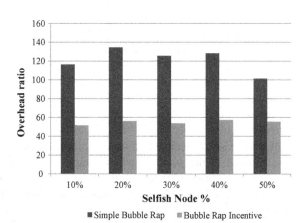

**Fig. 4** Comparison of overhead ratio for Bubble Rap routing algorithm with incentive scheme and without incentive scheme for Heterogeneous scenario

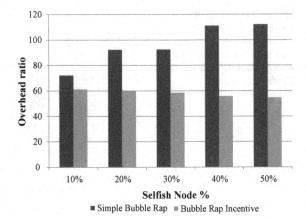

**Fig. 5** Comparison of average latency for Bubble Rap routing algorithm with incentive scheme and without incentive scheme for Cambridge scenario

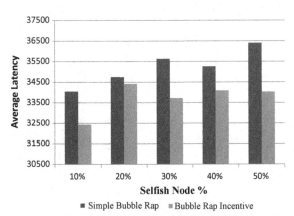

Figures 3 and 4 show the comparison of overhead ratio for the Bubble Rap routing algorithm in conjunction with incentive scheme and without incentive scheme for Cambridge and Heterogeneous scenario, respectively. In order to prevent resources, the selfish nodes do not forward the messages and prefer to drop the messages of other node. It is obvious that the overhead ratio will increase in the presence of selfish node. As the number of selfish nodes increases in the network, lesser number of messages are relayed and delivered in the network. Using an incentive scheme over bubble rap algorithm results in an increased number of delivered messages which automatically reduces the overhead ratio of the network as compared to the scenario where simple bubble rap algorithm is used in the presence of selfish nodes, but no incentives are provided.

Figures 5 and 6 illustrate the comparison of average latency for Bubble Rap routing algorithm with incentive scheme and without incentive scheme for Cambridge and heterogeneous scenarios, respectively. The average message latency tends to increase as the number of selfish nodes in the network increases in both the scenarios as selfish nodes drop the messages due to buffer scarcity. However when

**Fig. 6** Comparison of average latency for Bubble Rap routing algorithm with incentive scheme and without incentive scheme for Heterogeneous scenario

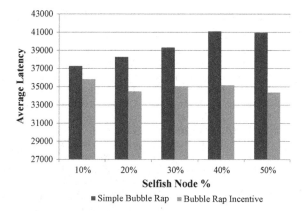

the incentive scheme is applied, nodes become ready to forward the messages of other nodes and hence latency reduces.

## 5 Conclusions

In this paper, a new incentive scheme is proposed to discourage selfish behaviour of nodes in delay-tolerant networks. To discourage selfish behaviour of nodes and to motivate them to forward the messages of other nodes, a credit-based scheme has been applied. The work proposed in this paper is a hybrid technique which combines both the incentive- and reputation-based scheme. The nodes behaviour is monitored by observing the number of messages forwarded and dropped by them to detect when a node starts acting selfishly. A credit-based incentive scheme is then applied to motivate nodes to avoid their selfish behaviour and cooperate in message forwarding. The credit rewarded is a function of the message size and the remaining TTL of the message. As a selfish node is motivated to cooperate even in the situation of resource crisis, an extra credit is rewarded which is inversely proportional to its available free buffer size. Performance evaluation has been done by comparing Bubble Rap routing algorithm without incentive scheme and with incentive scheme in the presence of a different number of selfish nodes. Some of the important conclusions that we observed were:

- As nodes are provided incentives to forward the message, more messages in the network get circulated and resulted in increased delivery probability.
- When nodes act selfishly, they drop and do not forward other node's messages and hence messages are delivered to the destination in more time resulting in increased message latency. But as nodes are incentivized, message delivery latency gets reduced.

# References

1. K. Fall, "A Delay-Tolerant Network Architecture for Challenged Internets," *in ACM SIGCOMM, Germany, pp. 244–251, August 2003.*
2. T. Spyropoulos, K. Psounis, and C. S. Raghavendra, "Efficient routing in intermittently connected mobile networks: The single-copy cast" *IEEE/ACM Transaction Networking, volume 16, no. 1, pp. 63–76, February 2008.*
3. S. Jain, K. Fall, and R. Patra, "Routing in a delay tolerant network," *in Proceedings of ACM SIGCOMM, 2004, pp. 145–158.*
4. A. Vahdat and D. Becker, "Epidemic routing for partially connected ad hoc networks," *Duke University, Durham, NC, Technical Report CS-200006, 2000.*
5. A. Lindgren, A. Doria, and O. Schelen, "Probabilistic routing in intermittently connected networks," *in Proceedings of SAPIR, volume 3126, pp. 239–254, 2004.*
6. S. Marti, T. Giuli, K. Lai, and M. Baker, "Mitigating routing misbehavior in mobile ad hoc networks," *in Proceedings of ACM MobiCom, Boston, Massachusetts, USA. pp. 255–265, August 2000.*
7. Q. He, D. Wu, and P. Khosla, "SORI: A secure and objective reputation based incentive scheme for ad hoc networks," *in Proceedings of WCNC, Atlanta, GA, volume 2, pp. 825–830, March 2004.*
8. Y. Zhang and Y. Fang, "A fine-grained reputation system for reliable service selection in peer-to-peer networks," *IEEE Transactions Parallel Distributed Systems, volume 18, no. 8, pp. 1134–1145, August 2007.*
9. S. Buchegger and J. Le Boudec, "Performance analysis of the CONFIDANT protocol: Cooperation of nodes-fairness in distributed ad-hoc networks," *in Proceedings of IEEE/ACM Workshop on Mobile Ad Hoc Networking and Computing, Lausanne, Switzerland, pp. 226–236, June 2002.*
10. Y. Zhang, W. Lou, W. Liu, and Y. Fang, "A secure incentive protocol for mobile ad hoc networks," *Wireless Network., volume 13, no. 5, pp. 569–582, October 2007.*
11. S. B. Lee, G. Pan, J.-S. Park, M. Gerla, and S. Lu, "Secure incentives for commercial ad dissemination in vehicular networks," *in Proceedings of MobiHoc, pp. 150–159, September 2007.*
12. R. Lu, X. Lin, H. Zhu, C. Zhang, P. H. Ho, and X. Shen, "A Novel Fair Incentive Protocol for Mobile Ad Hoc Networks," in Proceedings Of IEEE WCNC, pp. 3237–3242, Las Vegas, Nevada, USA, March 31-April 3, 2008.
13. L. Buttyan and J. P. Hubaux, "Enforcing Service Availability in Mobile Ad-hoc WANs," in Proceedings of ACM International Symposium on Mobile Ad Hoc Networking and Computing, pp. 87–96, 2000.
14. S. Zhong, J. Chen, and Y. R. Yang, "Sprite, A Simple, Cheat-proof, Credit-based System for Mobile Ad-hoc Networks," *in Proceedings of INFOCOM, San Francisco, IEEE Societies volume 3, pp. 1987–1997, April 2003.*
15. N. B. Salem, L. Buttyan, J. P. Hubaux, and M. Jakobsson, "A Charging and Rewarding Scheme for Packet Forwarding in Multi-hop Cellular Networks," *in Proceedings of MoBiHoc, pp. 13–24, June 1–3, 2003.*
16. M. Jakobsson, J. P. Hubaux, and L. Buttyan, "A Micropayment Scheme Encouraging Collaboration in Multi-hop Cellular Networks," *in Proceedings Of Financial Cryptography, volume 2742, pp. 15–33, 2003.*
17. V. Srinivasan, P. Nuggehalli, and C. Chiasserini, "Cooperation in Wireless Ad Hoc Networks," *in Proceedings of IEEE INFOCOM, San Francisco, volume 2, pp. 808–817, April 2003.*
18. X. Xie, H. Chen, and H. Wu, "Bargain-based Stimulation Mechanism for Selfish Mobile Nodes in Participatory Sensing Network," *in Proceedings of IEEESECON, NewOrleans, USA, pp. 1–9, 2009.*

19. A. Mei and J. Stefa, "Give2Get: Forwarding in Social Mobile Wireless Networks of Selfish Individuals," *in Proceedings of ICDCS, volume 8, pp. 1–14, june2010.*
20. U. Shevade, H. Song, L. Qiu, and Y. Zhang, "Incentive-aware Routing in DTNs," *in Proceedings of IEEE ICNP, no. 5, pp. 238–247, 2008.*
21. B. Chen and M. C. Chan, "MobiCent: a Credit-Based Incentive System for Disruption Tolerant Network," *in Proceedings of INFOCOM, pp. 1–9, 2010.*
22. Q. Li, S. Zhu, and G. Cao, "Routing in Socially Selfish Delay Tolerant Networks," *in Proceedings of INFOCOM, volume 38, pp. 857–865, 2010.*

# Security Analysis on Cognitive Radio Network

Fahia Nasnin, Mst. Najia Islam and Amitabha Chakrabarty

**Abstract** In traditional networking system, there is spectrum shortage problem. Therefore, cognitive radio (CR) is introduced to get the unlicensed users along with licensed users to maximize the bandwidth utilization. In that case, ensuring security is one major challenge and security issues are classified different types of attack. One major attack is known as primary user emulation attack (PUEA) that can affect the bandwidth utilization in CR network. CR network performance can be increased by mitigating the common security threats. The performance analysis based on miss detection and false alarm for primary user emulation attack in CR network has been observed. With the proposed model, the possibility of miss detection is successfully minimized with the increment of distance from the primary transmitter to primary exclusive region.

**Keywords** Cognitive radio · Cognitive radio network · Primary user emulation attack (PUEA) · False alarm · Miss detection · Primary exclusive region Probability density function

## 1 Introduction

Cognitive, the word was precisely used by Mitola and Maguire [1] in 1999. Due to increase spectrum demand day by day, it leads to the problem of demanding more available spectrum. Dynamic spectrum sharing and spectrum access methodology were proposed to eradicate the spectrum scarcity problems. Cognitive radio is an

F. Nasnin · Mst. Najia Islam (✉) · A. Chakrabarty
BRAC University, 66 Mohakhali, Dhaka 1212, Bangladesh
e-mail: naziatrina30@gmail.com

F. Nasnin
e-mail: fahiamim.hc@gmail.com

A. Chakrabarty
e-mail: amitabha@bracu.ac.bd

© Springer Nature Singapore Pte Ltd. 2019
B. K. Panigrahi et al. (eds.), *Smart Innovations in Communication and Computational Sciences*, Advances in Intelligent Systems and Computing 669, https://doi.org/10.1007/978-981-10-8968-8_26

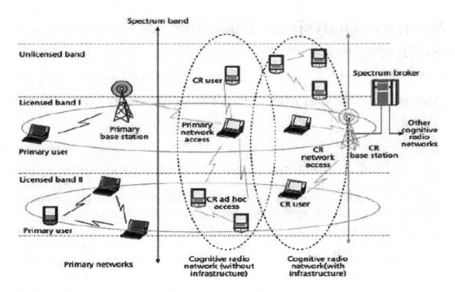

**Fig. 1** Basic architecture of cognitive radio network [3]

advanced intelligent wireless networking architecture that allows the secondary users (SU) to use the transmission channel. CR network provides high bandwidth to mobile users by its wireless architecture [2]. To overcome different challenges due to the change in the available transmission channel by the secondary user, and the quality of service (QoS) requirements of different implementations. There are four spectrum management features in CR, one of them is spectrum sensing, the process of identifying the free spectrum available at the primary user end. A CR user must have to sense the free channel to successfully detect the licensed users' presence in the network. The second feature is spectrum decision where the SU should access the spectrum according to bandwidth availability. Other CR users keep impact to take the decision of spectrum. Third one is spectrum sharing where all unlicensed user must have to share bandwidth along with other licensed users to maximize the bandwidth utilization. The last feature is spectrum mobility. During the spectrum using time if an unlicensed user (is also called secondary user) detects the primary user's (is also called licensed user) signal, then it is recommended to switch some other available channel (Fig. 1).

Hence, the opportunity of spectrum sensing, sharing, and mobility paves the way to CR users, and it is also difficult to detect malicious user who wants to use the whole spectrum. When unlicensed secondary user gets access in available spectrum, they may behave like malicious user [MU]. Malicious user can mitigate primary user's signal to occupy the full spectrum and sends false alarm to other secondary users. After getting the false alarm, secondary user cannot detect that this signal has sent by malicious user. Therefore, SU cannot utilize spectrum. It is a big challenge to mitigate malicious user attack.

## 2 Common Security Issues to Cognitive Radio Network

Among all security attacks, layer-based threats have been shortly described in this section. One of them is falsification attack which is a physical layer intrusion. In this intrusion, the malicious can cause jamming and it also can hide the primary user's presence. Following the cooperation of spectrum sensing process, malicious user can occupy the full spectrum by itself inside cognitive radio network. Malicious user sends false signal to other unlicensed user [4]. When malicious user affects the network performance creating collisions and controls the spectrum frequency, it is called DoS attack. The spectrum can allow a certain number of user at a same time. Therefore, when many CR users try to get accessed the channel, at the same time it faces huge spectrum scarcity problem [5]. This type of attack also occurs in physical layer on which we have mainly focused in this paper. Malicious users (MU) want to occupy full spectrum band by mitigating primary user (PU) signal; therefore, secondary user (SU) in the CR network gets the channel busy and such kind of attack restricts the access of SU. PUEA can fail the network to be utilized. Therefore, it is essential to mitigate PUEA. Mitigating PUEA can improve network performance. In this paper, performance for primary user emulation attack is studied from Neyman–Pearson criterion point of view. A new model with different configuration of the primary users and secondary users has been proposed and analyzed.

## 3 Literature Review

Primary user emulation attack (PUEA) has attracted considerable research attention. Researchers proposed an analytic model to detect PUEA in CR network. In this analytical model, energy detection helps to calculate receive power as a lognormally distributed random variable. Using Markov inequality, it is possible to track the wicked user and keep the low rate of PUEA [6]. So many methods have been approached to detect and mitigate PUEA. Chen et al. proposed transmitter verification scheme using received signal strength (RSS) to detect the source of primary user signal [7]. Another researcher proposed a presence of PU authentication process assisted by helper users deployed in near to the PUs [8]. Researchers approached the Neyman–Pearson criterion to mitigate the malicious primary user emulation attack and also used an energy detection algorithm for cooperative spectrum sensing [9]. Some researchers used the Neyman–Pearson test to discriminate between the main end users and wicked users. They proposed a channel-based detection for the noncooperative detection [10]. Z. Jin, S. Anand, K. P. Subbalakshmi analyzed primary user emulation attack. They claimed to keep the low rate of PUEA successfully, if there still a chance of missing primary signal [11]. Di Pu, Alexander M. Wyglinski proposed a way to recognize intrusion of wicked user in cognitive radio networks based on their action and techniques. They initiated the approach by recognizing power to locate the mimicry of wicked users [12]. Number of malicious user can affect the performance of CR network. Identifying wicked users and mitigation of primary emulation attack can maximize the bandwidth utilization.

## 4  Methodology and Design

Following assumptions are made for this ideology [11]. The system consists of N wicked users, and they relay *Pa* power. The distance between main transmitter and all the users is *Dp* and transmits at power *Pt*. The position of secondary user is at the center of the exclusive region. At the circle of radius *R*0r, there is no malicious user present; therefore, it is named primary exclusive region. There is no cooperation between the secondary users [11]. In the outer circle of exclusive region, malicious users are scattered evenly and they are not depended with each other. To the all secondary nodes, the position of primary transmitter is fixed and it is at $(r_{pt}, \theta_{pt})$ (Fig. 2).

### 4.1  Analytical Model

We estimated the collected signal at the subsidiary user from the transmission by the primary and wicked user is done to find out probability density function. It is acknowledged that N malicious users at $(r_k, \theta_k)$ where $1 \leq k \leq N$. Value of PDF which is $r_i$ is given as [11],

$$P_{(r_j)} = \frac{2r_j}{R^2 - R_0^2} \text{ Where } R_0 \leq r \leq R \qquad (1)$$

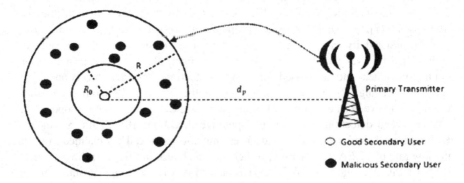

**Fig. 2** System model of cognitive radio network [3]

Here, R = radius of exterior circle and $R_0$ = radius of primary exclusive region. $\theta_k$ is evenly diffused in $(-\pi, \pi)$. The obtained energy from main transmitter is

$$P_r^{(p)} = P_t d_p^{-2} G_p^2 \text{ Where } G_p^2 = 10^{\frac{\epsilon p}{10}} \tag{2}$$

$d_p$ = distance from main transmitter to exclusive region and $P_t$ = power transference by main transmitter. Since $d_p$ and $P_t$ are fixed the probability density function of $P_r^p$ is

$$P^{(Pr)}(\gamma) = \frac{1}{\gamma A \sigma_p \sqrt{2\pi}} \exp\left\{ -\frac{\left(10 \log_{10} \gamma - \mu_p\right)^2}{2\sigma_p^2} \right\} \text{ Where } A = \frac{\ln 10}{10} \tag{3}$$

And $\mu_p = 10 \log_{10} P_t - 20 \log_{10} d_p$.

### Use of Neyman–Pearson Criterion for detecting PUEA

On the basis of deliberated values from the obtained signal, couple of assumptions can be made: $N_1$—main user resides with the analyzed signal and $N_2$—wicked user resides with the obtained signal or there is a chance that intrusion is ongoing. The subsidiary user acquaintance couple of menaces depending on the supervision of the assumptions [13].

## 4.2 Analysis

In this analysis, the chosen threshold value is 2, i.e., $\gamma = 2$. We have selected the radius value for primary exclusive region $R_0 = 30$ m, transmitting power by the primary node is $P_t = 100$ KW, transmitting power by the malicious attackers is $P_m = 4$ W and $\sigma_m = 3.5$ dB, $\sigma_p = 4$ dB.

Figures 3 and 4 are showing right continues graphs that can successfully detect the false alarm rate and miss detection rate. Here the two graphs are generated over five hundred times simulation. In both cases, the probability of sending false signal and miss detection rate is always close to 1–1.5, if we increase the malicious user number, then the false alarm probability will increase along with miss detection rate.

## 5 Proposed New Model

In our proposed model, there are N wicked users in this network and they are evenly scattered in the circular region. Pt1 and Pt2 are two primary transmitters, separated by a rigid area. The distance between secondary user and Pt1 is Dp1, and the distance between secondary user and Pt2 is Dp2. Our proposed CR network

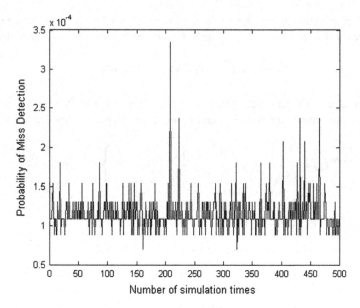

**Fig. 3** When secondary user misinterprets the transmission of primary user

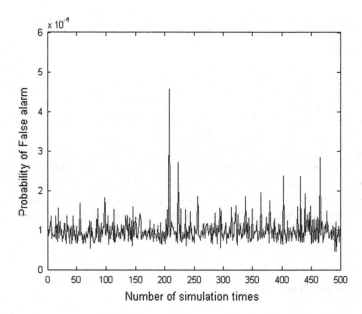

**Fig. 4** When secondary user thinks primary user is transmitting but it is generated by wicked user

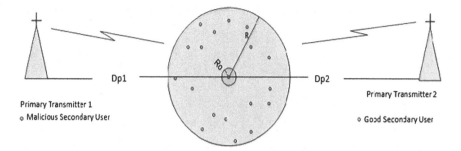

**Fig. 5** When there are two primary transmitters in CR model

model is shown in Fig. 5. We have deployed 10 malicious users that can transmit signal, and the value of transmitting signal is less than the value of primary transmitter's signal. There are two primary transmitters, and they are allocated in a certain distance from the primary exclusive region.

## 5.1 New Model Analysis

If we observe Fig. 6, we can see that false alarm and miss detection graphs are right continues. The presence of malicious user creates false alarm, and the secondary

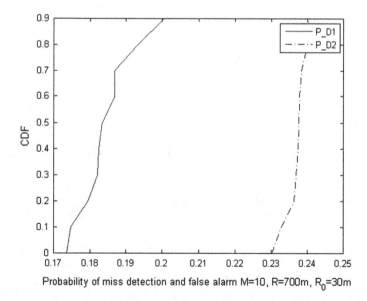

Probability of miss detection and false alarm M=10, R=700m, $R_0$=30m

**Fig. 6** Detection of false alarm and miss signal using one transmitter

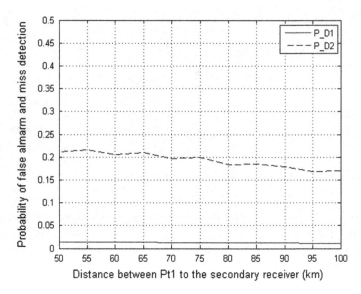

**Fig. 7** Detection of false alarm and miss signal using two transmitters

users miss the original signal transmitted by PU. In Fig. 6, false alarm rate is denoted by P_D1 and miss detection rate is denoted P_D2.

In Fig. 7, the change in generated probably of false alarm is not noticeable but the generated probability of miss detection rate is decreased by increasing the area value. Changing the distance between primary transmitter and primary exclusive region is affecting the miss detection rate but it has less impact on false signal generation by malicious intrusion.

## 6 Conclusion

In this paper, we have shown the analytical simulations by using theoretical value to plot probability of two different cases, first one is to detect the probability of miss signal and second one is to detect the probability of false alarm by wicked malicious user.

N = 10 Number of malicious users are generating false alarm and miss detections based on threshold value $\gamma = 2$. If false alarm and miss detection are generated, then it is obvious that there is malicious user. Identifying and knowing the presence of malicious user are the first step to mitigate PUEA. From our simulated result, we can conclude that miss detection can be decreed by increasing distance between P_t1 to the receiver. Our future work will be to secure CR network by implementing an efficient algorithm to verify secondary user using encryption techniques. Encrypted key will be generated by primary user and to occupy the channel by good secondary user, and they must have to match the key first. Comparative analysis of our work with existing model will be included in our further research.

# References

1. J. Mitola and G. Q. Maguire, Cognitive radio: making software radios more personal, IEEE Personal Communications, vol. 6, no. 4, pp. 1318, Aug 1999.
2. I. F. Akyildiz, W. y. Lee, M. C. Vuran and S. Mohanty, "A survey on spectrum management in cognitive radio networks," in IEEE Communications Magazine, vol. 46, no. 4, pp. 40–48, April 2008. https://doi.org/10.1109/mcom.2008.4481339.
3. D. A. Khare and M. Saxena, "Attacks and preventions of cognitive radio network-a-survey," International Journal of Advanced Research in Computer Engineering and Technology (IJARCET), vol. 2, March 2013.
4. A. A. Sharifi and M. J. M. Niya, "Defense against ssdf attack in cognitive radio networks: Attack-aware collaborative spectrum sensing approach," IEEE Communications Letters, vol. 20, no. 1, pp. 93–96, Jan 2016.
5. T. C. Clancy and N. Goergen, "Security in cognitive radio networks: Threats and mitigation," in 2008 3rd International Conference on Cognitive Radio Oriented Wireless Networks and Communications (Crown-Com 2008), May 2008, pp. 1–8.
6. S. Anand, Z. Jin, and K. P. Subbalakshmi, "An analytical model for primary user emulation attacks in cognitive radio networks," in 2008 3rd IEEE Symposium on New Frontiers in Dynamic Spectrum Access Networks, Oct 2008, pp. 1–6.
7. R. Chen, J. M. Park, and J. H. Reed, "Defense against primary user emulation attacks in cognitive radio networks," IEEE Journal on Selected Areas in Communications, vol. 26, no. 1, pp. 25–37, Jan 2008.
8. Y. Liu, P. Ning, and H. Dai, "Authenticating primary users' signals in cognitive radio networks via integrated cryptographic and wireless link signatures," in 2010 IEEE Symposium on Security and Privacy, May 2010, pp. 286–301.
9. Y. Zheng, Y. Chen, C. Xing, J. Chen, and T. Zheng, "A scheme against primary user emulation attack based on improved energy detection," in 2016 IEEE International Conference on Information and Automation (ICIA), Aug 2016, pp. 2056–2060.
10. T. N. Le, W. L. Chin, and Y. H. Lin, "Non-cooperative and cooperative puea detection using physical layer in mobile ofdm-based cognitive radio networks," in 2016 International Conference on Computing, Networking and Communications (ICNC), Feb 2016, pp. 1–5.
11. Z. Jin, S. Anand, and K. P. Subbalakshmi, "Detecting primary user emulation attacks in dynamic spectrum access networks," in 2009 IEEE International Conference on Communications, June 2009, pp. 1–5.
12. D. Pu and A. M. Wyglinski, "Primary-user emulation detection using database-assisted frequency-domain action recognition," IEEE Transactions on Vehicular Technology, vol. 63, no. 9, pp. 4372–4382, Nov 2014.
13. Z. Jin, S. Anand, and K. P. Subbalakshmi, "Mitigating primary user emulation attacks in dynamic spectrum access networks using hypothesis testing," SIGMOBILE Mob. Computer. Community. Rev., vol. 13, no. 2, pp. 74–85, Sep. 2009. [Online]. Available: http://doi.acm.org/10.1145/1621076.1621084.

# Android Anti-malware Techniques and Its Vulnerabilities: A Survey

Deepak Thakur, Tanya Gera and Jaiteg Singh

**Abstract** Android was initially presented in 2008, and luckily it has pulled in countless customers inside to make use of its services exhaustively. Utilization of Android gadgets is consistently expanding at colossal rate. As per Gartner's report, in the cell phone working framework (OS) market, Google's Android broadened its lead by catching 82% of the aggregate market in the final quarter of 2016. As per a report, the quantity of mobile applications in the Google Play Store had now outperformed a check of 2.5 million applications and all are accessible for download. Advanced features have been giving the clients simplicity of person-to-person communication, managing an account, i.e., making user to transfer money around the globe. As versatile applications are increasing, thereby expanding prominence among clients, and at the same time protection and security of smartphone clients turn into a worry, its open-source nature has turned into a fascination point for spammers and malware creators to perform unintended errands. Scholastic analysts and anti-malware organizations have understood that the static investigation strategies are helpless. Specifically, the common stealth strategies, for example, encryption and code obfuscation, are equipped for creating variations of known malware. This has prompted the utilization of dynamic examination-based strategies. Since a solitary approach might be incapable against the propelled procedures, numerous corresponding methodologies can be utilized as a part of pair for powerful malware discovery. This article gives an understanding into the qualities and inadequacies of the known anti-malware techniques from year 2009 to 2017. Our contribution to this paper is we have framed comparative study of various techniques and have provide major research gaps in summarized way which gives a

D. Thakur (✉) · T. Gera · J. Singh
Department of Computer Science and Engineering, Chitkara University,
Punjab, India
e-mail: deepak.thakur@chitkara.edu.in

T. Gera
e-mail: tanya.gera@chitkara.edu.in

J. Singh
e-mail: jaiteg.singh@chitkara.edu.in

© Springer Nature Singapore Pte Ltd. 2019                                           315
B. K. Panigrahi et al. (eds.), *Smart Innovations in Communication
and Computational Sciences*, Advances in Intelligent Systems
and Computing 669, https://doi.org/10.1007/978-981-10-8968-8_27

stage, to the scientists and specialists, toward proposing the cutting-edge Android security, examination, and malware discovery methods.

**Keywords** Android malware · Obfuscation · Static examination Dynamic examination

# 1  Introduction

Cell phones have turned out to be inescapable because of the accessibility of office applications, Internet, recreations, vehicle direction utilizing area-based services separated from traditional services, for example, voice calls, SMSes, and interactive media services. Expanded notoriety of the Android gadgets and related fiscal advantages pulled in the malware designers, bringing about huge ascent of the Android malware applications in the vicinity of 2010 and 2016. Scholarly analysts and commercial anti-malware organizations have understood that the signature matching approaches and static investigation techniques are powerless.

Universal Internet availability and accessibility of individual data, for example, contacts, messages, interpersonal organization data, web browser history, and keeping money-oriented credentials, have pulled in the consideration of malware engineers toward the cell phones and Android specifically. Android malware, for example, premium-rate SMS Trojans, spyware, botnets, forceful adware took an exponential ascent, as seen in the protected Google Play Store and third-party commercial mobile application stores [1–3].

Google Play has more than a million applications with countless downloads every day [4]. Not at all like the Apple App Store, Google Play checks the transferred applications without manual analysis. Rather, official market relies upon Bouncer tool [5, 6] a dynamic-based mechanism to control and shield the commercial center from the noxious application dangers. Malware creators exploit such defenseless applications and reveal the private client data to incidentally damage the application store and the mobile developer status. Besides, Android open-source rationality allows the establishment of third-party market applications, blending up many provincial and universal application stores [7–13]. In any case, the satisfactory assurance techniques and application quality at third-party mobile application stores is a worry [3]. The quantity of malevolent applications transferred on VirusTotal [8, 14] is expanding exponentially. Malware creators utilize stealth procedures and dynamic executions [15, 16] to sidestep the current insure systems provided by the Android platform and other vendors. Existing malware engenders by utilizing the above strategies and thrashing the signature-based methodologies.

Malware application engineers take the phone control by misusing device vulnerabilities [17], taking sensitive client data [15], to extricate money-related advantages by misusing the device communication services [18] or making phone a botnet [19]. The current companies against malware utilize signature-based algorithms because of its execution simplicity [20].

Signature-based strategies can be effortlessly evaded by utilizing code obfuscation techniques, requiring another signature for each malware variation [21], driving the victim to frequently refresh its signature database. Because of the restricted processing capacity and obliged battery accessibility, cloud-based techniques for examination and recognition have appeared [22, 23]. Manual analysis of an application and malware signature extraction requires adequate time and mastery. It can likewise produce false negatives (FN) while creating signature for the variations of known families. Because of the exponential expanded malware variations, there is a need to utilize programmed signature era strategies that bring about low false negatives. Off-gadget malware investigation techniques are expected to comprehend the malware techniques. Applications can be dissected to extricate the malware traces. Given the quick ascent of malware, there is an earnest need of the investigation strategies requiring least human interaction. Automatic detection helps the malware expert to distinguish the concealed malware. Static investigation can rapidly and accurately distinguish malware patterns. However, it comes up short against code obfuscation and Java reflections [24]. Accordingly, dynamic examination approaches, however tedious, is a contrasting option to extricate vindictive conduct of a stealth malware by executing them in a sandbox situation. This survey paper is organized in the following manner.

Section 2 discusses the notable Android anti-malware techniques between 2009 and 2017 and categorizes them according to their methodology, benefits and limitations and summarized as per their working in Table 1. In this section, comparison of various well-known anti-malware tools and techniques proposed by the researchers is also characterized as per their property, in Table 2. Section 3 provides the recommendations for future work. Finally, Sect. 4 concludes this paper.

# 2 Related Work

Practitioners and researchers have framed out many solutions to analyze and detect the Android malware threats. To the best of our knowledge, we have tried to cover significant malware analysis and detection methodologies from 2009 to 2017 in chronological order. This section discusses strengths and limitations of existing frameworks and also throws light in direction of future research possibilities.

A very interesting framework was implemented by practitioners in 2017 called Monet [44]. Monet is a suitable example of a hybrid approach for identifying malware variants. In their work, authors performed static as well as dynamic analysis to detect similar malicious applications from sample of around 3900 apps. Their investigation starts from on-device, for collecting useful behavioral information and generating signatures in form of graphs called static behavioral graphs. Then they combined static behavior graph with runtime information to produce runtime behavior graphs and send to server for further analysis. At server, signature matching algorithm plays its role to classify each app as malicious or not. The main benefits of this approach lie in capability of identification of malware variants with

**Table 1** Characterization of anti-malware tools based on their objectives, technique used, and implementation criteria

| R.No | Framework | Objective | | | Technique | | Implementation | |
|---|---|---|---|---|---|---|---|---|
| | | Assessment | Analysis | Detection | Static | Dynamic | On-device | Off-device |
| [25] | Kirin | ✓ | | | ✓ | | ✓ | |
| [26] | Scandroid | | ✓ | ✓ | | ✓ | ✓ | |
| [27] | TaintDroid | | | ✓ | ✓ | ✓ | | ✓ |
| [28] | AA Sandbox | | | ✓ | ✓ | ✓ | | ✓ |
| [29] | Stowaway | | ✓ | | ✓ | | ✓ | |
| [30] | Tang Wei | ✓ | ✓ | | ✓ | | ✓ | |
| [31] | ComDroid | | ✓ | | ✓ | | ✓ | |
| [32] | CrowDroid | | ✓ | ✓ | | ✓ | ✓ | ✓ |
| [33] | DroidMOSS | | | ✓ | | ✓ | ✓ | ✓ |
| [34] | Andromaly | | | ✓ | | ✓ | | ✓ |
| [35] | DroidScope | | ✓ | | | ✓ | | ✓ |
| [36] | AndroSimilar | | | ✓ | ✓ | | | ✓ |
| [37] | DroidAnalytics | ✓ | ✓ | | | ✓ | | ✓ |
| [38] | PUMA | | | ✓ | ✓ | | | ✓ |
| [39] | Androguard | ✓ | ✓ | ✓ | ✓ | | | ✓ |
| [40] | Andrubis | | ✓ | ✓ | ✓ | ✓ | | ✓ |
| [41] | APK Inspector | | ✓ | | ✓ | | | ✓ |
| [42] | CopperDroid | | ✓ | ✓ | | ✓ | | ✓ |
| [43] | SmartDroid | | | ✓ | ✓ | | | ✓ |

**Table 2** Summary of anti-malware techniques and its associated peculiarities

| Peculiarities | AA Sandbox [7] | Andrubis [83] | CopperDroid [85] | DroidAnalytics [17] | SmartDroid [86] |
|---|---|---|---|---|---|
| GUI interaction | ✗ | ✗ | ✗ | ✓ | ✓ |
| System call analysis | ✓ | ✗ | ✗ | ✓ | ✓ |
| Risk prediction | ✓ | ✓ | ✓ | ✓ | ✗ |
| Identifying data leakage | ✗ | ✓ | ✗ | ✓ | ✓ |
| Identifying SMS/call misuse | ✗ | ✗ | ✗ | ✓ | ✗ |
| Network traffic analysis | ✗ | ✓ | ✗ | ✓ | ✗ |
| File monitoring | ✗ | ✓ | ✗ | ✓ | ✗ |
| Native code analysis | ✗ | ✗ | ✗ | ✗ | ✓ |
| On-device analysis | ✗ | ✓ | ✗ | ✗ | ✗ |

only 7% performance overhead and 3% battery overhead. This technique also poses a challenging problem of defending against transformation attacks while detecting malware variants.

In 2016 [45], authors performed analysis in an isolated virtualized environment using emulators to provide increased security. The mechanism is VM-based analyses. Complete control and oversight of the environment achieved through the use of emulators. They also added that sandboxing non-Java code compiled when run with Android CPU in the future may add further protection to Android devices [45]. In 2016, researchers generated another traditional approach and performed static analysis over 2000 applications [46]; they extracted permissions of all the 2,000 applications. Results showed 93% of applications were found malicious due to requested network connectivity.

In 2015 [47, 48], authors used decompiling methods after applying amended post-processing to extract variety of features like API calls, Java classes and methods, structure sequences. They also included dependency graphs as a special feature to be extracted. Similarly, in 2014, DREBIN [49] collected static information which tends to extract API calls, network addresses, dangerous permissions, explicit and implicit intents from malware applications and performed analysis using Support Vector Machine (SVM). Their results showed that they were able to achieve low FP rate. Andrubis [40] is a online portal for malware analysis in which we can feed malware samples. Andrubis generates reports after performing static as well as dynamic analysis on Web itself. Its strength lies in detection of zero-day malware.

Further in 2013, researchers extracted API-level features and applied multiple classifiers for performing API-level malware detection. This mechanism is named as DroidAPIMiner [50] and showed acceptable accuracy. The results conclude that API feature set with package-level and parameter information is more accurate than permission-based feature set. The limitations of DroidAPIMiner are high rate of false positive (FP) and also resulted in wrong classification of APKs. Also in 2013, authors aimed to formalize Dalvik bytecode language with unique feature of reflection, experimented on Dalvik bytecode [51], and provided analysis of its control flow. The main benefits of this method are that it can trace the API calls, work on extraordinary feature of reflection and dynamic dispatch properties. Its limitation lies in concurrency handling. In 2013 [38], practitioners focused on analyzing extracted permissions and followed by taking K-fold cross validation in which value of K is taken to be 10 by the researchers and incorporating various machine learning algorithms to extract the permissions and features for the classification of an applications. Though this experiment achieves a great rise in detection rate, but at the same time it also observed increased FP rate. An another smart mechanism was used by researchers in 2013 [37] for automating the process of collecting samples, identifying important information, and performing analysis over them. Their target was to match suspected applications with large malicious database. For this, they make use of API calls for generating class-level signatures and performed analysis on the basis of op-code. The limitation lies in that it may often classify benign apps as malicious due to similarity score. Also it is not capable of identifying unknown malware types. The main features were that it is effective against mutations and repackaged apps, and also capable of identifying associate malwares at op-code level and dynamic malware payloads. CopperDroid [42], a dynamic off-device technique for analysis and detection of malicious applications, was experimented on virtual machine.

AndroSimilar [36] has been created keeping in view the rise in concealed and zero-day known malware samples. Authors in 2013 firstly created variable length signature to perform comparison with the large database of signatures followed by applying the procedure of fuzzy hashing for classifying each application as legitimate and harmful apps on the basis of matching score. This technique is very prominent against identifying encrypted code and repackaged code but having limited signature database is its lacking point. It has Androguard [39], a static analysis and detection tool for reverse engineering the Android applications. It first generates control flow graph for all the methods and then finds similarity and differences between the two clones of apps.

Another very exciting approach of static analysis was implemented in 2013 [41], named APK Inspector. Its main charm lies in its rich GUI for analysis of malwares, and also it focused in generating detailed meta data information of applications. This tool analyzes sensitive permission and generates call graph, control flow graphs, call structures to scrutinize suspicious events. The main loophole of all the tools which pursue static approaches is resource consumption, and they usually get deprived of apps behavior when it is actually executing. Also static approaches fail when code is obfuscated.

DroidScope [35] is also built by researchers for Android malware analysis in 2012. It tracks the system calls and helps in monitoring the operating system activities as it was created on quick emulator. It also performed dynamic analysis through introspection of virtual machine. It has limited code coverage but was able to identify administrative root attacks on the kernel layer. In Andromaly [34], detection of malware involves monitoring continuously, the processes being running in the device, the CPU usage, the amount of battery being consumed, the data being sent off the device. Then they applied machine learning algorithms and classifiers to segregate benign and malicious applications. They experimented their system by self-created malware variants. This method lacked in major points as it has issue of battery drainage issue. In DroidMOSS [33], they generated the fingerprint by making use of META-INF file and classes.dex file, and they extracted the author information and full developer certificate from META-INF file and mapped into 32-bit identifier. On the other hand, they extracted the actual Dalvik bytecode using Baksmali tool and they condense each instruction into much smaller instruction using fuzzy hashing technique. For this purpose, they set a sliding window of size $= 7$, later they applied the similarity scoring algorithm which tends to compute the edit distance between two fingerprints. They set threshold value to be 70. If the score exceeds the prescribed threshold value, the system will declare it to be as repackaged app. It has constraint that original application is must to be present in the database for detection of its malicious version. Some researchers omit the process of reverse engineering of APKs and directly they extracted bytecode of apps to transform to an efficient language for analysis of privacy leaks [52]. This process has many drawbacks: As it enhances the memory consumption, improvising performance is required. It does not support java native code and reflections. SmartDroid [42] is an automated effective approach for detection of UI-triggered conditions for assessment of several Android malwares which were evaded from eminent approach like TaintDroid [27].

Stowaway [29] makes use of static tools for keeping trace of API calls and its related permissions which are required for executing those API calls. It notifies the user about the over privileged applications. It cannot resolve complex reflective calls. Further, one more static approach [30] was implemented for assessment and analysis of applications. In this approach, combination of various permissions was analyzed and a distance model was utilized to measure the dangerous permissions. The main strength of this method was that it has the ability to identify the unknown malwares too and it can detect malware at the time of installation. The only limitation was that it apps with danger level 50–100 were tough to identify. ComDroid [31], a static approach in 2011, is used for inter-application communication threats by keeping logs of DEX files. Its strength lies in its capability of issuing warning to users. The manual investigation was must to ensure the presence of malicious doings in an application. In CrowDroid [32], dynamic analysis of an application had been done by the researchers. It used a tool called Strace to trace the system calls of device. The goodware version and malware version interactions are thus formed and collected in a file. The file is being sent to the server for further processing of system calls, thus system call vectors are formed. K-means algorithm in combination with Euclidean distance is used to find the similar system calls. Clusters are generated to distinguish good and malicious applications.

TaintDroid [27] automatically labels the data, and whenever the data flows outside the mobile device, a notification is sent to the user. It keeps track the data flow from source to destination. It only analyzes the data flow and does not analyzes the control flow. It is also not capable of tracking information related with network data flow. Kirin [25] builds risk assessment mechanism for checking certification of apps at install time. They used their own security rules for comparing the security configuration of application, and if an application does not follow the security rules, declared by them, their system declared that application as malicious. Their system has the ability to discard certain malicious applications. At the time of installation, application is checked against security rules. In Scandroid [26], researchers concentrated on analyzing data flow of an application, to conclude whether an Android application is malicious or safe. Malware detection's approach includes AA Sandbox [28], in which class.dex files are extracted, analyzed statically, and implemented the application in sandbox for its runtime analysis using monkey tool. The weakness of this method is its incapability in identifying new malware, but it can be used for enhancing efficiency of anti-malware programs.

# 3   Our Contribution

Our major contribution to this paper is that after performing intense survey of many anti-malware techniques, we are able to frame comparative study of various techniques (Table 1) and have provided major research gaps in summarized way (Table 2) which gives a great opportunity to researcher to work on those areas.

## 3.1   Comparative Analysis and Its Summary

In the comparative study, we have characterized the anti-malware approaches on the basis of its objectives, technique, and implementation which are clearly described in Table 1. Then followed by Table 2 in which we have synthesized present state of art, research gaps, and future recommendation for work. Based on comparative study, we generated a summary table, to the best of our knowledge, and most of the researchers have identified malwares using graphical user interaction, system calls, API. But as malware authors have started using much smarter ways than before and due to which it has becomes cumbersome for analysts to identify malware in obfuscated code, encrypted code, and dynamic code loading. The major research challenges lie in analyzing code of apps when present in other formats like hexadecimal. This provides a clear glance on scope of research over above-mentioned concepts.

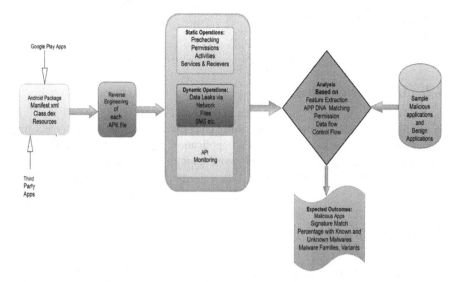

**Fig. 1** Proposed model

## 3.2   Proposed Model

See Fig. 1.

# 4   Future Directions

Numerous straightforward techniques do give security against the majority of malware; however, strategies like bytecode signatures are powerless against the developing measure of cutting-edge and contemporary malware [53]. We accordingly propose the accompanying areas for future research.

## 4.1   Hybrid-Based Analysis

Static detections are not useful in trivial code transformations [54], yet numerous applications, and most malware, are now utilizing larger amounts of obscurity [55]. In some situations where transformations (e.g., reflection, encryption, and native) are distinguished, dynamic-based analysis can be utilized as a part of conjunction for culmination. On the other hand, dynamic-based analysis intrinsically has less code scope yet it can utilize static examination of an application to guide analysis strategies in more ways [56–58] or utilize applications exercisers like MonkeyRunner.

## 4.2  Multileveled Analysis

In multilevel-based systems, the numerous layers will work parallel to detect the malicious activities, concealed within an application. Parallel handling could likewise significantly upgrade multilevel investigations and give efficient detection frameworks [59].

## 4.3  Code Coverage

Code coverage mechanism is key fundamental for powerful malware analysis. This may lead to troublesome task when managing progressively runtime code, native code, and activities related with network. There are a few advantages to dynamic like in ART [60], such as having the capacity to bar malware dodging investigations with reflection or native code. For instance, system call based analysis is good approach yet can in any case dissect Android-level and dynamic level practices behavior [61] and can be utilized to stop certain rooting of an Android phone [62]. While hybrid systems and more quick simulations (e.g., IntelliDroid, a static and API-based information generator) [63] would enormously build code coverage, distinctive methodologies ought to be additionally inquired about in light of malware patterns. Code coverage likewise presents an intriguing inquiry on whether malware tends to utilize "effectively" picked ways to execute more vindictive conduct or harder ways to keep away from being detected. Therefore, this can be an intriguing zone for future research directions, to distinguish malware patterns and, in that, increment the adequacy of future examinations.

## 4.4  Detection Place

Given the plenty of security infringement and malware recognition systems, there is a point of concern that where the malware detection system should be installed. On-device malware detection mechanism is limited by CPU processing, less space/memory, and mAh battery, while cloud-based systems may experience the ill effects of high correspondence overhead and more response time. Concocting a hybrid system, combination of static and dynamic analysis techniques will be useful to the community.

# 5 Conclusion

The distinction between our lives and our gadgets gets obscured with each passing day. We are bit by bit getting concealed smart device, for example, mobile-based shopping and banking transactions, learning, and even controlling home appliances. Cybercriminals have expanded their attack edge to trap each sort of client who is associated with the Internet. Information still remains the most valuable prize for cybercriminals—take it, offer it, and profit out of it.

This article concentrated an extensive survey on variety of anti-malware methodologies, delineating changing patterns in their techniques. This article likewise examined Android malware's capacity to surpass investigation. By understanding the malware dangers and solutions proposed, this article contributes to frame comparative study of most prominent techniques from year 2009 to 2017, and also we are able to come to fact that maximum of approaches work for GUI interactions, system calls, risk predictions, file monitoring, etc., but their system fails to defend against java native code, dynamic loading, java reflections. We conclude that there is a requirement of cutting-edge methodology which should be able to handle aforementioned factors to provide better security to Android device.

# References

1. http://www.gartner.com/newsroom/id/3609817, last accessed 30 March, 2017.
2. C. A. Castillo, "Android malware past, present, future," Mobile Working Security Group McAfee, Santa Clara, CA, USA, Tech. Rep., 2012.
3. W. Zhou, Y. Zhou, X. Jiang, and P. Ning, "Detecting repackaged smartphone applications in third-party android marketplaces," in Proc. 2$^{nd}$ ACM CODASPY, New York, NY, USA, 2012, pp. 317–326. [Online]. Available: http://doi.acm.org/10.1145/2133601.2133640.
4. AppBrain, Number of applications available on Google Play, (Online; Last accessed March. 18, 2017). [Online]. Available: http://www.appbrain.com/stats/number-of-android-apps.
5. Google Bouncer: Protecting the Google Play market, (Online; Last Accessed March. 18, 2017). [Online]. Available: http://blog.trendmicro.com/trendlabs-security-intelligence/a-look-at-google-bouncer/.
6. Android and security: Official mobile google blog, (Online; Last Accessed March. 18, 2018). [Online]. Available: http://googlemobile.blogspot.in/2012/02/android-and-security.html.
7. AndroidAPKSFree, http://www.androidapksfree.com/, last accessed on March 18, 2017.
8. APKMirror, http://www.apkmirror.com/, last accessed on March 18, 2017.
9. APK4Fun, https://www.apk4fun.com/, last accessed on March 18, 2017.
10. APKPure, https://apkpure.com/, last accessed on March 18, 2017.
11. PandaApp, http://android.pandaapp.com/, last accessed on March 18, 2017.
12. AppChina, http://www.appchina.com/, last accessed on March 18, 2017.
13. Baidu, http://shouji.baidu.com/?from=as, last accessed on March 18, 2017.
14. VirusTotal, https://www.virustotal.com/, last accessed on March 11, 2017.
15. Android.Bgserv, http://www.symantec.com/security_response/writeup.jsp?docid=2011-031005-2918-99, last accessed on Feb. 12, 2017.
16. Backdoor.AndroidOS.Obad.a, http://contagiominidump.blogspot.in/2013/06/, last accessed on Feb. 25, 2017.

17. RageAgainstTheCage, https://github.com/bibanon/android-developmentcodex/wiki/rageaga-instthecage, last accessed on Feb. 11, 2017.
18. Android Hipposms, http://www.csc.ncsu.edu/faculty/jiang/HippoSMS/, last accessed on Jan. 11, 2017.
19. Android/NotCompatible Looks Like Piece of PC Botnet, http://blogs.mcafee.com/mcafee-labs/androidnotcompatible-looks-like-piece-of-pc-botnet, last accessed on Dec. 25, 2016.
20. E. Fernandes, B. Crispo, and M. Conti, "FM 99.9, radio virus: Exploiting FM radio broadcasts for malware deployment," IEEE Trans. Inf. Forensics Security, vol. 8, no. 6, pp. 1027–1037, Jun. 2013. [Online]. Available: http://dblp.uni-trier.de/db/journals/tifs/tifs8.html#FernandesCC13.
21. R. Fedler, J. Schütte, and M. Kulicke, "On the Effectiveness of Malware Protection on Android," Fraunhofer AISEC, Berlin, Germany, Tech. Rep., 2013.
22. C. Jarabek, D. Barrera, and J. Aycock, "ThinAV: Truly lightweight Mobile Cloud-based Anti-malware," in Proc. 28th Annu. Comput. Security Appl. Conf., 2012, pp. 209–218.
23. Kaspersky Internet Security for Android, (Online; Last Accessed Feb. 11, 2017). [Online]. Available: https://www.kaspersky.co.in/android-security.
24. M. Grace, Y. Zhou, Q. Zhang, S. Zou, and X. Jiang, "RiskRanker: Scalable and accurate zero-day Android malware detection," in Proc. 10th Int. Conf. MobiSys, New York, NY, USA, 2012, pp. 281–294. [Online]. Available: http://doi.acm.org/10.1145/2307636.2307663.
25. W. Enck, M. Ongtang, and P. McDaniel 2009. "On lightweight mobile phone application certification,". In Proceedings of 16th ACM Conf. Comput. Commun. Secur. - CCS "09, pages.
26. Fuchs, A. P., Chaudhuri, A., & Foster, J. S. 2009. "Scandroid: Automated security certification of android".
27. W. Enck, P. Gilbert, B.-G. Chun, L. P. Cox, J. Jung, P. McDaniel, and A. N. Sheth 2010. "TaintDroid: An Information-Flow Tracking System for Realtime Privacy Monitoring on Smartphones," Osdi "10, vol. 49, pages 1–6.
28. T. Bläsing, L. Batyuk, A. D. Schmidt, S. A. Camtepe, and S. Albayrak 2010. "An android application sandbox system for suspicious software detection," In Proc. of 5th IEEE Int. Conf. Malicious Unwanted Software, Malware 2010, pages 55–62.
29. Felt, A. P., Chin, E., Hanna, S., Song, D., & Wagner, D. 2011. "Android permissions demystified", In Proceedings of the 18th ACM conference on Computer and communications security, pages 627–638.
30. Tang, W., Jin, G., He, J., & Jiang, X. 2011. "Extending Android security enforcement with a security distance model", In Internet Technology and Applications (iTAP), International Conference on, pages 1–4.
31. E. Chin, A. Felt, K. Greenwood, and D. Wagner 2011. "Analyzing inter-application communication in Android," In Proceedings of the 9th international conference on Mobile systems, applications, and services, pages 239–252.
32. I. Burguera, U. Zurutuza, and S. Nadjm-Tehrani 2011. "Crowdroid: Behavior-Based Malware Detection System for Android," In Proceedings of 1st ACM Workshop on Security and Privacy in smartphones and mobile devices, pages 15–26.
33. W. Zhou, Y. Zhou, X. Jiang, and P. Ning 2012. "Detecting repackaged smartphone applications in third-party android marketplaces," In Proceedings of 2nd ACM CODASPY, New York, NY, USA, pages 317–326.
34. A. Shabtai, U. Kanonov, Y. Elovici, C. Glezer, and Y. Weiss 2012. "'Andromaly': A behavioral malware detection framework for android devices," J. Intell. Inf. Syst., vol. 38, no. 1, pages 161–190.
35. L. K. Yan and H. Yin 2012. "DroidScope: Seamlessly reconstructing the OS and Dalvik semantic views for dynamic Android malware analysis," In Proceedings of 21st USENIX Security Symp., p. 29.
36. P. Faruki, V. Ganmoor, V. Laxmi, M. S. Gaur, and A. Bharmal 2013, "AndroSimilar: Robust statistical feature signature for Android malware detection," In Proc. SIN, A. Eli, M. S. Gaur, M. A. Orgun, and O. B. Makarevich, Eds., pages 152–159.

37. M. Zheng, M. Sun, and J. C. S. Lui 2013. "DroidAnalytics: A signature based analytic system to collect, extract, analyze and associate android malware," In Proceedings of 12th IEEE Int. Conf. TrustCom, pages 163–171.
38. B. Sanz, I. Santos, C. Laorden, X. Ugarte-Pedrero, P. G. Bringas, and G. Álvarez 2013. "PUMA: Permission usage to detect malware in android," Adv. Intell. Syst. Comput., vol. 189 AISC, pages 289–298.
39. BlackHat, Reverse Engineering with Androguard, (Online; Accessed Mar. 29, 2013). [Online]. Available: https://code.google.com/androguard.
40. Lindorfer, M., Neugschwandtner, M., Weichselbaum, L., Fratantonio, Y., Van Der Veen, V., & Platzer, C. (2014, September). Andrubis–1,000,000 apps later: A view on current Android malware behaviors. In Building Analysis Datasets and Gathering Experience Returns for Security (BADGERS), 2014 Third International Workshop on (pp. 3–17). IEEE.
41. APKInspector, 2013. [Online]. http://www.kitploit.com/2013/12/apkinspector-powerful-gui-tool-to.html.
42. Reina, A., Fattori, A., & Cavallaro, L. (2013). A system call-centric analysis and stimulation technique to automatically reconstruct android malware behaviors. EuroSec.
43. Zheng, C., Zhu, S., Dai, S., Gu, G., Gong, X., Han, X., & Zou, W. (2012, October). Smartdroid: an automatic system for revealing ui-based trigger conditions in android applications. In Proceedings of the second ACM workshop on Security and privacy in smartphones and mobile devices (pp. 93–104). ACM.
44. Sun, M., Li, X., Lui, J. C., Ma, R. T., & Liang, Z. (2017). Monet: A User-Oriented Behavior-Based Malware Variants Detection System for Android. IEEE Transactions on Information Forensics and Security, 12(5), 1103–1112.
45. Afonso, V., Bianchi, A., Fratantonio, Y., Doupé, A., Polino, M., de Geus, P. & Vigna, G. (2016). Going native: Using a large-scale analysis of android apps to create a practical native-code sandboxing policy. In Proceedings of the Annual Symposium on Network and Distributed System Security (NDSS).
46. Hein, C. L. P. M., & Myo, K. M. (2016). Characterization of Malware Detection on Android Application. In Genetic and Evolutionary Computing (pp. 113–124). Springer International Publishing.
47. Yang, W., Xiao, X., Andow, B., Li, S., Xie, T., & Enck, W. (2015, May). Appcontext: Differentiating malicious and benign mobile app behaviors using context. In Software Engineering (ICSE), 2015 IEEE/ACM 37th IEEE International Conference on (Vol. 1, pp. 303–313). IEEE.
48. Li, L., Bartel, A., Bissyandé, T. F., Klein, J., Le Traon, Y., Arzt, S., ...& McDaniel, P. (2015, May). Iccta: Detecting inter-component privacy leaks in android apps. In Proceedings of the 37th International Conference on Software Engineering-Volume 1 (pp. 280–291). IEEE Press.
49. Arp, D., Spreitzenbarth, M., Hubner, M., Gascon, H., Rieck, K., & Siemens, C. E. R. T. (2014, February). DREBIN: Effective and Explainable Detection of Android Malware in Your Pocket. In NDSS.
50. Y. Aafer, W. Du, and H. Yin 2013. "DroidAPIMiner: Mining API-Level Features for Robust Malware Detection in Android," Secur. Priv. Commun. Networks, vol. 127, pages 86–103.
51. E. R. Wognsen, H. S. Karlsen, M. C. Olesen, and R. R. Hansen 2014. "Formalisation and analysis of Dalvikbytecode," Sci. Comput. Program., vol. 92, pages 25–55.
52. J. Kim, Y. Yoon, K. Yi, J. Shin, and S. Center 2012. "ScanDal: Static analyzer for detecting privacy leaks in Android applications," In Proc. Workshop MoST, 2012, in conjunction with the IEEE Symposium on Security and Privacy.
53. Securelist. 2013. Mobile malware evolution: 2013. Retrieved from https://securelist.com/analysis/kaspersky-security-bulletin/58335/mobile-malware-evolution-2013/.
54. Feng, Y., Anand, S., Dillig, I., & Aiken, A. (2014, November). Apposcopy: Semantics-based detection of android malware through static analysis. In Proceedings of the 22nd ACM SIGSOFT International Symposium on Foundations of Software Engineering (pp. 576–587). ACM.

55. Securelist. 2013. The most sophisticated Android Trojan. Retrieved from https://securelist. com/blog/research/35929/the-most-sophisticated-android-trojan/.

56. Zheng, C., Zhu, S., Dai, S., Gu, G., Gong, X., Han, X., &Zou, W. (2012, October). Smartdroid: an automatic system for revealing ui-based trigger conditions in android applications. In Proceedings of the second ACM workshop on Security and privacy in smartphones and mobile devices (pp. 93–104). ACM.

57. Spreitzenbarth, M., Freiling, F., Echtler, F., Schreck, T., & Hoffmann, J. (2013, March). Mobile-sandbox: having a deeper look into android applications. In Proceedings of the 28th Annual ACM Symposium on Applied Computing (pp. 1808–1815). ACM.

58. Mahmood, R., Mirzaei, N., & Malek, S. (2014, November). Evodroid: Segmented evolutionary testing of android apps. In Proceedings of the 22nd ACM SIGSOFT International Symposium on Foundations of Software Engineering (pp. 599–609). ACM.

59. Dini, G., Martinelli, F., Saracino, A., & Sgandurra, D. (2012, October). MADAM: a multi-level anomaly detector for android malware. In International Conference on Mathematical Methods, Models, and Architectures for Computer Network Security (pp. 240–253). Springer Berlin Heidelberg.

60. Marko Vitas. 2013. ART vs Dalvik. Retrieved from http://www.infinum.co/the-capsized-eight/articles/art-vsdalvik-introducing-the-new-android-runtime-in-kit-kat, last accessed on 29th March, 2017.

61. Reina, A., Fattori, A., & Cavallaro, L. (2013). A system call-centric analysis and stimulation technique to automatically reconstruct android malware behaviors. EuroSec, April.

62. Ho, T. H., Dean, D., Gu, X., & Enck, W. (2014, March). PREC: practical root exploit containment for android devices. In Proceedings of the 4th ACM conference on Data and application security and privacy (pp. 187–198). ACM.

63. Wong, M. Y., & Lie, D. (2016). Intellidroid: A targeted input generator for the dynamic analysis of android malware. In Proceedings of the Annual Symposium on Network and Distributed System Security (NDSS).

# Simple Transmit Diversity Techniques for Wireless Communications

**Kommabatla Mahender, Tipparti Anil Kumar and K. S. Ramesh**

**Abstract** Multipath fading encountered in time-varying channel renders wireless communications highly non-reliable. To obtain an average bit-error rate of $10^{-3}$ using BPSK modulation with coherent detection, performance degradation due to Rayleigh fading can account for a SNR of 15 dB higher than in AWGN [1]. This paper is limited to the review of several study cases of MISO configuration where suitable coding or signal processing techniques are exploited to allow the extraction of transmit diversity without channel knowledge at the receiver. If the subchannels associated with transmit antennas have independent fades, the order of diversity is proven to be equal to the number of transmit antennas. This approach is attractive in public broadcasting systems such as cellular (for voice) or broadband wireless access (for data communications) to keep the subscriber-side equipment cost down with simpler hardware requirement and more compact form factor by avoiding the implementation of several receive antennas.

**Keywords** Diversity · Fading channels · MIMO · MRC and SNR

## 1 Introduction

### 1.1 Mitigating Rayleigh Fading

Theoretically, transmitter power control technique known as water-filling is the most effective way to mitigate multipath fading [2–5]. However, the transmitted

K. Mahender (✉) · K. S. Ramesh
Department of Electronics and Communication Engineering,
KL University, Vaddeswaram, Guntur, Andhra Pradesh, India
e-mail: kmsharma2@yahoo.co.in

T. A. Kumar
Department of Electronics and Communication Engineering,
CMR Institute of Technology, Kandlakoya, Medchal road,
Hyderabad, Telangana, India

© Springer Nature Singapore Pte Ltd. 2019
B. K. Panigrahi et al. (eds.), *Smart Innovations in Communication and Computational Sciences*, Advances in Intelligent Systems and Computing 669, https://doi.org/10.1007/978-981-10-8968-8_28

power adaptation requires knowledge of channel signal-to-noise ratio (SNR) to be evaluated at receiver end and send feedback to transmitter that inevitably results in throughput reduction and higher complexity in both transmitter and receiver. Dynamic ranges of transmitter amplifier necessary to accommodate power back-off represent another disadvantage in using water-filling technique.

Diversity is a powerful technique which is more practical and therefore widely used in combination with space-time coding to combat signal fading. Diversity is characterized by the number of independently fading subchannels being created by multiple antennas at transmitter and/or at receiver. Depending on antenna configurations, space-time wireless systems can be categorized into single-input single-output (SISO) being the traditional channel, single-input multiple-output (SIMO) having one single transmit antenna and multiple receive antennas, multiple-input single-output (MISO) using multiple transmit antennas and a single receive antenna, and multiple-input multiple-output (MIMO) having multiple transmit antennas and multiple receive antennas [6, 7].

## 1.2 Overview of the Paper

This paper is limited to the review of several study cases of MISO configuration where suitable coding or signal processing techniques are exploited to allow the extraction of transmit diversity without channel knowledge at the receiver. If the sub channels associated to transmit antennas have independent fades, the order of diversity is proven to be equal to the number of transmit antennas. This approach is attractive in public broadcasting systems such as cellular (for voice) or broadband wireless access (for data communications) to keep the subscriber-side equipment cost down with simpler hardware requirement and more compact form factor by avoiding the implementation of several receive antennas. This paper is organized as follows:

- Alamouti's simple but efficient transmit diversity technique [3] is introduced. This approach uses space-time coding to achieve diversity on one single receive antenna in flat-fading wireless channels. Extension to multiple receive antennas is also reviewed.
- Winters' approach [8] using signal processing to create diversity from multiple transmit antennas is briefly reviewed.
- An extension to frequency selective fading for Alamouti's technique in the OFDM.

## 2 Alamouti's Simple Transmit Diversity Techniques

This technique is described to be applicable to two transmit antennas and one receive antenna configuration [3]. A simple diversity technique is Alamouti's scheme; it is a simple transmit diversity that can be used in 2 × 1 MISO mode or in a 2 × 2 MIMO mode. Transmitted symbols can be transmitted through each antenna using an M-ary modulation technique, then these transmitted signals encoded by using space-time block code (STBC). In this paper, we assumed that the transmitter does not know the channel state information (CSI) whereas receiver has full knowledge of channels.

## 2.1   Flat-Frequency-Fading Channels

We present the received signals following separate propagation paths as identical except for a complex scalar. This implies flat-frequency-fading channel, i.e., $BT_m \ll 1$ where $B$ is the signal bandwidth and $T_m$ the channel delay spread. In addition, it is necessary that $BT_z \ll 1$, where $T_Z$ is transit time of the wave front across antenna array or difference in propagation delay from transmit antennas to receive antenna. For distance between transmit antennas in the order of 10–20 wavelengths and for carrier frequencies in the range 2–5 GHz, the transit time $T_Z \approx 5$ ns, i.e., much smaller than practical values of delay spread or symbol period.

Under the aforementioned conditions, transmitted symbols can be sent through N number of transmit antennas and receiver antennas as shown in Fig. 1. The channels can be modeled as the as a row vector $h = [\, h_0 \quad h_1 \quad \ldots \quad h_{N-1} \,]$ and the received signal at time $k$ as

$$y[k] = \sqrt{\frac{E_s}{N}} h x[k] + n[k] \tag{1}$$

where $\mathbf{x}[k] = [\mathbf{x}_0[k] \quad \mathbf{x}_1[k] \quad \ldots \quad \mathbf{x}_{N-1}[k]]^T$ is the transmitted signal vector at time $k$ and $n[k]$ additive noise [5].

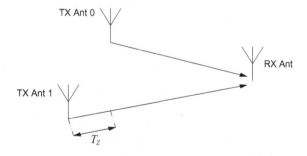

**Fig. 1** Transit time across antenna array

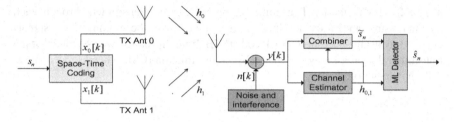

**Fig. 2** Alamouti's simple transmit diversity scheme

## 2.2 Two Branch Transmit Diversity with One Receiver

This technique performs the following three functions

- The encoding and transmission sequence of information symbols at the transmitter
- The combining scheme at the receiver
- The decision rule for maximum likelihood detection [2] as shown in Fig. 2.

## 2.3 The Encoding and Transmission Sequence

The space-time coding scheme is based on pairs of symbols that are transmitted in two consecutive symbol intervals. During the first (even-numbered) interval, the two symbols are transmitted unaltered and simultaneously by two antennas. During the following (odd-numbered) interval, the complex conjugates are transmitted as shown in Fig. 3.

The transmitted symbol sequence is denoted as $s_n$, each transmitted signals sent through antenna zero is $x_0[k]$ and antenna one is $x_1[k]$, respectively, at time $k$. Assume that the symbol time $k$ is introduced to highlight the sequential aspect in transmitted and received signals. However, the symbol index $n$ is retained to avoid eventual confusion that apparent non-causality in the following equations may

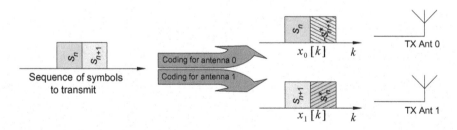

**Fig. 3** Alamouti's space-time coding for TX antennas

cause. Two consecutive symbols are transmitted simultaneously over two antennas during the symbol interval $k$, where $k$ is even, as

$$\begin{cases} x_0[k] = s_n \\ x_1[k] = s_{n+1} \end{cases} \tag{2}$$

During the odd-numbered symbol interval, the same symbols are re-transmitted, in complex conjugate form and inverted polarity on antenna zero, as

$$\begin{cases} x_0[k+1] = -s_{n+1}^* \\ x_1[k+1] = s_n^* \end{cases} \tag{3}$$

While the fading is assumed to remain unchanged across two consecutive symbol intervals, the channel can be modeled by complex multiplicative distortions for transmit antenna zero and one as

$$\begin{cases} h_0[k+1] = h_0[k] = h_0 = \alpha_0 e^{j\varphi_0} \\ h_1[k+1] = h_1[k] = h_1 = \alpha_0 e^{j\varphi_1} \end{cases} \tag{4}$$

And since the time delay between receptions of $x_0[k]$ and $x_1[k]$ caused by spatial distance between transmit antenna locations (order of several wavelengths) is assumed negligible compared to the symbol interval, the received signal can be expressed as

$$y[k] = h_0 x_0[k] + h_1 x_1[k] + n[k] \tag{5}$$

## 3 The Combining Scheme

The received signal is combined across two consecutive symbol intervals following manner [4] as shown in Fig. 4.

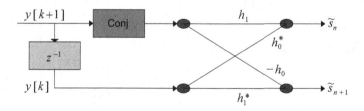

**Fig. 4** Alamouti's receive combining scheme

$$\begin{cases} \tilde{s}_n = h_0^* y[k] + h_1^* y[k+1] \\ \tilde{s}_{n+1} = h_1^* y[k] - h_0^* y[k+1] \end{cases} \tag{6}$$

where the channel complex multiplicative factors $h_{0,1}$ are estimated and made available by channel estimator. Substituting Eqs. (2) through (3) into (6) and simplifying produce

$$\begin{cases} \tilde{s}_n = \left(\alpha_0^2 + \alpha_1^2\right)s_n + h_0^* n[k] + h_1 n^*[k+1] \\ \tilde{s}_{n+1} = \left(\alpha_0^2 + \alpha_1^2\right)s_{n+1} + h_1^* n[k] - h_0 n^*[k+1] \end{cases} \tag{7}$$

Eq. (7) shows the combined signals are in fact transmitted symbols being scaled and received in additive noise.

## 3.1  The Maximum Likelihood Decision Rule

The detector makes decision of which symbol being transmitted based on the Euclidean distance to all possible symbols

$$\hat{s}_n = s_j, j = \arg \min_i |\tilde{s}_n - s_i| \tag{8}$$

## 3.2  Comparison with Maximal Ratio Combining

To illustrate the comparable performance between the simple transmit diversity scheme presented in previous sections and maximal ratio combining diversity scheme, let us consider the one transmit antenna and two receive antennas diversity system as shown by Fig. 5 [9, 10].

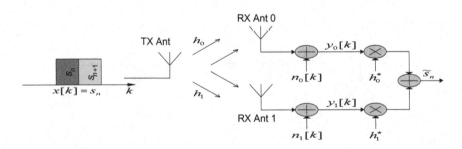

**Fig. 5** Maximal ratio combining diversity

The received signals are

$$
\begin{aligned}
y_0[k] &= h_0 x[k] + n_0[k] \\
y_1[k] &= h_1 x[k] + n_1[k]
\end{aligned}
\tag{9}
$$

where each symbol interval transmitted signal is $x[k] = s_n$, channel response $h_{0,1} = \alpha_{0,1} e^{j\varphi_{0,1}}$ is assumed to remain unchanged across consecutive symbol intervals and $n_{0,1}[k]$ additive noise terms.

The MRC is given as

$$
\tilde{s}_n = h_0^* y_0[k] + h_1^* y_1[k] = \left(\alpha_0^2 + \alpha_1^2\right) s_n + h_0^* n_0[k] + h_1^* n_1[k]
\tag{10}
$$

That is indeed very similar to the expressions in Eq. (7).

## 3.3 Extension to M Receive Antennas

Alamouti's technique can be applicable to MIMO configuration of two transmit antennas and $M$ receive antennas to provide a diversity order of $2M$. While the transmission coding remain as specified by Eq. (4), the channel matrix becomes

$$
H = \begin{bmatrix} h_{0,0} & h_{0,1} \\ \vdots & \vdots \\ h_{M-1,0} & h_{M-1,1} \end{bmatrix}
$$

where $h_{l,m}$ is the complex scalar associated with transmission path between transmit antenna $l$, $l \in \{0, 1\}$, and receive antenna $m$, $m \in \{0, 1, \ldots, M-1\}$. The received signals are

$$
\mathbf{y}[k] = \begin{bmatrix} y_0[k] \\ \vdots \\ y_{M-1}[k] \end{bmatrix} = H \begin{bmatrix} x_0[k] \\ x_1[k] \end{bmatrix} + \begin{bmatrix} n_0[k] \\ \vdots \\ n_{M-1}[k] \end{bmatrix}
\tag{11}
$$

And the combining scheme is

$$
\begin{cases}
\tilde{s}_n = [h_{0,0} & \cdots & h_{M-1,0}]^* \mathbf{y}[k] + [h_{0,1} & \cdots & h_{M-1,1}] \mathbf{y}^*[k+1] \\
\tilde{s}_{n+1} = [h_{0,1} & \cdots & h_{M-1,1}]^* \mathbf{y}[k] + [h_{0,0} & \cdots & h_{M-1,0}] \mathbf{y}^*[k+1]
\end{cases}
\tag{12}
$$

Replacing Eqs. (11) and (12) into Eq. (7) yields

$$
\begin{cases}
\tilde{s}_n = s_n \sum_{m=0}^{M-1} \left(\alpha_{m,0}^2 + \alpha_{m,1}^2\right) + \sum_{m=0}^{M-1} \left(h_{m,0}^* n_m[k] + h_{m,1} n_m^*[k+1]\right) \\
\tilde{s}_{n+1} = s_{n+1} \sum_{m=0}^{M-1} \left(\alpha_{m,0}^2 + \alpha_{m,1}^2\right) + \sum_{m=0}^{M-1} \left(h_{m,1}^* n_m[k] + h_{m,0} n_m^*[k+1]\right)
\end{cases}
\tag{13}
$$

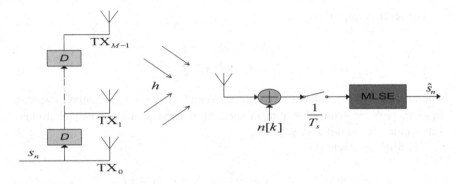

**Fig. 6** Winters' transmit diversity

## 4   Winters' Transmit Diversity Technique

Unlike Alamouti's, the technique being briefly reviewed here operates based on artificial introduction of inter-symbol interference achieved by multiple antennas sending delayed versions of transmit symbols [8] as shown in Fig. 6.

The channels between transmit antennas and receive antenna are assumed to be independent Rayleigh. In order to obtain uncorrelated signals over transmit antennas, the delay $D$ must be larger than $\max(T_s, T_m)$, where $T_s$ is the symbol interval and $T_m$ the multipath delay spread. Aggregate noise is assumed additive white Gaussian. At the receiver side, the signal is sampled at the symbol rate before maximum likelihood sequence estimation (MLSE) with perfect channel knowledge is employed to determine the transmitted symbols.

We consider the a tree pruning algorithm, modified to handle multilevel signal in complex channel, to determine the minimum Euclidean distance in MLSE implementation, and to produce the performance curves presented in his paper. The paper reports simulated results that are identical to the theoretical matched filter bound performance for the case of $M = 2$ and have negligible degradation, i.e., less than 0.1 dB, for larger $M$.

One significant difference Winters' approach has from Alamouti's resides in the matched filter bound reference. In contrast to MRC receive diversity, the matched filter has only one noise source that is unable to provide array gain. This results in a gain reduction by $M$ as discussed in this paper.

# 5 Extension of Alamouti's to Wideband Fading

Alamouti's proposed technique is based on narrowband fading assumption where the channel is determined by a set of complex scalars that remains unchanged across several symbol intervals. This section summarizes an extension to wideband fading channel.

## 5.1 OFDM Overview

The basic idea being exploited here is to divide the symbol stream into substreams to be transmitted over a set of orthogonal subcarriers where resulted symbol period in each substream is much larger than the multipath channel delay spread to efficiently mitigate the ISI.

In practice, OFDM can be efficiently implemented using the well-known FFT and IFFT functions as shown in Fig. 7.

In the context of OFDM where each subchannel centered on a subcarrier experiences flat fading, Alamouti's technique can be applied using a coding scheme spanning multiple antennas and adjacent OFDM subcarriers, instead of timeslots, to obtain diversity over a single receive antenna.

## 5.2 Space-Frequency Coding

Let consider transmitted symbol is denoted as $s_n$. The transmit encoding scheme consists of mapping two consecutive symbols on to same subcarriers $f$ to be transmitted over antennas zero and one as

$$\begin{cases} X_0(f) = s_n \\ X_1(f) = s_{n+1} \end{cases}$$

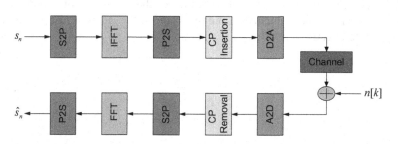

**Fig. 7** OFDM modem block diagram

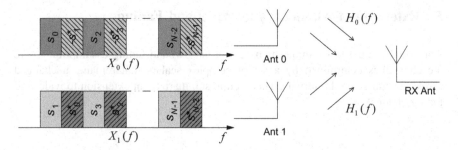

**Fig. 8** Space-frequency coding for OFDM

and complex conjugates of same symbols on to adjacent subcarrier $f + \Delta f$ as

$$\begin{cases} X_0(f + \Delta f) = -s_{n+1}^* \\ \quad X_1(f + \Delta f) = s_n^* \end{cases}$$

Figure 8 shows the formation of the OFDM frames, from $N = N_{FFT}/2$ symbols, to be transmitted over two antennas.

Assuming fading is similar across adjacent subchannels that can then be represented by complex factors $H_{0,1}(f) = H_{0,1}(f + \Delta f) = H_{0,1} = \alpha_{0,1} e^{j\varphi_{0,1}}$, the received signals, for the transmitted symbols $s_n$ and $s_{n+1}$, are

$$Y(f) = H_0 x_0(f) + H_1 x_1(f) + n(f)$$

$$Y(f + \Delta f) = H_0 x_0(f + \Delta f) + H_1 x_1(f + \Delta f) + n(f + \Delta f)$$

A combining scheme similar to Eq. (9) yields the following estimates

$$\begin{cases} \tilde{s}_n = \left(\alpha_0^2 + \alpha_1^2\right) s_n + H_0^* n(f) + H_1 n^*(f + \Delta f) \\ \tilde{s}_{n+1} = \left(\alpha_0^2 + \alpha_1^2\right) s_{n+1} + H_1^* n(f) - H_0 n^*(f + \Delta f) \end{cases}$$

The simulation results with QPSK-OFDM over multipath channels, respectively, 4- and 16-tap long, to be very close to narrowband performance for FFT length $N_{FFT} \geq 512$. For $N_{FFT} = 256$, the SNR degradation is about 2 dB for $BER = 10^{-3}$. This reduced performance is explained by the fact that as number of bins decreases, adjacent channel spacing increases, and the assumed flat fading across subchannels ceases to be effective.

# 6 Simulation Results and Discussions

This paper reviewed the performance of two transmit diversity techniques proposed by Alamouti [2, 3] and Winters [7]. Compared to equivalent receive diversity, the presented techniques observe the advantage of having simpler, smaller, and less expensive subscriber equipments while providing efficient remedy to Rayleigh fading channel.

These techniques are reported based on different benchmarks such as MRC and matched ched filter bound criteria, and direct comparison is impractical. However, in general, Alamouti's and MRC provides an array gain due to different noise sources, but the matched filter bound doesn't, instead suffers a gain reduction proportional to diversity order M. Therefore, considering same TX power equally distributed among transmits' antennas, Alamouti's technique would be expected to outperform Winters' due to array gain. Besides, Alamouti's approach appears to be more popular thanks to its simple formulation, efficiency, and flexible adaptability to other configurations (MISO and MIMO) and coding scheme (space-time vs. space-frequency). And finally, following are several issues that need to be properly addressed and made provisions for in order to take full advantage of the reviewed techniques.

## 6.1 Power Requirements

As per Alamouti's paper, the simulated cases (2-by-1 or 2-by-2 antennas) report 3 dB performance degradation from the equivalent MRC (1-by-2 or 1-by-4 antennas). This performance loss is due to the fact only half of total power is radiated from each transmitting antenna. It can also be seen as one receive antenna can pick up only half the equivalent power compared to the MRC cases. Of course, MRC comparable performance can be achieved whenever total power is not a constraint and, hence, can be doubled to accommodate both transmit antennas. In Winters' report, it is unclear how power repartition would affect the overall performance.

## 6.2 Sensitivity to Channel Estimation Errors

Both techniques are built upon the assumption of perfect knowledge of the channels. Pilot tones or training symbols are commonly used to assist the receiver in estimating channel information. In the case of transmit diversity, same pilot symbols need to be alternately transmitted from every antenna or orthogonal symbols from all antennas simultaneously, i.e., more overhead than receive diversity is needed for channel estimation purpose. Some constraint is also imposed on received signal sampling rate that must meet both Nyquist requirement and exceed twice the maximum Doppler frequency.

## 6.3 Decoding Latency

Both methods are introducing delayed symbols to allow diversity extraction at receiver side. Detection function must wait until all transmitted symbols are received to make optimal decision. Instead of space-time coding, Alamouti's technique can also be combined with coding across space-frequency dimensions. If so, the decoding delay can be eliminated.

## 6.4 Soft Failure

If one transmitted signal fails to reach the receive antenna, e.g., due to transmitter failure or deep fade, the diversity gain is reduced but the transmitted symbols can still be decoded successfully.

Fig. 9 illustrated that various Alamouti coding comparisons with MRC expressed in terms of BER performance obtained. In this paper, we assumed that perfect channel estimation at receiver is available and independent Rayleigh fading channels used. Same diversity order is achieved by Alamouti coding as compared to 1 × 2 MRC schemes, which implies that same BER obtained due to transmit power constraint. However, MRC performs better than Alamouti's technique.

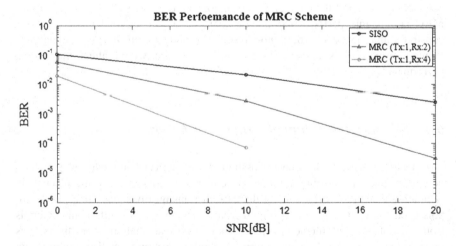

**Fig. 9** BER performance of different Alamouti's schemes

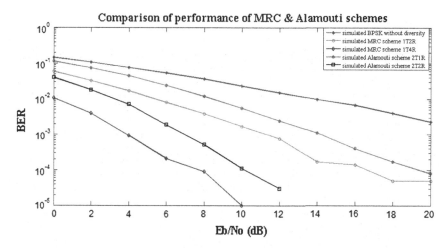

**Fig. 10** Comparison of performance of MRC and Alamouti's scheme

Figure 9 shows that 2 × 2 Alamouti achieves same diversity where 1 × 4 MRC schemes.

From the Fig. 10 we can clearly understand the BER performance between transmit (Alamouti) and receive diversity (MRC) scheme using QPSK modulation. Also achieved BER of the MRC with transmit antenna and two receive antenna. We need around 16 dB of SNR and for the Alamouti scheme with two transmit antenna and one receive antenna required SNR is 18 dB. This shows us that the Alamouti scheme 2 dB (practically 3 dB) is worse than MRC due to system radiation limitations.

# 7 Conclusion

In Alamouti's coding technique, we get better error probability performance than SISO. However, MRRC with the same diversity order with transmit diversity gives the same error performance if there is no power limitation. In transmit diversity using different kind of modulation and coding technique, we can achieve almost zero probability error whereas in receiver diversity, we can achieve it by increasing SNR which can be improved by increasing the number of antenna element at receiver side.

# References

1. Winters, J. H.; "The diversity gain of transmit diversity in wireless systems with Rayleigh fading," *IEEE Transactions on Vehicular Technology*, Feb 1998, vol. 47, no. 1, pp. 119–23.
2. N.S. Murthy, S. Sri Gowri and JNV Saileela" Transmit Diversity Techniques for wireless Communication over Rayliegh Channels using 16-QAM Modulation Schemes", Communications in Computer and Information Science-197, pp: 127–137, 2011 Springer.
3. Ahmed, N.; Baraniuk, R.G; "Asymptotic performance of transmit diversity via OFDM for multipath channels," *IEEE Conference on Global Telecommunications*, Nov 2002, vol. 1, pp. 691–95.
4. Alamouti, S. M.; "A simple transmit diversity technique for wireless communications," *IEEE Journal on Select Areas in Communications*, Oct 1998, vol. 16, no. 8, pp. 1451–58.
5. Paulraj, A. J.; Papadias, C. B.; "Space-time processing for wireless communications," *IEEE Signal Processing Magazine*, Nov 1997, pp. 49–83.
6. Lindskog, E.; Paulraj, A. J.; "A transmit diversity scheme for channels with intersymbol interference," *IEEE Conference on International Communications*, Jun 2000, vol. 1, pp. 307–11.
7. Paulraj, A.; Nabar, R.; Gore, D.; "*Introduction to Space-Time Wireless Communications,*" Cambridge University Press 2003.
8. Tarokh, V.; Alamouti, S. M.; Poon, P.; "New detection schemes for transmit diversity with no channel estimation," *IEEE International Conference on Universal Personal communications*, Oct 1998, vol. 2, pp. 917–20.
9. Goldsmith, A.; "EE359 Wireless Communications, Course Reader," 2003.
10. Stuber, G.; "*Principles of Mobile Communications,*" Kluwer Academic Publishers, 2nd Edition, 2001.

# Optimal Threshold Detection and Energy Detection in Cognitive Radio

Mahua Bhowmik, Nilambika Jawalkote and P. Malathi

**Abstract** Cognitive radio is an emerging technology that is an advanced version of the radio network. The goal of cognitive radio is to efficiently use all empty spaces that are not fully utilized for communication. It is an intelligent radio that fits dynamically and changes its transmitter parameters according to the environment. In the proposed work, the quality of services has been optimized through detection threshold. We get closed-form optimum threshold expression for local equilibrium detection, probability of failure detection, and probability of false alarm.

**Keywords** Cognitive radio · Spectrum sensing · Threshold detection
Energy detection

## 1 Introduction

Although intensive research is done for cognitive radio network technology that is revolutionary. In the cognitive network, there are two types of users: primary and secondary; primary users are authorized to use the spectrum at a given frequency band, while secondary users use the space available for primary users always. But there are some unused spaces are called white holes so the main task of the radio network is to effectively use spaces that have not been. Cognitive radio network aims to provide an optimal solution. It aims to improve spectrum utilization and maintains a high-quality service. Its efficiency can be improved by minimizing

M. Bhowmik (✉) · N. Jawalkote
Department of Electronics and Telecommunication, Dr. D.Y. Patil Institute
of Technology, Pimpri, Pune, India
e-mail: m_ghorai2001@yahoo.com

N. Jawalkote
e-mail: nilambikajawalkote@gmail.com

P. Malathi
Dr. D.Y. Patil College of Engineering, Akurdi, Pune, India
e-mail: malathijesudason@ymail.com

© Springer Nature Singapore Pte Ltd. 2019
B. K. Panigrahi et al. (eds.), *Smart Innovations in Communication and Computational Sciences*, Advances in Intelligent Systems and Computing 669, https://doi.org/10.1007/978-981-10-8968-8_29

343

interference. Unused user frequencies can not be taken by authorized users (secondary users) so efficiently using unused frequency spectrum. Which can actively allocate change parameters without degrading (QoS) unused ASIGNA band parameters [1]. In this paper the two different form of detection is considered that is using the concept of transmitted power by primary user and secondly the closed loop equation using the center square distribution is used to fet the optimal value of QoS.

But the spectrum detection procedure is generally alienated into basic two categories: rapid detection and thin line detection [2, 3]. Quick detection uses an energy detection system that only provides information of the existence of the signal wherein it measures the amount of the energy present in it. Therefore, it is possible to run the system with very few microseconds. Thin line detection uses a developed detection technique that can be administered as a filter method or cyclostationary adaption method wherein we can easily able to recognize the type of system and it takes ample of times for its cycle to operate.

## 2 Proposed Sensing by Power Transmission

If the power of the received signal exceeds the predetermined power, Energy threshold detection determines that a particular primary signal. The IEEE 802.22 system already stipulates the threshold detection for some authorized users which are given as follows: −94–107 dBm for PAL and SECAM television transmission and −116 dBm for digital television transmission over terrestrial, cable, and satellite networks [4]. The threshold value to is very poor in these above cases as the noise level is varied and unknown to the input parameters of the signals. Therefore, for the uncertain event the possibility of exchanging energy detector noise of the primary user is high.

The false alarm probability Pf and the detection probability play a major role in the detection of the spectrum. The missed detection probability indicates that the secondary user fails in detecting the signal of primary user. In this case, the secondary to the primary cause of serious interference due to the presence of both signals simultaneously. For improved sensing, the false alarm probability should be minimized as it gives the forged alarm of the existence of the primary signal even though the signal is absent. One can say that there is a negotiation between the probability of detection which is also miss detection and probability of false alarm, regarding the concept of threshold detection [2, 5, 6].

If we increase the detection threshold, missed detection probability also increases, which further is the cause of decrease in the false alarm probability. Therefore, the determination of the threshold value plays an important role in detection methodology. In the cognitive Radio system by considering the various environmental parameters such as channel capacity, signal strength, signal to noise ratio, the bandwidth on which the detector has to work demands a appreciable choice of value detection which can improve detection performance so that an efficient exchange of possible spectrum happens in cognitive system.

## 2.1 System Algorithm

This paper emphasis basically on the calculation of energy threshold calculation. The transmission power of secondary user is adjusted in the cognitive network region by considering the distance between the two nodes. If the SU transmission power is smaller, then the interference of its processing unit is lower. As the interference in primary user is reduced if the transmitted power from secondary user is minimized, thus, the false alarm probability decreases by increasing the threshold value. Nowadays, many researches are going on in the power control of the transmitter and receiver for spectrum issues in the cognitive radio network [2, 7–11]. It is common that the SU (secondary user) first checks the hole of primary user and then it controls the power of its own transmission level. At the same time, it also checks that there should not be any interference of the two signals. But the secondary user could not able to reduce its power below to a certain level which is essential to maintain its quality of services. To maintain the speed of the signal, its power plays the major role in transmission [2, 12–15]. In such circumstances, the SU must jump from one channel to another so that it can avoid primary user interference. The channel for transmission shared by primary and secondary users is the same; thus, SU signals interfere with PU. The SINR interferences received are compared by the primary receiver, and if it is greater than the minimum requirement of the receiver to decode the primary signal, then the primary user easily accepts the interference caused by the secondary user. To understand the above concept, consider the worst scenario where the processing system has less SINR. Normally, the worst case means that PR is present has less SINR at the limit of the primary coverage area and the closest possible position with SU.

Or building the cognitive scenario some input parameters are considered as follows:

$d$      distance between secondary user and primary receiver.
$Rs$      radius coverage of secondary user.
$Qp$      power transmission of PT.
$Rp$      radius coverage of *primary user*.
$Qs$      power transmission of secondary user. (Let the transmission power be shared by two secondary users).
$T_n$      noise power due to heat/thermal.
$\gamma_p$      signal-to-noise interference received.
$\gamma_{min}$      minimum decodable interference for primary receiver.
$\lambda$      threshold level.

Then, $Rp$ is defined as [2],

$$R_p = \frac{Q_p - f(R_p)}{T_n + (Q_s - f(d)) \times n} \tag{1}$$

where v $f(\cdot)$ is a path loss function and $n$ is the number of secondary user pairs that is interfering with primary receiver. The path loss function is given by [1]

$$f(d) = 10p \times \log(d) \text{ (DB)} \tag{2}$$

where $d$ =distance between the two secondary users and $P =$ the power loss exponent,
whereas the thermal noise power is determined by [2]

$$T_n = -174 + nf + \log(bw) \text{ (dBm)} \tag{3}$$

From the same frequency band, if the SINR is greater than the minimum SINR, then the spectrum is identified. The condition Rp > γ min must be given by below equation which later will be satisfied by the different user and frequency channel. Then, the following inequality is given by (1) [2].

$$\frac{Q_p - f(R_p)}{T_n + (Q_s - f(d) \times n)} > \gamma_{min} \tag{4}$$

By calculating the values of $Qp$, $Rp$, and $\gamma min$, the SU gets the information pertaining to the PU which it wants to sense. To determine the path loss function and also know other SUs using the same channel, $d$ is the only unknown parameter in above equation. From Eqs. (2) and (4), the minimum $d$ is obtained

$$d_{min} = exp\left(\frac{1}{k} ln\left(\frac{nQ_s \gamma_{min}}{Q_p/R_p^n - T_n \gamma_{min}}\right)\right) \tag{5}$$

which is the distance between the secondary user and the primary receiver for simultaneous detection of the channels. It can be noted that the value of *dmin* is related to the broadcast of secondary user power; *dmin* is becoming more with increase in $Qs$. But at the time of the sharing of spectrum, *Qr represents* the signal power of the receiver which is found by

$$Q_r = Q_p - f(d_{min} + R_p)$$
$$= Q_p - 10 \log(d_{min} + R_p) \tag{6}$$

Once if Qr is less than the power of the received signal, then we can say that PR may be closer to SU Dmin, so at this time it is not possible to share the spectrum.

However, if the power of the detected signal is lower than Qr, then it can be said that dmin is less than the distance between the SU and the PU, and hence, the spectrum allocation can be done by the process of the energy threshold. Consequently, the power received Qr power can be used as a norm to evaluate the presence of primary user. The threshold $\lambda$ then can be obtained by determining the value of Qr by keeping some input parameters constant.

Figure 1 depicts that there is a settlement between false alarm and lost detection. In the same adaptive condition method, the probability of false alarm is slightly

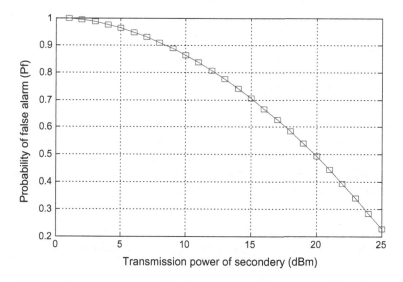

**Fig. 1** Probability of false alarm versus secondary transmission power

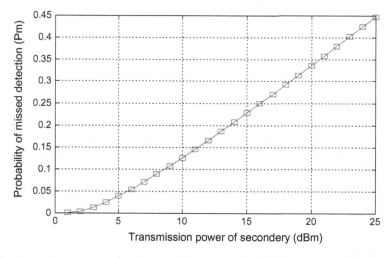

**Fig. 2** Probability of missed detection versus secondary power transmission

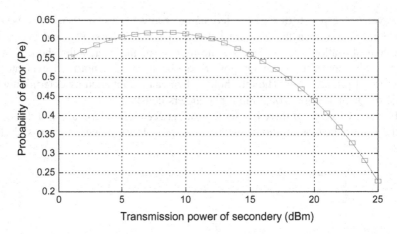

**Fig. 3** Probability of errors versus transmission power of secondary

greater than the probability of failure detection. Since the fading effect dominates to make the probability of false alarm greater than the probability of detection [2, 16].

Figures 2 and 3 show that as the interferences increase due to the power of SU, the detection capability loses its chances. According to the SU power transmission, the adaptive threshold method is adapted in the work done.

## 3  Proposed Sensing by Central Square Distribution

The paper also proposed the energy detection where spectrum performance parameters are altered by means of optimum detection technique and optimal threshold. The optimal cooperative threshold detection is found to be greater than local optimum only when the signal-to-noise ratio starts decreasing [17, 18].

$$P_f = P(k > 1/Z_0) \tag{7}$$

$$P_d = P(k < 1/Z_1) \tag{8}$$

where $Z_0$ indicates the absence of primary user and Pf indicates the presence of primary user with the probability of false alarming. Pf should be kept minimum as it determines the missing chance of the spectrum. $Z_1$ indicates the presence of main primary user, and Pd is the probability of detection by supporting the main user and must be kept high enough to shun the interference. The two scenarios are given by

$$Y_1(t) = \begin{cases} a_1(t)Z_0 \\ a_2 m(t) + a_1 Z_1 \end{cases} \tag{9}$$

where $a_1(t)$ is the pure additive white Gaussian noise (AWGN), and $a_2$ is channel gain of the spectrum used. Energy which is given as $Y_1(t)$ and can give the following equation:

$$A_1(t) = \begin{cases} a_{2u}Z_1 \\ a_2(2\gamma) \end{cases} \tag{10}$$

where $(\alpha 2u)$ indicates a central square distribution (CSD) which gives freedom of $2u$ degrees and is equal to the $2TW$ (product of two time bandwidth) and $(\alpha 2\ (2\gamma))$ indicates a distribution of non-central square with the same degrees of freedom [1].

For cognitive radio, the mean false probability, the average detection probability, and probability of false detection of AWGN channels are determined by [1]

$$P_{f,i} = \frac{r\left(u, \frac{\lambda_i}{2}\right)}{\Gamma(u)} \tag{11}$$

$$P_{d,i} = Q_u\left(\sqrt{2\gamma_i}, \sqrt{\lambda_i}\right) \tag{12}$$

$$P_{m,i} = 1 - P_{d,i} \tag{13}$$

where $\lambda_i$ is the energy threshold in the cognitive radio; for example, $\Gamma(x, y)$ is the gamma function of the range of $\Gamma(x)$ and $Q_u(a, b)$ is the generalized Marcum Q function.

In the work proposed, the sensing time of the network is quite greater than the normal one. Thus, the probability of two hypotheses is given below:

$$P_{Z0} + P_{Z1} = 1 \tag{14}$$

In this algorithm, optimum threshold is calculated by the following formula where $N$ = product bandwidth

$$\lambda_i = \frac{N\gamma_i(\gamma_i + 1) + (\gamma_i + 1)\sqrt{N^2\gamma_i^2 + 2N\gamma_i(\gamma_i + 2)\left[\ln(1 + \gamma_i)^2 - \ln 4\right]}}{\gamma_i(\gamma_i + 2)} \tag{15}$$

## 3.1 Results

The parameter setups for the algorithm are as follows:

$N = 2TW = 4000$, SNR = $-15$ to $30$ dB.

With above setup, the algorithm was implemented in MATLAB and the following results are obtained.

We get the number of Pm and Pd against various SNRs. Figure 4 shows that Pm decreases with the increase in the probability of false alarm $P_f$, and thus, if false alarm is more, then the miss detection is less. Figure 5 shows Pd versus SNR. It increases the threshold as the noise starts dying and the signal strength becomes more stronger by increasing the SNR. We are able to adjust the threshold according to the different needs of the cognitive radio environment. The optimal value of threshold is preferred in such a way that for low value of Pm and $P_f$, there must be a trade-off between the two factors. But it is not easy to detect for the low SNR.

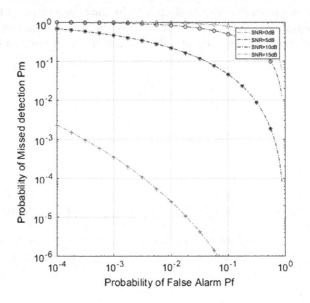

**Fig. 4** Missed detection probability Pm versus $P_f$

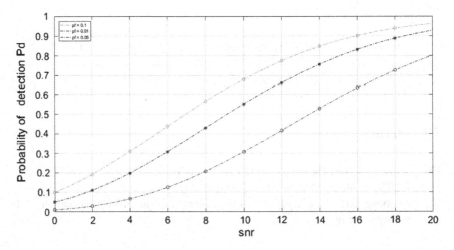

**Fig. 5** Probability of detection Pd versus SNR

# 4 Conclusion

In energy detection, the threshold value has been adjusted at various different situations accordingly. Once the optimal threshold value is detected, and both for low value of $P_m$ and $P_f$, the $P_m$ is chosen above the threshold level inversely the $P_f$ is also chosen below the threshold level. Once for minimizing the probability of error $P_e$, there is sharp trade-off between the optimal threshold and SNR. At the same time to maintain the trade-offs, the probability of miss detection $P_{md}$ is also minimized to compensate the detection method. At last, the rule of the half outcome voting number showed that the probability of total system error can be minimized with a higher threshold when the good SNR locale is minimum. Thus the same concept is applied for the $PH_o$ and $PH_1$ if it is been changed. From the simulation result, the threshold calculated by using the different equations shows the result to be the most optimal output keeping all the QoS under strict observations. For low SNR value, the optimal sensing implies at half rate rule. The future work definitely scores the work on the absence of the parameters of the signal with and without the closed form of equations.

# References

1. Shujing Xie, Lianfeng Shen and Jishun Liu, "Optimal Threshold of Energy Detection for Spectrum Sensing in Cognitive Radio", IEEE Internal conference on Wireless Communications and Signal Processing, 978-1-4244-5668-November 13–15, 2009.
2. Hyun–Ho Choi, Kyunghun Jang, Yoonchae Cheong, "Adaptive Sensing Threshold Control based on Transmission power in Cognitive Radio Systems", 3rd International Conference on Cognitive Radio oriented Wireless Networks and Communications (Crowncom 2008), May 15–17, 2008.
3. D. Cabric, S. M. Mishra, and R. W. Brodersen, "Implementation issues in spectrum sensing for cognitive radios," Asilomar Conference on Signals Systems and Computers, vol. 1, pp. 772–776, Nov. 2004.
4. H. Urkowitz, "Energy detection of unknown deterministic signals," Proceedings of the IEEE, vol. 55, issue 4, pp. 523–531, Apr. 1967.
5. F. F. Digham, M.-S. Alouini, and M. K. Simon, "On the Energy Detection of Unknown Signals Over Fading Channels," IEEE Transactions on Communications, vol. 55, issue 1, pp. 21–24, Jan. 2007.
6. N. Hoven and A. Sahai, "Power scaling for cognitive radio," International Conference on Wireless Networks, Communications and Mobile Computing, vol. 1, pp. 250–255, June 2005.
7. Xia Wang and Qi Zhu, "Power Control for Cognitive Radio Base on Game Theory," Wireless Communications, Networking and Mobile Computing, 2007. WiCom 2007. International Conference on 21–25 Sept. 2007 Page(s):1256–1259.
8. Wang Wei, Peng Tao, and Wang Wenbo, "Optimal Power Control Under Interference Temperature Constraints in Cognitive Radio Network," Wireless Communications and Networking Conference, 2007. WCNC 2007. IEEE 11–15 March 2007 Page(s):116–120.
9. ERC Report 42, "Handbook on radio equipment and systems radio microphones and simple wide band audio links".
10. Sun. C, Zhang. W, Letaief. K. B, "Cluster-based cooperative spectrum sensing in cognitive radio systems", in Proc. IEEE ICC'07, pp. 2511–2515, 2007.

11. Ganesan. G and Li. Y, "Cooperative spectrum sensing in cognitive radio, Part I: Two User Networks", IEEE Transactions on Wireless Communications, vol. 6, no. 6, 2007, pp. 2204–2213.

12. Ian F. Akyildiz, et al., "NeXt generation/dynamic spectrum access/cognitive radio wireless networks: A survey," Elsevier Computer Networks, pp. 2127–2159, May 2006.

13. Carlos Cordeiro, MonishaGhosh, Dave Cavalcanti, and KiranChallapali, "Spectrum Sensing for Dynamic Spectrum Access of TV Bands (Invited Paper)," ICST CrownCom 2007, Aug. 2007.

14. K. Hamdi, W. Zhang, and K. B. Letaief, "Power Control in Cognitive Radio Systems Based on Spectrum Sensing Side Information," IEEE International Conference on Communications, pp. 5161–5165, June 2007.

15. Wei Zhang, Ranjan K. M, Letaief K. B "Cooperative Spectrum Sensing Optimization in Cognitive Radio Networks" in Proc. IEEE ICC'08, pp. 3411–3415, 2008.

16. F. F. Digham, M.-S. Alouini, and M. K. Simon, "On the energy detection of unknown signals over fading channels", IEEE Transactions on Communications, vol. 55, no. 1, Jan 2007, pp. 21–24.

17. E. Peh and Y.-C. Liang, "Optimization for cooperative sensing in cognitive radio networks," in Proc. IEEE Int. Wireless Commun. Networking Conf., Hong Kong, Mar. 11–15, 2007, pp. 27–32.

18. Hyun–Ho Choi, "Transmission Power Based Spectrum Sensing for Cognitive Ad Hoc Networks", journal of Information and Communication Convergence Engineering 2014.

# Energy Consumption Analysis of Multi-channel CSMA Protocol Based on Dual Clock Probability Detection

Zhi Xu, Hongwei Ding, Zhijun Yang, Liyong Bao, Longjun Liu
and Shaojing Zhou

**Abstract** In the era of Internet of Things, it is necessary to have a strong ability to transmit information, but the transmission capacity needs to have enough energy resources to support. The Multichannel dual clock probability detection CSMA protocol is proposed. The paper use average cycle analysis method and Newton solution of nonlinear equation to build mathematical model, and the system throughput and energy consuption can be obtained. The final experiment shows that the simulation value is consistent with the theoretical value. The new CSMA protocol has higher throughput; while the network energy consumption is low, it is an excellent communication protocol.

**Keywords** Internet of Things · Average period analysis · Newton's solution of nonlinear equation · Throughput · Network energy consumption

## 1 Introduction

In wireless sensor networks [1], there are many nodes; however, the channel resources are limited and the energy consumption is limited. If the nodes send information without regularity, then the probability of packet conflict will increase greatly. However, if the CSMA protocol is used to intercept the channel and then transmit the information, the utilization of the channel will be greatly improved, the transmission efficiency of the communication system will be improved, and the energy consumption of the network will be reduced.

Z. Xu (✉) · H. Ding · Z. Yang · L. Bao · L. Liu
Information College, Yunnan University, Kunming 653000, Yunnan, China
e-mail: 775862901@qq.com

S. Zhou
School of Computer and Information, Kunming College of Metallurgy,
Kunming 650300, Yunnan, China

© Springer Nature Singapore Pte Ltd. 2019
B. K. Panigrahi et al. (eds.), *Smart Innovations in Communication and Computational Sciences*, Advances in Intelligent Systems and Computing 669, https://doi.org/10.1007/978-981-10-8968-8_30

There are three traditional CSMA protocols: 1-adherence to CSMA protocol [2], no- adherence to CSMA protocol [3], and P-persistent CSMA protocol [4]. But they all have some disadvantages; they are only single policy control protocol and cannot effectively use idle period in communication channel, which makes their system throughput lower and system energy consumption higher. Based on the theory of traditional CSMA protocol, proposed a new protocol that is dual clock probability detection of multi-channel CSMA protocol, which is a hybrid control strategy, not only can improve the utilization rate of the system idle period, and can according to different priorities, control network resources flexibly, to meet various requirements. This article carries on the analysis in two aspects: the first part of the system throughput and energy consumption, firstly combines the advantages of discrete continuous hybrid system [5], get the accurate mathematical formula by the average cycle analysis method, then the concept of multi-channel, through the Newton's method for solving nonlinear equations by mathematical expressions of the power consumption of the system [6]; the second part uses the MATLAB software simulation. The study protocol throughput and energy consumption compared with other traditional communication protocols, obtained results show that this protocol not only has a high throughput, load capacity, and channel access efficiency, more important is its low energy consumption, is a respected excellent communication network protocol.

## 2 Model Introduction

In the wireless communication system, there are N channels for all the end users, and the terminal uses new protocol technology. A multi-channel CSMA load balancing system model is shown in Fig. 1. The model has N priority, and priority is ascending order, from priority 1 to priority n; and in these priorities, each priority user number is not limited. Namely, if a user is in priority x, then his business channel can cover to channel 1 to x. And priority y in x channels of the arrival rate for $\lambda_y = \frac{G_x}{N-x+1}$ ($y \leq x$). At this time, the system load balancing and utilization rate of each channel are for $G_x = G(x = 1, 2, \ldots, N)$.

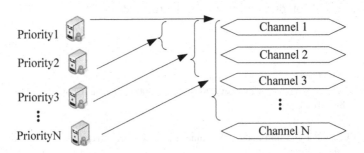

**Fig. 1** Multichannel load balancing system model

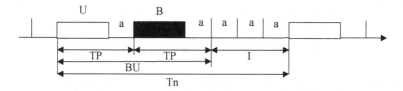

**Fig. 2** the function model of multi-channel CSMA based on double clock probability3

There are three events in the proposed dual clock probability detection multi-channel CSMA protocol: U event: information packets to send a successful event; B event: collision event in the information group; I event: a free event; the BU event that represents the joint event of a successful and sent collision. The function model of multi-channel CSMA based on double clock probability is shown in Fig. 2.

## 2.1 Throughput Analysis of the New Protocol

Firstly, we define the average length of the event $U_x$ as the message packet in section x successfully. The average length of the event is defined as $E[U_x]$.

In a communication system that accesses the multi-channel dual clock probability detection CSMA protocol, the X channel is taken as an example, if you want to send your message to success, there will be two cases:

(1) The last slot in the idle period is only one message to arrive, then the next time slot is sent successfully, which is denoted as $U_{x1}$; i and j, respectively, represent the number of consecutive idle and busy events on the channel, the average length of $U_{x1}$:

$$E(U_{x1}) = \frac{aG_x e^{-aG_x}}{1 - e^{-aG_x}} \tag{1}$$

Information is grouped in a busy period (TP), and this information is the only one in the current TP P probability to listen to the information packet; the information is sent to the next time slot success, and this event is recorded as $U_2$. After calculation:

$$E(U_{x2}) = \sum_{i=1}^{\infty} \sum_{j=1}^{\infty} \sum_{k=0}^{i-1} kP(i,j) = P(1+a)G_x \tag{2}$$

$$E(U_x) = E(U_{x1}) + E(U_{x2}) = \frac{aG_x e^{-aG_x}}{1 - e^{-aG_x}} + P(1+a)G_x \tag{3}$$

Secondly, the length of the busy period $BU_x$ of the channel x is analyzed:

$$E(BU_x) = \sum_{i=1}^{\infty} \sum_{j=1}^{\infty} (1+a) * i * P(i,j) = \frac{1+a}{e^{-P(1+a)G_x}} \qquad (4)$$

Finally, the average length of the idle period $E(I_x)$ of the X channel is analyzed:

$$E(I_x) = a\left(\frac{1}{1-e^{-aG_x}} - 1\right) + \frac{aG_x e^{-aG_x}}{1-e^{-aG_x}} * \frac{a}{2} + \frac{1-aG_x e^{-aG_x} - e^{-aG_x}}{1-e^{-aG_x}} * \frac{a}{2} \qquad (5)$$

As the system load balances, all channels have the same packet arrival rate: the $G_1 = G_2 = .. = G_N = G$. Therefore, according to the above analysis system throughput can all channel on the throughput of sum; if the set $E\left(U_x^{(p_y)}\right)$ for the average length of y packets in priority transmission channel x in the average period, because of channel load equalization, so the priority x in the channel y on the arrival rate is: $\lambda_x^{p_y} = G_x/(N-x+1)$. Then the same way:

$$E\left(U_x^{(p_y)}\right) = \frac{\lambda_x^{p_y}}{G_x} E(U_j) = \frac{1}{N-x+1}\left(\frac{aG_x e^{-aG_x}}{1-e^{-aG_x}} + P(1+a)G_x\right) \qquad (6)$$

According to the arrival rate of G equals, the throughput of priority y can be obtained:

$$S_{Py} = \left(\sum_{x=1}^{y} \frac{1}{N-x+1}\right) \frac{\frac{aG_x e^{-aG_x}}{1-e^{-aG_x}} + P(1+a)G_x}{a\left(\frac{1}{1-e^{-aG_x}} - 1\right) + \frac{aG_x e^{-aG_x}}{1-e^{-aG_x}} * \frac{a}{2} + \frac{1-aG_x e^{-aG_x} - e^{-aG_x}}{1-e^{-aG_x}} * \frac{a}{2} + \frac{1+a}{e^{-P(1+a)G_x}}} \qquad (7)$$

## 2.2 Energy Consumption Analysis of the New Protocol

Because the equation is nonlinear, we cannot get an exact analytical solution. Therefore, a series of approximate solutions are obtained by solving the nonlinear equations by Newton's method. When the load is lighter, P to maintain 1; when $G \geq 1$, P gradually reduced. On the basis of the above analysis, the approximate analytic solution of P can be obtained by using the idea of interval division and the maximum value of system throughput. At that time, P increased with G and reduced; so that the relationship between P and G is as follows:

$$p|_{MAX(s)} = \frac{1}{\beta \times G} \qquad (8)$$

In order to obtain the mathematical expression of the system power consumption, the power of the terminal station is set in three different states: ① when the power is in the transmission state $P_s$; ② when the power is in the listening state $P_l$; ③ the power of the receiving state is $P_r$. The system power is divided into the following situations:

the number of terminals in the system is M, and when the number of terminals in the idle slot a is 1, the system power is:

$$P_s + (M-1)P_r \tag{9}$$

In an average transmission cycle, the power required to transmit the packet successfully is:

$$P_U = \frac{E(U)}{E(T)} * [P_s + (M-1)P_r] \tag{10}$$

When the idle time slot a in the non-terminal to decide to send a packet, all terminal stations are listening to the channel; the system power is $MP_l$; then in an average transmission cycle, the channel is idle state of the system power is:

$$P_I = \frac{E(I)}{E(T)} * MP_l \tag{11}$$

Based on the previous analysis, the system is divided into (I, BU, Tn) average cycle period, so that the system power of the channel collision occurs in an average transmission cycle:

$$P_B = \left[ E(B)P_s + \left( M - \frac{1}{E(B)} \right) P_r \right] * \frac{1}{E(T)} \tag{12}$$

$$E(B) = E(BU) - E(U) * (1 + a) \tag{13}$$

Therefore, in an average transmission cycle, the average power of the system is:

$$P_A = P_U + P_B + P_I \tag{14}$$

## 3   Simulation Experiment and Result Analysis

Based on the above analysis, the mathematical expressions of the throughput and the average power of the system have been obtained. On this basis, the use of MATLAB 2012b simulates the results of the above experiments. Set node power consumption in different states in the sending state power is: $P_s = 70$ mw. In the listening state power: $P_l = 16$ mw. In the receiving state power: $P_r = 40$ mw.

The energy consumption and throughput of the new protocol and other random multiple access protocols are shown in Figs. 3 and 4. The following points can be obtained from the graph:

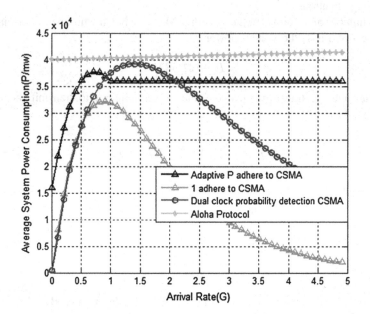

**Fig. 3** Comparison of the power consumption

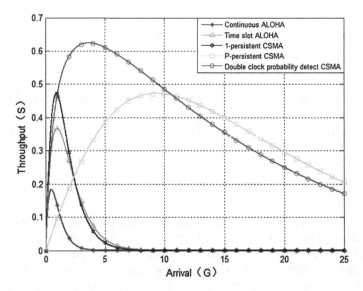

**Fig. 4** Comparison of throughput

1. The new protocol shows a good low energy consumption performance in the low arrival rate range.
2. Compared with the common single policy protocol (CSMA), the energy consumption of the new protocol is equal to that of the new protocol. However, in the case of the same energy consumption, the new protocol can have a higher throughput.
3. The power consumption of the new protocol has slightly higher because which is a hybrid control protocol strategy, combined with the continuous discrete system a bit, can greatly improve the throughput of the system, and the introduction of the multi-channel, to the reasonable allocation of the system channel, for the user to provide a variety of services.

# 4 Conclusion

The paper using Newton's method for solving nonlinear equations, the approximate solution is obtained and the energy consumption of the system, through mathematical model analysis, accurate mathematical expression is obtained, and then through the computer software simulation, a series of parameters of dual clock probability detection the channel CSMA protocol and other random multiple access protocol comparison; finally comes to the conclusion that this agreement is put forward in this paper, not only has a high system throughput, and can meet the demand of different information users, distribution communication network resources reasonably, so that the load capacity of the system improves the efficiency of information access. Greatly enhance, the most valuable is the double clock probability detection of multi-channel CSMA protocol that has a lower power consumption of the system. Protocol is very effective in this efficient energy saving and emission reduction of the communication era.

**Acknowledgements** Supported by the National Natural Science Foundation of China (61461053, 61461054, 61072079).

# References

1. Yuhong Zhang, Wei Li. Modeling and energy consumption evaluation of a stochastic wireless sensor network[J]. EURASIP Journal on Wireless Communications and Networking, 2012(1): 79–82.
2. Ding Hongwei, Zhao Dongfeng, Zhao Yifan. Monitoring of the Internet of things in the discrete 1-adhere to the CSMA protocol [C] International on Cellular Conference, Molecular Biology, Biophysics & Bioengineering.
3. Li Yingtao, Hao Chuan. Performance analysis of multichannel non persistent CSMA protocol [J]. Journal of Harbin Institute of Technology, 1995 (4): 61–67.

4. He Wei, Nan Jing Chang, Pan Feng. Improved dynamic p-adhere to CSMA protocol, [J].computer engineering, 2010, 36 (21): 118–120.
5. Guo Yingying. Analysis and research of BTMC-CSMA protocol with in wireless Ad Hoc network [D]. Yunnan University, China.
6. Xiaofei Zeng. Energy consumption of the communication network analysis and Research on energy saving technology of [J]. information technology, 2012, 10: 17.

# Detecting Primary User Emulation Attack Based on Multipath Delay in Cognitive Radio Network

Yongcheng Li, Xiangrong Ma, Manxi Wang, Huifang Chen and Lei Xie

**Abstract** Primary user emulation attack (PUEA) is a typical threat in the cognitive radio networks (CRN). In this paper, in order to detect the PUEA, we propose a detection method according to the characteristics of wireless channel. Using the multipath delay characteristics of the fading channels, PUEA can be detected by the secondary users (SUs). The performance of the proposed PUEA detection method is theoretically analyzed, and the closed-form expression of detection performance is derived. Simulation results illustrate that the proposed PUEA detection method achieves high detection probability in terms of low false positive probability.

**Keywords** Cognitive radio network · Primary user emulation attack (PUEA) Channel characteristic

## 1 Introduction

In order to mitigate the increasingly problem of spectrum resource scarcity, cognitive radio (CR) has been proposed to efficiently access the spectrum occupied by the primary user (PU). In the cognitive radio network (CRN), secondary users (SUs) can access the spare spectrum dynamically. The prerequisite of this mechanism is that the SUs cannot interfere with the primary network.

The cognitive nature of the CRN introduces new security issues, where the primary user emulation attack (PUEA) is a typical denial-of-service threat [1].

Y. Li · M. Wang · H. Chen (✉)
State Key Laboratory of Complex Electromagnetic Environment Effects
on Electronics and Information System, Luoyang 471003, China
e-mail: chenhf@zju.edu.cn

X. Ma · H. Chen · L. Xie
College of Information Science and Electronic Engineering, Zhejiang University,
Hangzhou 310027, China

© Springer Nature Singapore Pte Ltd. 2019
B. K. Panigrahi et al. (eds.), *Smart Innovations in Communication and Computational Sciences*, Advances in Intelligent Systems and Computing 669, https://doi.org/10.1007/978-981-10-8968-8_31

In the PUEA, the attacker emulates the PU to transmit the signal in order to deterring SUs from accessing the unoccupied spectrum band. There are two types of PUEA, selfish and malicious. The selfish PUEA aims to occupy the unoccupied spectrum band, while the malicious PUEA is to obstruct the normal operation of the CRN.

Many researchers have come up with different kinds of approaches to defense the PUEA [2–11]. In [2], a localization-based authentication scheme is proposed to mitigate the PUEA in TV spare spectrum. If the received signal is similar to the TV signal, a received signal strength measurement based on localization is performed. PUEA is declared to be presented if the location of the transmitter is different from that of the TV tower. In [3], an analytical model is presented to analyze the successful detection probability of the PUEA detection method based on the energy detection mechanism. However, authors in [4] proposed a strategy to invalidate the method proposed in [3]. To defense this PUEA strategy presented in [4], an advanced defense method using the variance of the power of the received signal is proposed. In [5, 6], authors assume that the CRN has multiple channels, an honest SU senses and intelligently selects one channel to access at each time slot in order to avoid the PUEA. In [7], a passive and nonparametric classification method is proposed to decide whether an emulated signal exists or not using the frequency deviation of the transmitter. In [8, 9], a PUEA defense method is proposed by permitting a shared key between the PU and the SU, which is based on the cross-correlation between the received signal and locally generated signal. In [10], a game theory-based approach is used to defense the PUEA, and the equilibrium point is determined for the game. In [11], authors proposed the non-cooperative and cooperative detection methods in terms of the channel-tap power between the transmitter and the receiver.

In the PUEA defense methods mentioned above, the detection of the PUEA is mainly using the mean and/or variance of the received signal energy, the transmitter fingerprint, or the channel-tap power. When these characteristics are obtained by the attacker, a PU signal can be emulated by reconfiguring the parameters. Although the variance characteristic or channel-tap power is difficult to emulate for an attacker, the variance characteristic needs a long-time observation to obtain the well-performed estimation, and the channel-tap power is time variant.

In this paper, a novel detection method is proposed to detect the PUEA, where the characteristic of the fading channel is utilized to determine the received signal is sent by a PU or an attacker. According to the fact that the signal received by an SU experiences different propagation paths, the multipath delay can be estimated as the intrinsic feature of fading channel for distinguishing the PU from the attacker. The performance of the PUEA detection method is analyzed and validated using Monte Carlo simulations and experimental hardware implementations.

## 2 System Model

In this section, we introduce the system model, including the network model, the channel model, and a channel estimation method.

### 2.1 Network Model

Figure 1 shows a simplified network model including a PU, an attacker, and multiple SUs. If the PU is working, SUs and the attacker cannot access the spectrum occupied by the PU network. When the PU does not transmit the signal, the attacker can transmit an emulated PU signal in order to deter the SU to access the unoccupied spectrum. Moreover, it is assumed that time-domain synchronous Orthogonal Frequency Division Multiplexing (TDS-OFDM) technology is used in the PU network.

In this work, it is assumed that the attacker and SUs have a priori knowledge of the PU signal in terms of the frequency band and the frame structure. The attacker can reconfigure the emission parameters after the spectrum sensing process, which nullifies the PUEA detection methods using the signal energy or the transmitter fingerprint.

### 2.2 Channel Model

We assume that the CRN operates in an urban environment, which means that the wireless signal transmitted from the PU or the attacker should consider not only the path loss and shadowing, but also the multipath fading. Hence, the signal received at an SU has different strength and propagation delay. Compared with the channel-tap power, the multipath delay is relatively stationary. Hence, the multipath delay is chosen as a channel characteristic for the PUEA detection.

**Fig. 1** A simplified network model

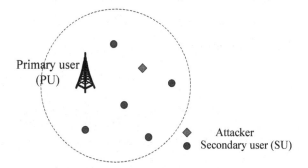

As we know, the propagation delay is spatially irrelevant between different geographic locations [12]. Hence, the multipath delay between the PU and the SU is independent of that between the attacker and the SU.

The multipath channel can be modeled as an impulse response filter. That is,

$$h(t) = \sum_{l=0}^{L-1} h_l \delta(t - t_l), \tag{1}$$

where $h_l$ and $t_l$ denote the amplitude and the delay of the $l$th path, respectively; $L$ denotes the number of propagation paths.

In [13], the propagation delay is modeled by a Poisson process that models the arrival times of multipath components. The time interval between a pair of adjacent paths, which is described as multipath delay, $\tau_l = t_{l+1} - t_l$, $l = 0, 1, \ldots, L-2$, follows an exponential distribution with parameter $\Lambda$.

It is well known that the distribution of the square root of an exponential random variable is Rayleigh. The Rayleigh distribution can be expressed as a square root of the sum of squares of two independent Gaussian random variables. Hence, $\tau_l$ can be mathematically expressed by

$$\tau_l = (\tau_R)^2 + (\tau_I)^2, \tag{2}$$

where $\tau_R$ and $\tau_I$ are independent Gaussian variables, i.e., $\tau_R \sim \mathcal{N}(0, \frac{1}{2\Lambda})$ and $\tau_I \sim \mathcal{N}(0, \frac{1}{2\Lambda})$.

## 2.3 Channel Estimation

In this work, Fig. 2 shows the frame structure of the PU signal. Many channel estimation approaches have been proposed based on the PN sequence [14, 15].

As shown in Fig. 2, each frame has two parts, the head with length $L_{PN}$ and the body with length $L_d$. The head consists of a preamble with length $L_{pre}$, a pseudo-noise (PN) sequence with length $L_m$, and a post-amble with length $L_{post}$ [16]. The PN sequence is generated by an m-sequence, and the preamble and post-amble are the cyclical extensions of m-sequence.

**Fig. 2** Frame structure of the PU signal

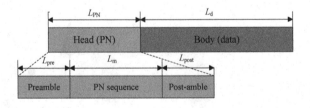

According to the frame structure, the $n$th sample of signal $s(t)$, $s(n)$, is denoted as

$$s(n) = PN(n) + d(n), \tag{3}$$

where $$PN(n) = \begin{cases} p(n), & 0 \le n < L_{PN} \\ 0, & \text{otherwise} \end{cases}$$ and

$$d(n) = \begin{cases} s(n), & L_{PN} \le n \le L_{PN} + L_d - 1 \\ 0, & \text{otherwise} \end{cases}.$$

Based on the priori knowledge of the frame head, the SU and the attacker can regenerate the m-sequence denoted as $C(n)$. The cross-correlation between the received signal and the local generated m-sequence can be computed as

$$R_{yC}(n) = [s(n) * h(n) + w(n)] \otimes C(n) = s(n) \otimes C(n) * h(n) + w'(n), \tag{4}$$

where $*$ denotes the linear convolution operation, $\otimes$ denotes the correlation operation, and $w'(n)$ is the cross-correlation between $w(n)$ and $C(n)$.

Substituting (3) into (4), the linear cross-correlation between $s(n)$ and $C(n)$, $R_{sC}(n)$, is

$$R_{sC}(n) = s(n) \otimes C(n) = PN(n) \otimes C(n) + d(n) \otimes C(n) = R_{pC}(n) + R_{dC}(n), \tag{5}$$

where $R_{pC}(n)$ is the cross-correlation between $PN(n)$ and $C(n)$, and $R_{dC}(n)$ is the cross-correlation between $d(n)$ and $C(n)$.

The auto-correlation of $C(n)$ is denoted as $$R_C(n) = \sum_{i=0}^{L_m - 1} C(i) C\left((n+i)_{L_m}\right) = \begin{cases} L_m, & n = 0 \\ -1, & n \ne 0 \end{cases}, n = 0, 1, \ldots, L_m - 1,$$ where $C\left((n+i)_{L_m}\right)$ is the periodic extension sequence of $C(n)$ with length $L_m$. Hence, $R_{pC}(n)$ can be denoted as

$$R_{pC}(n) = \sum_{i=0}^{L_m - 1} C(i) PN(n+i) = \begin{cases} -1, & 0 \le n < L_{pre} \\ L_m, & n = L_{pre} \\ -1, & L_{pre} < n \le L_{pre} + L_{post} \\ noise, & \text{otherwise} \end{cases}, \tag{6}$$

Hence, $R_{pC}(n)$ can be represented as an ideal impulse response with an additive sequence as,

$$R_{pC}(n + L_{pre}) = G\delta(n) + I(n), n = -L_{pre}, -L_{pre} + 1, \ldots, L_{post}, \tag{7}$$

where $G = R_{pC}(L_{pre})$, and $I(n) = \begin{cases} 0, & n = 0 \\ R_{pC}(n + L_{pre}), & \text{otherwise} \end{cases}.$

When the sampling rate $f_s$ at the receiver is high enough, $R_{yC}(n)$ can be rewritten as

$$R_{yC}(n) = \underbrace{\sum_{l=0}^{L-1} Gh_l\delta(n - L_{\text{pre}} - t_l f_s)}_{\text{term 1}} + \underbrace{\sum_{l=0}^{L-1} h_l I(n - L_{\text{pre}} - t_l f_s)}_{\text{term 2}} + \underbrace{R_{dC}(n)*h(n)}_{\text{term 3}} + \underbrace{w'(n)}_{\text{term 4}},$$

(8)

where term 1 includes the correlation peaks for estimating $h_l$; term 2 contains the inter-path interference caused by the additive sequence $I(n)$; term 3 introduces an interference resulting from the time spread of random data; and term 4 is the AWGN noise.

When the maximum channel delay is smaller than $L_{\text{post}}$, the amplitude of each path can be estimated as

$$\hat{h}_l \approx \frac{R_{yC}(L_{\text{pre}} + t_l f_s)}{G},$$

(9)

That is, the positions of peak in the range from $L_{\text{pre}}$ to $L_{\text{pre}} + L_{\text{post}}$ are the corresponding delay, $\{t_l\}_{l=0}^{L-1}$. Hence, using the cross-correlation in (8), the multipath delay can be estimated.

## 3  Proposed PUEA Detection Method

From the network model introduced in Sect. 2.1, the location of the PU is different from that of the attacker. Therefore, the propagation delay can be exploited as an intrinsic characteristic of the wireless channel to distinguish different signal transmitters.

According to the channel model introduced in Sect. 2.2, the multipath delay of fading channel from the PU to the SU can be expressed as

$$\tau_l^P = \left(\tau_R^P\right)^2 + \left(\tau_I^P\right)^2,$$

(10)

where $\tau_R^P$ and $\tau_I^P$ are two independent Gaussian variables, i.e., $\tau_R^P \sim \mathcal{N}(0, \sigma_P^2)$ and $\tau_I^P \sim \mathcal{N}(0, \sigma_P^2)$.

Similarly, the time interval between first two paths of the fading channel from the MU to the SU can be expressed as

$$\tau_l^M = \left(\tau_R^M\right)^2 + \left(\tau_I^M\right)^2,$$

(11)

where $\tau_R^M$ and $\tau_I^M$ are two independent Gaussian variables, i.e., $\tau_R^M \sim \mathcal{N}(0, \sigma_M^2)$ and $\tau_I^M \sim \mathcal{N}(0, \sigma_M^2)$.

Using the channel estimation method described in Sect. 2.3, the multipath delay at time slot $k$, $\hat{\tau}_l(k)$, can be estimated. And two Gaussian random variables $\tau_R$ and $\tau_I$ in (2) are modeled, respectively, by

$$\begin{cases} \hat{\tau}_R(k) \triangleq \tau_R + \omega_R(k) \\ \hat{\tau}_I(k) \triangleq \tau_I + \omega_I(k) \end{cases},$$

(12)

where $\omega_R(k)$ and $\omega_I(k)$ are two independent estimation error. It is assumed that $\omega_R$ and $\omega_I$ are Gaussian random variables, where $\omega_R \sim \mathcal{N}(0, \sigma_\omega^2)$ and $\omega_I \sim \mathcal{N}(0, \sigma_\omega^2)$. And the estimated time interval still satisfies the exponential distribution with parameter $\Lambda^{'} = \frac{1}{2\sigma^2 + 2\sigma_\omega^2}$, where $\sigma^2 = \sigma_P^2$ if the received signal is transmitted by the PU; $\sigma^2 = \sigma_M^2$ if the attacker is transmitting. That is, $\hat{\tau}_l(k) \sim \exp\left(\frac{1}{2\sigma^2 + 2\sigma_\omega^2}\right)$.

A binary hypothesis test is constructed to determine that the received signal is transmitted by the PU or the attacker. The binary hypothesis test can be expressed as

$$\begin{cases} array*20cH_0: \hat{\tau}_l(k) = \left(\tau_R^P + \omega_R(k)\right)^2 + \left(\tau_I^P + \omega_I(k)\right)^2 H_1: \hat{\tau}_l(k) = \left(\tau_R^M + \omega_R(k)\right)^2 + \left(\tau_I^M + \omega_I(k)\right)^2, \end{cases}$$

(13)

where $H_0$ and $H_1$ denote the hypotheses that the PU is transmitting and the attacker is transmitting, respectively.

In the proposed method, the time interval between first two paths is used as the feature to detect the PUEA. The difference between the current estimated propagation delay $\hat{\tau}_0(k)$ and the predicted propagation delay $\tau_0^P$ is defined as the test statistic. Hence, SU performs the binary hypothesis test given as

$$\left|\hat{\tau}_0(k) - \tau_0^P\right| \underset{\mathcal{H}_0}{\overset{\mathcal{H}_1}{\gtrless}} \lambda,$$

(14)

where $\lambda$ is the pre-defined threshold of the PUEA detection method; $\mathcal{H}_0$ and $\mathcal{H}_1$ denote the inferences that the PU is transmitting and the attacker is transmitting, respectively.

The performance of the proposed PUEA detection method is commonly measured by two probabilities, the false positive probability and the false negative probability.

The false positive probability, $P_{FP}$, is defined as the probability conditioned $H_0$ that the received signal is inferred to be transmitted by an attacker. That is,

$$P_{FP} = \Pr\left(\left|\hat{\tau}_0(k) - \tau_0^P\right| \geq \lambda | H_0\right).$$

(15)

The false negative probability, $P_{FN}$, is defined as the probability conditioned $H_1$ that the received signal is inferred to be transmitted by the PU. That is,

$$P_{FN} = \Pr\left(\left|\hat{\tau}_0(k) - \tau_0^P\right| \le \lambda | H_1\right). \tag{16}$$

For the sake of obtaining the closed-form expressions of $P_{FP}$ and $P_{FN}$, the distribution of the test statistic is needed. If the PU exists, the test statistic in (16) is further rewritten as

$$\hat{\tau}_0(k) - \tau_0^P = (\tau_R^P + \omega_R(k))^2 + (\tau_I^P + \omega_I(k))^2 - (\tau_R^P)^2 - (\tau_I^P)^2$$
$$= \underbrace{(\omega_R(k))}_{P_1} \cdot \underbrace{(2\tau_R^P + \omega_R(k))}_{P_2} + \underbrace{(\omega_I(k))}_{P_3} \cdot \underbrace{(2\tau_I^P + \omega_I(k))}_{P_4}. \tag{17}$$

In (17), since $\omega_R(k)$, $\omega_I(k)$, $\tau_R^P$, and $\tau_I^P$ are independent Gaussian distributions, the random variables, $P_1$, $P_2$, $P_3$, and $P_4$, are also Gaussian variables with mean zero and variances $\sigma_\omega^2$, $\sigma_\omega^2 + 4\sigma_P^2$, $\sigma_\omega^2$ and $\sigma_\omega^2 + 4\sigma_P^2$.

Obviously, $P_1$ and $P_2$ are independent from $P_3$ and $P_4$. The correlation coefficient between $P_1$ and $P_2$ equals to that between $P_3$ and $P_4$. That is,

$$\rho_P = \frac{\sigma_\omega}{\sqrt{\sigma_\omega^2 + 4\sigma_P^2}}. \tag{18}$$

Based on the distribution given in [17, Eq. (6.19)], the probability distribution function of the test statistic in (17) is denoted as

$$F_{P_1 P_2 + P_3 P_4}(x) = \begin{cases} \frac{1-\rho_P}{2} \exp\left(\frac{x}{\sqrt{\sigma_\omega^2(\sigma_\omega^2 + 4\sigma_P^2)}(1-\rho_P)}\right), & x < 0 \\ 1 - \frac{1+\rho_P}{2} \exp\left(-\frac{x}{\sqrt{\sigma_\omega^2(\sigma_\omega^2 + 4\sigma_P^2)}(1+\rho_P)}\right), & x \ge 0 \end{cases}. \tag{19}$$

With (19), the closed-form expression of $P_{FP}$ can be expressed as

$$P_{FP} = 1 - F_{P_1 P_2 + P_3 P_4}(\lambda) + F_{P_1 P_2 + P_3 P_4}(-\lambda)$$
$$= \left(\frac{1+\rho_P}{2}\right) \exp\left(-\frac{\lambda}{\sqrt{\sigma_\omega^2(\sigma_\omega^2 + 4\sigma_P^2)} + \sigma_\omega^2}\right) + \frac{1-\rho_P}{2} \exp\left(-\frac{\lambda}{\sqrt{\sigma_\omega^2(\sigma_\omega^2 + 4\sigma_P^2)} - \sigma_\omega^2}\right). \tag{20}$$

Similarly, when the attacker exists, the distribution of the test statistic can be denoted by

$$
\begin{aligned}
\hat{\tau}_0(k) - \tau_0^P &= \left(\tau_R^M + \omega_R(k)\right)^2 + \left(\tau_I^M + \omega_I(k)\right)^2 - \left(\tau_R^P\right)^2 - \left(\tau_I^P(k)\right)^2 \\
&= \underbrace{\left(\tau_R^M - \tau_R^P + \omega_R(k)\right)}_{M_1} \cdot \underbrace{\left(\tau_R^M + \tau_R^P + \omega_R(k)\right)}_{M_2} + \underbrace{\left(\tau_I^M - \tau_I^P + \omega_I(k)\right)}_{M_3} \cdot \underbrace{\left(\tau_I^M + \tau_I^P + \omega_I(k)\right)}_{M_4}.
\end{aligned}
$$
(21)

Since $M_1$, $M_2$, $M_3$, and $M_4$ are Gaussian distribution with mean zero and variance $\sigma_P^2 + \sigma_M^2 + \sigma_\omega^2$. $M_1$ and $M_2$ are independent from $M_3$ and $M_4$. The probability distribution function of the test statistic in (21) is

$$
F_{M_1M_2+M_3M_4}(x) = \begin{cases} \dfrac{1-\rho_M}{2}\exp\left(\dfrac{x}{\left(\sigma_P^2+\sigma_M^2+\sigma_\omega^2\right)(1-\rho_M)}\right), & x < 0 \\ 1 - \dfrac{1+\rho_M}{2}\exp\left(-\dfrac{x}{\left(\sigma_P^2+\sigma_M^2+\sigma_\omega^2\right)(1+\rho_M)}\right), & x \geq 0 \end{cases},
$$
(22)

where $\rho_M$ is the correlation coefficient between random variables $M_1$ and $M_2$. Similarly, the correlation coefficient between $M_1$ and $M_2$ equals to that between $M_3$ and $M_4$. That is,

$$
\rho_M = = \frac{\sigma_\omega^2 + \sigma_M^2 - \sigma_P^2}{\sigma_\omega^2 + \sigma_P^2 + \sigma_M^2}.
$$
(23)

With (23), the closed-form expression of $P_{FN}$ can be expressed as

$$
\begin{aligned}
P_{FN} &= F_{M_1M_2+M_3M_4}(\lambda) - F_{M_1M_2+M_3M_4}(-\lambda) \\
&= 1 - \left(\frac{1+\rho_M}{2}\right)\exp\left(-\frac{\lambda}{2\sigma_\omega^2+2\sigma_M^2}\right) - \frac{1-\rho_M}{2}\exp\left(-\frac{\lambda}{2\sigma_P^2}\right).
\end{aligned}
$$
(24)

In order to find the optimal choice of $\lambda$ in (14), we adopt the Neyman-Pearson framework in the context of the PUEA detection, where the goal is to minimize $P_{FN}$ subject to the condition that $P_{FP} \leq \bar{P}_{FP}$. Hence, the pre-defined threshold can be approximately calculated as

$$
\lambda \approx -\ln(P_{FP})\sigma_\omega^2\left(\sqrt{1+4\frac{\sigma_P^2}{\sigma_\omega^2}}\right).
$$
(25)

# 4 Numerical Results

In this section, the performance of the proposed PUEA detection method based on the multipath delay is first evaluated by computer simulations. Then, the proposed PUEA detection method is validated on a software defined radio (SDR) platform.

According to the theoretical analysis of the performance of the PUEA detection method in Sect. 3, we plot the receiver operating characteristics (ROC) curves. We obtain the different values of $P_{FP}$ and $P_{FN}$ by varying threshold $\lambda$. According to the multipath delay estimated in the experiments of [13], $\Lambda_P = 10^6$.

Figure 3 shows the ROC curves, where Fig. 3a shows the impact of $\Lambda_M = 1/\sigma_M^2$ on the performance, and $\sigma_\omega^2 = 0.01\sigma_P^2$; Fig. 3b shows the impact of $\sigma_\omega^2$ on the performance, and $\Lambda_M = 8\Lambda_P$. From Fig. 3a, one finds that $P_{FN}$ decreases as $\Lambda_M$ increases. Moreover, from Fig. 3b, one finds that $P_{FN}$ increases as the estimation error decreases.

The performance of the multipath delay estimation method introduced in Sect. 2.3 is evaluated. The parameters of the frame structure are listed in Table 1 and the parameters of the fading channel from PU to SU are set as in Table 2. From Table 2, propagation paths 1 and 2 are two paths used for the multipath delay estimation. Let $N_s$ be the number of sampling times within an OFDM symbol.

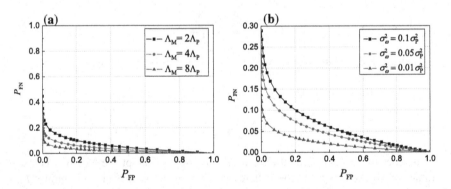

**Fig. 3** ROC curves. **a** ROC curves under different $\Lambda_M$. **b** ROC curves under different estimation errors

**Table 1** Parameters of frame structure

| Parameter | Value |
|---|---|
| Symbol period | 0.04 μs |
| FFT size, $L_d$ | 3780 |
| Frame head length, $L_{PN}$ | 945 |
| PN sequence length, $L_m$ | 511 |
| Preamble length, $L_{pre}$ | 217 |
| Post-amble length, $L_{post}$ | 217 |

**Table 2** Channel profile

| Path no. | Delay (μs) | Power (dB) |
|---|---|---|
| 1 | 0.0 | 0 |
| 2 | −1.8 | −18 |
| 3 | 0.15 | −20 |
| 4 | 1.8 | −20 |
| 5 | 5.7 | −10 |
| 6 | 8.0 | 0 |

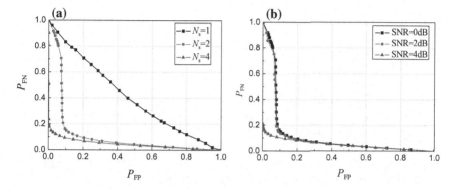

**Fig. 4** ROC curves. **a** ROC curves under different $N_s$. **b** ROC curves under different $SNR$

Figure 4 shows the ROC curves, where Fig. 4a shows the impact of $N_s$ on the performance, and $SNR = 0$ dB; Fig. 4b shows the impact of SNR on the performance, and $N_s = 2$. The time interval between first two paths of the attacker varies from 0 to 3.6 μs. From Fig. 4, we find that as the $P_{FP}$ increases, the $P_{FN}$ decreases obviously. Moreover, from Fig. 4a, we find that the $P_{FN}$ decreases as $N_s$ increases. For example, for $P_{FP} = 0.1$, $N_s = 4$ should be set at the SU to make $P_{FN} \leq 0.1$. However, for $P_{FP} = 0.2$, $N_s = 2$ should be set at the SU to make $P_{FN} \leq 0.1$. In addition, from Fig. 4b, the false negative probability decreases as $SNR$ increases.

From the discussions mentioned above, it can be concluded that the detection performance improves by increasing the sampling rate and the received SNR. The detection probability can achieve more than 0.9 when $N_s = 4$ and $SNR = 0$ dB.

Furthermore, the proposed PUEA detection method is validated a real wireless communication environment by using the Universal Software Radio Peripheral (USRP) SDR platform. The validation scenario is illustrated in Fig. 5. Three users, PU, attacker, and SU, are located at different places in a big room, as illustrated in Fig. 5b. The parameters of the signal are set as in Table 1, and $N_s = 4$.

The experimental results show that as the false positive probability is set as 0.1, the detection probability can achieve about 93.55% for the proposed PUEA detection based on multipath delay.

(a)                                          (b)

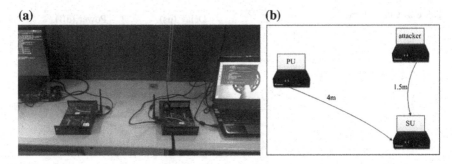

**Fig. 5** Validation scenario. **a** USRP-based user. **b** The locations of three users

## 5   Conclusion

A multipath delay-based PUEA detection method is proposed in our work. Since the intrinsic feature of the fading channel in the CRN cannot be emulated by reconfiguring, the proposed PUEA detection method uses the multipath delay of the fading channel as the feature to distinguish the PU from the attacker. Simulation and analytical results demonstrate that the proposed PUEA detection approach performs well with appropriate sampling rate and received SNR.

**Acknowledgements** This work was partly supported by fund of the State Key Laboratory of Complex Electromagnetic Environment Effects on Electronics and Information System (No. CEMEE2015Z0202A), National Natural Science Foundation of China (No. 61471318, 61671410) and the Zhejiang Provincial Natural Science Foundation of China (No. LY14F010014).

## References

1. R. Chen and J.-M. Park: Ensuring Trustworthy Spectrum Sensing in Cognitive Radio Networks. In: Proc. of IEEE Workshop Networking Technology and Software Defined Radio Networks, pp. 110–119, Reston, USA (2006)
2. R. Chen, J. M. Park and J. H. Reed: Defense against Primary User Emulation Attacks in Cognitive Radio Networks. IEEE Journal on Selected Areas in Communications, vol. 26, no. 1, pp. 25–37 (2008)
3. S. Anand, Z. Jin and K. P. Subbalakshmi: An Analytical Model for Primary User Emulation Attacks in Cognitive Radio Networks. In: Proc. of the 3rd IEEE Symposium New Frontiers Dynamic Spectrum Access Networks, pp. 1–6, Chicago, USA (2008)
4. Z. Chen, T. Cooklev, C. Chen and C. P. Raez: Modeling Primary User Emulation Attacks and Defenses in Cognitive Radio Networks. In: Proc. of the 28th International Performance Computing and Communications Conference, pp. 208–215, Scottsdale, USA (2016)
5. H. Li and Z. Han: Dogfight in Spectrum: Jamming and Anti-jamming in Cognitive Radio Systems. In Proc. of IEEE Conf. Global Commun., pp. 1–6, Hawaii, USA (2009)
6. H. Li and Z. Han: Dogfight in Spectrum: Combating Primary User Emulation Attacks in Cognitive Radio Systems, Part I: Known Channel Statistics. IEEE Transaction on Wireless Communications, vol. 9, no. 11, pp. 3566–3577 (2010)

7. T. Le, W. Chin and C. Tsai: Channel-Based Detection of Primary User Emulation Attacks in Cognitive Radios. In: Proc. of VTC Spring, pp. 1–5, Yokohama, Japan (2012)
8. N. Nguyen, R. Zheng, and Z. Han: On Identifying Primary User Emulation Attacks in Cognitive Radio Systems using Nonparametric Bayesian Classification. IEEE Transaction on Signal Processing, vol. 60, no. 3, pp. 1432–1445 (2012)
9. A. Alahmadi, Z. Fang, T. Song and T. Li: Subband PUEA Detection and Mitigation in OFDM-Based Cognitive Radio Networks. IEEE Transactions on Information Forensics and Security, vol. 10, no. 10, pp. 397–402 (2015)
10. N. Nguyen-Thanh, P. Ciblat, A T. Pham and V. Nguyen: Surveillance Strategies Against Primary User Emulation Attack in Cognitive Radio Networks. IEEE Transactions on Wireless Communications, vol. 14, no. 9, pp. 4981–4993 (2015)
11. T. Le, W. Chin and Y. Lin: Non-cooperative and Cooperative PUEA Detection using Physical Layer in Mobile OFDM-based Cognitive Radio Network". In Proc. of IEEE Conf. on Computing, Networking and Commun., pp. 1–5, Kauai, USA (2016)
12. W. C. Jakes: Microwave Mobile Communications. Wiley (1974)
13. A. A. Saleh and R. A. Valenzuela: A Statistical Model for Indoor Multipath Propagation. IEEE Journal on Selected Areas in Communications, vol. 5, no. 2, pp. 128–137 (1987)
14. B. Song, L. Gui, Y. Guan and W. Zhang: On Channel Estimation and Equalization in TDS-OFDM Based Terrestrial HDTV Broadcasting System. IEEE Transaction on Consumer Electronics, vol. 51, no. 3, pp. 790–797 (2005)
15. G. Liu and J. Zhang: ITD-DFE Based Channel Estimation and Equalization in TDS-OFDM Receivers, IEEE Transaction on Consumer Electronics, vol. 53, no. 2, pp. 304–309 (2007)
16. Framing Structure, Channel Coding and Modulation for Digital Television Terrestrial Broadcasting System. Chinese National Standard GB 20600–2006 (2006)
17. M. K. Simon: Probability Distributions Involving Gaussian Random Variables: A Handbook for Engineers and Scientists. Springer (2002)

# Part III
# Computational Sciences

# Optimization of Multi-objective Dynamic Optimization Problems with Front-Based Yin-Yang-Pair Optimization

Varun Punnathanam and Prakash Kotecha

**Abstract** Multi-objective Dynamic Optimization Problems (MDOP) are a set of challenging engineering problems in which one or more of the terms in the problem are dependent on independent variables. In this work, we employ a recently proposed stochastic multi-objective optimization algorithm, Front-based Yin-Yang-Pair Optimization, to solve such problems. The algorithm is applied on three Multi-objective Dynamic Optimization Problems (MDOP) from literature: (i) a batch reactor, (ii) a plug flow reactor and (iii) a fed-batch reactor problem. F-YYPO is able to determine efficient Pareto curves for the MDOP problems and shows competitive performance with literature results.

**Keywords** Multi-objective optimization · Stochastic optimization
Multi-objective dynamic optimization problems

## 1 Introduction

Dynamic optimization problems involve objectives which can vary with respect to independent variables, such as time or spatial coordinates. Optimal control problems are dynamic optimization problems, where the control variables at specific time intervals are to be set by the optimization algorithm. In many real-world cases, such problems are accompanied with multiple objectives, such as maximizing/minimizing the concentrations of various products/by-products in a reactor or minimizing the utilities consumed by a facility. These objectives frequently involve simultaneous ordinary or partial differential equations which render them to be challenging optimization problems as the value of the decision variable at a specific interval will be influenced by its value at the previous intervals. Stochastic optimization techniques are popularly employed on these problems due to their flexibility in handling such objectives [1–3].

V. Punnathanam · P. Kotecha (✉)
Indian Institute of Technology Guwahati, Guwahati 781039, Assam, India
e-mail: pkotecha@iitg.ernet.in

© Springer Nature Singapore Pte Ltd. 2019
B. K. Panigrahi et al. (eds.), *Smart Innovations in Communication and Computational Sciences*, Advances in Intelligent Systems and Computing 669, https://doi.org/10.1007/978-981-10-8968-8_32

There are various methods available to solve multi-objective problems with the aid of single objective optimization algorithms [4]. Techniques such as the weighted-sum method or the ε-constraint technique solve numerous single objective instances to obtain the final Pareto front. On the other hand, stochastic multi-objective optimization techniques [5, 6] such as Non-dominated Sorting Genetic Algorithm-II (NSGA-II) [7] and Multi-Objective Differential Evolution (MODE) [8] are capable of obtaining the Pareto front in a single run.

Front-based Yin-Yang-Pair Optimization (F-YYPO) [9] is a recently proposed multi-objective stochastic optimization algorithm which was shown to perform well on problems related to Stirling engine systems. The algorithm is based on Yin-Yang-Pair Optimization [10] and emphasizes on low computational complexity. In this work, we consider the optimization of three MDOP problems by utilizing F-YYPO and subsequently compare with the results in literature. This paper is structured in the following manner: the F-YYPO algorithm is briefly discussed in Sect. 2, followed by the problem description of the multi-objective dynamic optimization problems in Sect. 3. Subsequently, we present the results in Sect. 4 and conclude the work in Sect. 5.

## 2  Front-Based Yin-Yang-Pair Optimization

For the sake of brevity, we restrict the description of F-YYPO as a detailed description is available in literature [11, 12]. The algorithm utilizes two stages for search: (i) the Splitting Stage and (ii) the Archive Stage. The flowchart for the main algorithmic framework is provided in Fig. 1, and the pseudo-code for the Splitting Stage is provided in Fig. 2. Note that the parameters ($I_{min}$, $I_{max}$ and $\alpha$), maximum number of Pareto Points ($P_{max}$), the termination criteria and the lower and upper bounds of the decision variables are to be provided by the user.

The algorithm generates new points viz. the Splitting Stage, which is very similar to the Splitting Stage from YYPO. Here, $2D$ (where $D$ is the number of decision variables of the problem) new points are generated for each point entering the Splitting Stage. Subsequently, the fittest of the $2D$ point (with respect to its respective objective) replaces the entering point. Note that the input is a point $Q$, the pair number ($q$) to which $Q$ belongs and the corresponding search radius $\delta$. The output is the updated point $Q$ and its fitness values. The equations for Figs. 1 and 2 are as follows:

$$x^{'} = x_l + x(x_u - x_l) \tag{1}$$

$$\left.\begin{array}{l} S_j^j = S^j + r_j\delta \\ S_{D+j}^j = S^j - r_{D+j}\delta \end{array}\right\} \forall\, j = 1, 2, 3 \ldots D \tag{2}$$

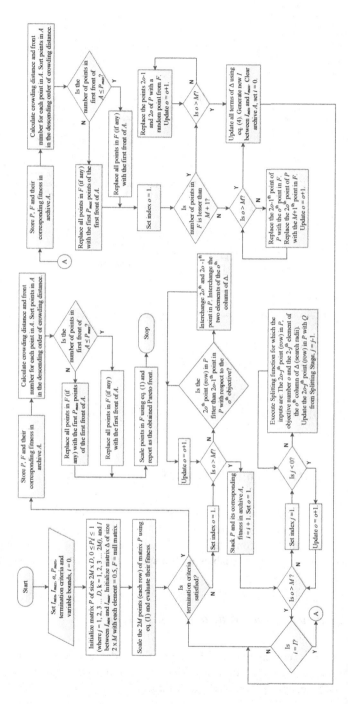

**Fig. 1** Flowchart for F-YYPO

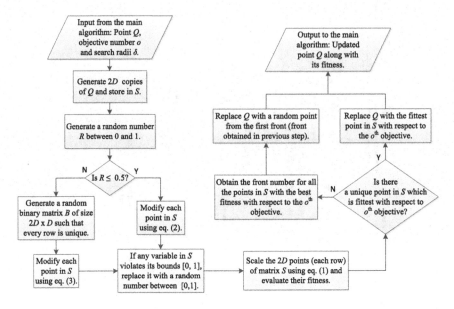

**Fig. 2** Flowchart for splitting stage

$$\left. \begin{array}{l} S_k^j = S^j + r_k\left(\delta/\sqrt{2}\right) \quad if \; B_k^j = 1, \\ S_k^j = S^j - r_k\left(\delta/\sqrt{2}\right) \quad otherwise \end{array} \right\} \forall \; k = 1,2,3\ldots 2D \; and \; j = 1,2,3\ldots D \qquad (3)$$

$$\left. \begin{array}{l} \Delta(1,o) = \Delta(1,o) - (\Delta(1,o)/\alpha) \\ \Delta(2,o) = \min(\Delta(2,o) + (\Delta(2,o)/\alpha), 0.75) \end{array} \right\} \forall o = 1,2,3\ldots M \qquad (4)$$

# 3  Multi-objective Dynamic Optimization Problems

In this section, we study the application of F-YYPO on solving three MDOPs which have been widely considered in literature [1, 13, 14]. The values of the constants and their units are consistent. The decision variables for each problem are the trajectories of the control variables with respect to time or a spatial variable.

## 3.1  Temperature Control of a Batch Reactor

A batch reactor is considered in which the first-order exothermic irreversible reactions $A k_1 B k_2 C$ occur. The dimensionless concentrations of component $A$ ($C_A$) and $B$ ($C_B$) vary according to the ODEs in Eq. (5)

$$\frac{dC_A}{dt} = -k_1 C_A^2; \quad \frac{dC_B}{dt} = k_1 C_A^2 - k_2 C_B;$$
$$C_A(0) = 1, C_B(0) = 0 \tag{5}$$

where $C_A(0)$ and $C_B(0)$ are the initial concentrations of $A$ and $B$. The rate constants are known to be functions of the reactor temperature and are given by Eq. (6)

$$k_1 = 4 \times 10^3 \times e^{-2500/T}; \quad k_2 = 6.2 \times 10^5 \times e^{-5000/T}; 298 \leq T \leq 398 \tag{6}$$

where $T$ is the reactor temperature (K). The objectives are to maximize the concentration of $B$ while minimizing that of $A$ at the final reaction time ($t_f = 1$) and are defined as

$$Min \, f_1 = -C_B(t_f); \quad Min \, f_2 = C_A(t_f) \tag{7}$$

The reactor temperatures at specified intervals of time form the decision variables of the problem.

## 3.2 Catalyst Mixing Problem in a Plug Flow Reactor

A steady-state plug flow reactor packed with two types of catalysts (catalyst $A$ and $B$) is considered in which the following reactions occur: $S_1 \leftrightarrow S_2 \rightarrow S_3$ (one reversible and the other irreversible). The concentrations of components $S_1$ ($x_1$) and $S_2$ ($x_2$) vary according to the ODEs in Eq. (8)

$$\frac{dx_1}{dz} = u(z)[10x_2(z) - x_1(z)];$$
$$\frac{dx_2}{dz} = -u(z)[10x_2(z) - x_1(z)] - [1 - u(z)]x_2(z); \tag{8}$$
$$x_1(0) = 1, \quad x_2(0) = 0$$

where $z$ is the spatial coordinate, $x_1(0)$ and $x_2(0)$ are the concentrations of $S_1$ and $S_2$ at $z = 0$, and $u(z)$ is the fraction of catalyst $A$ in the plug flow reactor at specific spatial coordinates. The bounds for $u(z)$ are given as $0 \leq u(z) \leq 1$, and the outlet of the plug flow reactor is at $z_f = 1$. As catalyst $A$ is the costlier of the two catalysts, the objectives of the problem are to minimize the quantity of catalyst $A$ utilized while maximizing the production of the product $S_3$ in the outlet of the plug flow reactor and are given as

$$Min \, f_1 = -\left(1 - x_1(z_f) - x_2(z_f)\right); \quad Min \, f_2 = \int_0^{z_f} u(z)dz \tag{9}$$

The fraction of the catalyst $A$ at specified locations forms the decision variables of the problem.

## 3.3 Optimal Operation of a Fed-Batch Reactor

The reactions given in Eq. (10) occur in a batch reactor which may be fed continuously

$$A + Bk_1C; \ B + Bk_2D \tag{10}$$

The ODEs in Eq. (11) describe the dynamics of the system

$$\frac{d[A]}{dt} = -k_1[A][B] - ([A]/V)u;$$

$$\frac{d[B]}{dt} = -k_1[A][B] - 2k_2[B]^2 + \left(\frac{b_{feed} - [B]}{V}\right)u$$

$$\frac{d[C]}{dt} = -k_1[A][B] - ([C]/V)u; \quad \frac{d[D]}{dt} = -2k_2[B]^2 - ([D]/V)u \tag{11}$$

$$\frac{dV}{dt} = u;$$

$$t = 0, \ b_{feed} = 0.2 \ mol/l, \ [A] = 0.2 \ mol/l, \ V = 0.5 \ l$$

where $[A]$, $[B]$, $[C]$ and $[D]$ are the concentrations of the components $A$, $B$, $C$ and $D$, and $V$ is the total reaction volume. The feed rate $u$ has the bounds $0 \leq u \leq 0.01$, and the concentration of all the components except $A$ is initially zero. The rate constants of the reactions are assumed to be constant and are given as $k_1 = 0.5$ and $k_2 = 0.5$. The feed rates at specified intervals of time form the decision variables and the objectives of the problem are to maximize the concentrations of product $C$ while minimizing that of the by-product $D$ at the final time ($t_f = 120$ min) and are defined as

$$Min \ f_1 = -[C](t_f)V(t_f); \ Min \ f_2 = [D](t_f)V(t_f) \tag{12}$$

## 4  Results on the MDOPs

For the convenience of comparing results with those available in literature [1], a uniform discretization of the time/spatial variable is employed by dividing the entire interval into 10 uniformly sized sub-intervals. The solvers in-built in MATLAB 2015a for ODEs (*ode45*) and integration (*integral*) are employed in this work as per requirement. Due to the stochastic nature of the algorithm, the three MDOPs were each solved 10 times. In order to obtain the best run corresponding to each problem, the non-dominated points obtained from the 10 runs corresponding to each problem are combined and sorted to obtain a single Pareto front corresponding to each problem. This is followed by the calculation of the IGD corresponding to each run and each problem; the run corresponding to the best IGD is selected as the best run for each problem. F-YYPO settings utilized for solving the MDOPs are: $I_{min} = 1$, $I_{max} = 3$ and $\alpha = 20$, and the maximum number of functional evaluations is set at 10,100 (corresponding to MODE-RMO with population size and number of generations equal to 100) for the purpose of providing a fair comparison with the MODE-RMO results available in literature [1]. The five non-dominated points obtained by MODE-RMO on each of the MDOPs are utilized here for comparison with F-YYPO.

The Pareto front obtained by F-YYPO in its best run and the five non-dominated points reported by MODE-RMO for each problem is presented in Fig. 3. On the batch reactor problem, the Pareto front obtained by F-YYPO is observed to be very close to the points obtained by MODE-RMO, while extending significantly farther towards the extremes of both the objectives, thus providing a larger resultant front. The front obtained by F-YYPO on the plug flow reactor problem is observed to be very similar to that obtained by MODE-RMO except for the region around the $f_1 = -0.02$ on the x-axis, where F-YYPO has not converged satisfactorily as compared to MODE-RMO. However, the front obtained by F-YYPO extends beyond that obtained by MODE-RMO towards both extremes in this case as well. F-YYPO shows a very strong performance with respect to MODE-RMO on the fed-batch problem, where the front obtained by F-YYPO extends well beyond that obtained by

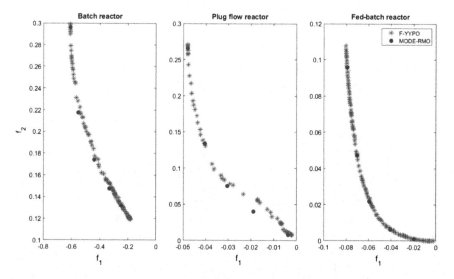

**Fig. 3** Pareto fronts obtained by F-YYPO and MODE-RMO

MODE-RMO towards both the extremes of the front. Thus, F-YYPO is observed to explore a significantly larger section of the Pareto front on all the three problems.

Table 1 presents the five non-dominated points selected corresponding to the batch reactor problem. It can be observed that maintaining temperatures close to the upper limit (Point 5) leads to the least value for $f_2$, i.e. minimization of the concentration of component $A$. On the other hand, maintaining a regularly decreasing temperature profile (Point 1) is observed to yield the maximum concentration of component $B$ (minimization of $f_1$). The temperatures of the intermediate points fluctuate between the temperature bounds at different intervals and do not adhere to any pattern.

Table 2 presents the five non-dominated points selected corresponding to the plug flow reactor problem. The objective $f_2$ corresponds to minimizing the catalyst

**Table 1** Selected Pareto optimal points for the batch reactor problem

| Interval | Point 1 | Point 2 | Point 3 | Point 4 | Point 5 |
|---|---|---|---|---|---|
| 1 | 364.74 | 363.28 | 376.46 | 390.14 | 397.84 |
| 2 | 343.89 | 362.84 | 397.61 | 391.40 | 397.70 |
| 3 | 337.68 | 365.45 | 387.59 | 390.37 | 397.60 |
| 4 | 334.51 | 351.69 | 385.23 | 394.16 | 397.82 |
| 5 | 333.00 | 362.76 | 390.58 | 377.24 | 397.94 |
| 6 | 331.04 | 356.60 | 363.34 | 388.50 | 397.90 |
| 7 | 328.07 | 363.66 | 366.19 | 391.51 | 396.88 |
| 8 | 328.41 | 356.89 | 352.96 | 382.86 | 395.66 |
| 9 | 326.83 | 365.98 | 387.11 | 387.30 | 396.68 |
| 10 | 327.40 | 311.68 | 344.45 | 387.53 | 396.41 |
| $f_1$ | −0.6102 | −0.5345 | −0.3720 | −0.2661 | −0.1827 |
| $f_2$ | 0.2989 | 0.2163 | 0.1612 | 0.1354 | 0.1192 |

**Table 2** Selected Pareto optimal points for the plug flow reactor problem

| Interval | Point 1 | Point 2 | Point 3 | Point 4 | Point 5 |
|---|---|---|---|---|---|
| 1 | 0.9964 | 0.9144 | 0.6337 | 0.0238 | 0.0032 |
| 2 | 0.4947 | 0.6311 | 0.0065 | 0.0065 | 0.0022 |
| 3 | 0.3302 | 0.0066 | 0.0066 | 0.0066 | 0.0010 |
| 4 | 0.1032 | 0.0553 | 0.0128 | 0.0128 | 0.0128 |
| 5 | 0.3404 | 0.0894 | 0.0048 | 0.0048 | 0.0048 |
| 6 | 0.1515 | 0.0479 | 0.0396 | 0.0479 | 0.0039 |
| 7 | 0.2390 | 0.0688 | 0.0184 | 0.0184 | 0.0064 |
| 8 | 0.0475 | 0.0285 | 0.0285 | 0.0285 | 0.0142 |
| 9 | 0.0056 | 0.0765 | 0.0154 | 0.0765 | 0.0154 |
| 10 | 0.0000 | 0.0131 | 0.0131 | 0.0131 | 0.0131 |
| $f_1$ | −0.0480 | −0.0457 | −0.0295 | −0.0066 | −0.0021 |
| $f_2$ | 0.2708 | 0.1932 | 0.0779 | 0.0239 | 0.0077 |

**Table 3** Selected Pareto optimal points for the fed-batch reactor problem

| Interval | Point 1 | Point 2 | Point 3 | Point 4 | Point 5 |
|----------|---------|---------|---------|---------|---------|
| 1 | 0.0099 | 0.0098 | 0.0082 | 0.0042 | 0.0002 |
| 2 | 0.0100 | 0.0081 | 0.0027 | 0.0018 | 0.0001 |
| 3 | 0.0099 | 0.0025 | 0.0091 | 0.0028 | 0.0001 |
| 4 | 0.0099 | 0.0088 | 0.0020 | 0.0013 | 0.0001 |
| 5 | 0.0100 | 0.0073 | 0.0054 | 0.0018 | 0.0000 |
| 6 | 0.0098 | 0.0094 | 0.0038 | 0.0013 | 0.0001 |
| 7 | 0.0100 | 0.0069 | 0.0037 | 0.0033 | 0.0000 |
| 8 | 0.0099 | 0.0098 | 0.0078 | 0.0003 | 0.0000 |
| 9 | 0.0099 | 0.0088 | 0.0041 | 0.0050 | 0.0001 |
| 10 | 0.0100 | 0.0088 | 0.0016 | 0.0002 | 0.0001 |
| $f_1$ | −0.0806 | −0.0762 | −0.0649 | −0.0411 | −0.0019 |
| $f_2$ | 0.1077 | 0.0748 | 0.0320 | 0.0065 | 0.0000 |

$A$ which is required by the reactor, which is the profile obtained by Point 5. A general trend of a gradually decreasing fraction of catalyst $A$ with respect to the reactor length is observed among the rest of the points, with the Point 1 (corresponding to the minimum obtained $f_1$) having the highest initial fraction of catalyst $A$. Point 4 and Point 5 correspond to maintaining a negligible fraction of catalyst $A$ throughout the reactor.

Table 3 presents the five non-dominated points selected corresponding to the fed-batch reactor problem. The extreme points corresponding to the maximum concentration of product $C$ is Point 1, and that corresponding to the minimum concentration of the by-product $D$ is Point 5. These points correspond to maintaining the feed rate almost constantly at its upper and lower limits, respectively. The feed-rate profiles for the remaining three points are observed to be distributed within these extremes and have a tendency to fluctuate significantly.

# 5  Conclusion

Real-life engineering problems frequently deal with independent variables such as time or spatial coordinates which can be dynamic in nature. These dynamic optimization problems can involve multiple objectives and ordinary or partial differential equations, making them challenging to solve. Stochastic optimization algorithms have been utilized to solve such problems in literature.

In this work, three multi-objective dynamic optimization problems considering a (i) batch reactor, (ii) plug flow reactor and (iii) fed-batch reactor were solved using a recently proposed stochastic optimization algorithm termed as F-YYPO. This algorithm is shown to perform well on all the three problems and is able to obtain well distributed Pareto fronts on all cases. In particular, it shows a strong performance compared to the literature for the batch reactor and fed-batch reactor problems. A sample Pareto optimal point has been selected for each problem and analysed. This work may be extended to include multiple competing algorithms to determine the more suitable technique for MDOPs.

# References

1. Chen, X., W. Du, and F. Qian, *Multi-objective differential evolution with ranking-based mutation operator and its application in chemical process optimization*. Chemometrics and Intelligent Laboratory Systems, 2014. **136**: p. 85–96.
2. Babu, B.V. and R. Angira, *Modified differential evolution (MDE) for optimization of non-linear chemical processes*. Computers & Chemical Engineering, 2006. **30**(6–7): p. 989–1002.
3. Campos, M. and R.A. Krohling, *Entropy-based bare bones particle swarm for dynamic constrained optimization*. Knowledge-Based Systems, 2016. **97**: p. 203–223.
4. Deb, K., *Multi-objective optimization using evolutionary algorithms*. Vol. 16. 2001: John Wiley & Sons.
5. Marler, R.T. and J.S. Arora, *Survey of multi-objective optimization methods for engineering*. Structural and multidisciplinary optimization, 2004. **26**(6): p. 369–395.
6. Reyes-Sierra, M. and C.C. Coello, *Multi-objective particle swarm optimizers: A survey of the state-of-the-art*. International Journal of Computational Intelligence Research, 2006. **2**(3): p. 287–308.
7. Deb, K., et al., *A fast and elitist multiobjective genetic algorithm: NSGA-II*. IEEE Transactions on Evolutionary Computation, 2002. **6**(2): p. 182–197.
8. Xue, F., A.C. Sanderson, and R.J. Graves. *Pareto-based multi-objective differential evolution*. in *IEEE Congress on Evolutionary Computation*. 2003. Canberra, Australia.
9. Punnathanam, V. and P. Kotecha, *Multi-objective optimization of Stirling engine systems using Front-based Yin-Yang-Pair Optimization*. Energy Conversion and Management, 2017. **133**: p. 332–348.
10. Punnathanam, V. and P. Kotecha, *Yin-Yang-pair Optimization: A novel lightweight optimization algorithm*. Engineering Applications of Artificial Intelligence, 2016. **54**: p. 62–79.
11. Punnathanam, V., *Yin-Yang-Pair Optimization: A novel lightweight optimization algorithm (Unpublished Master's thesis)*, 2016, Indian Institute of Technology Guwahati: Guwahati, India.
12. Punnathanam, V. and P. Kotecha. *Front-based Yin-Yang-Pair Optimization and its performance on CEC2009 benchmark problems*. in *International Conference on Smart Innovations in Communications and Computational Sciences*. 2017. Punjab, India.
13. Logist, F., et al., *Multi-objective optimal control of chemical processes using ACADO toolkit*. Computers & Chemical Engineering, 2012. **37**: p. 191–199.
14. Herrera, F. and J. Zhang, *Optimal control of batch processes using particle swam optimisation with stacked neural network models*. Computers & Chemical Engineering, 2009. **33**(10): p. 1593–1601.

# Front-Based Yin-Yang-Pair Optimization and Its Performance on CEC2009 Benchmark Problems

**Varun Punnathanam and Prakash Kotecha**

**Abstract** The Front-based Yin-Yang-Pair Optimization (F-YYPO) is a novel meta-heuristic for multi-objective problems that works with multiple Yin-Yang pair of points to handle conflicting objectives and utilizes the non-dominance-based approach to sort points into fronts. The performance of F-YYPO along with that of Multi-Objective Differential Evolution with Ranking-based Mutation Operator, Multi-Objective Grey Wolf Optimizer, and Non-dominance Sorting Genetic Algorithm-II is benchmarked on a suite of ten multi-objective problems. A comparative analysis between the performances of these algorithms is obtained by utilizing nonparametric statistical tests for multiple comparisons. F-YYPO is observed to outperform the competing algorithms on the benchmark problems with regard to the quality of solutions and computational time.

**Keywords** Multi-objective optimization · Stochastic optimization
Congress on evolutionary computation 2009

## 1 Introduction

Real-world optimization problems frequently require multiple objectives of conflicting natures to be taken into account. The solution of these problems is usually characterized by a set of trade-off points commonly known as the Pareto front [1]. Every point in this front is better than the others with respect to at least one of the objectives of the problem. Stochastic optimization procedures can be used to solve problems with multi-modal, non-differentiable, and discontinuous functions. Additionally, their inherent nature enables them to solve complex combinatorial problems [2] as well as problems involving simultaneous ordinary or partial differential equations. Hence, they have been extensively utilized in the literature [3] for both single- and multi-objective applications.

V. Punnathanam · P. Kotecha (✉)
Indian Institute of Technology Guwahati, Guwahati 781039, Assam, India
e-mail: pkotecha@iitg.ernet.in

© Springer Nature Singapore Pte Ltd. 2019
B. K. Panigrahi et al. (eds.), *Smart Innovations in Communication and Computational Sciences*, Advances in Intelligent Systems and Computing 669, https://doi.org/10.1007/978-981-10-8968-8_33

387

There exist a number of stochastic multi-objective optimization techniques which can obtain the Pareto front with a single execution of the algorithm such as Non-dominated Sorting Genetic Algorithm-II (NSGA-II) [4] and Multi-Objective Grey Wolf Optimizer (MOGWO) [5]. For discussions on various such techniques, the reader may refer to surveys available in the literature [6]. A vast majority of the aforementioned algorithms are population-based techniques which are computationally intensive. Yin-Yang-Pair Optimization (YYPO) is a recently proposed meta-heuristic which is shown to be significantly faster than other meta-heuristics while performing just as well [7]. We build upon the YYPO framework of optimization to design an effective multi-objective optimization algorithm which can provide quality solutions in a relatively shorter time. The performance of F-YYPO on Stirling engine systems with multiple objectives was showcased in the literature [8]. The objective of this work is to describe the formulation of F-YYPO in its entirety (which was not done in [8, 9] for the sake of brevity) such that an interested reader may independently implement the algorithm. Additionally, we evaluate its performance by providing a comparative analysis against three other optimization algorithms from the literature on benchmark problems. For this purpose, three statistical comparison tests were utilized [10]. The description of the multi-objective F-YYPO is discussed in Sect. 2, followed by the performance evaluation of the algorithms on the benchmark problems in Sect. 3. Subsequently, we conclude the work in Sect. 4.

## 2  Formulation of F-YYPO

The formulation of F-YYPO is based on the concept of sorting points into its respective non-dominance fronts, hence the name Front-based Yin-Yang-Pair Optimization. The pseudo-code for the main algorithmic framework is provided in Algorithm 1 and the pseudo-code for the Splitting Stage is provided in Algorithm 2. F-YYPO utilizes scaled variables, and hence explicitly overcomes issues which could arise from varying magnitudes of the decision variables. Each variable is maintained within the range 0 to 1 and is scaled to its actual bounds prior to every functional evaluation using Eq. (1).

$$x^{'} = x_l + x(x_u - x_l) \tag{1}$$

where $x$ is a point vector (within the range 0 to 1) in the algorithm whose fitness is to be evaluated, $x_l$ and $x_u$ are the lower and upper bounds of the variables and $x'$ is $x$ scaled to its original bounds. In addition to the maximum number of Pareto Points ($P_{max}$), the user-defined parameters are given below.

- $I_{min}$ and $I_{max}$—Integer. The number of iterations for Archive Stage is generated between these two numbers.
- $a$—Real number. This parameter governs the rate of increase or decrease in the region searched around each point.

**Algorithm 1**. Main algorithm for F-YYPO

| | |
|---|---|
| Step 1 | Initialize matrix $P$ of size $2M$ x $D$, where each element is a random number between 0 and 1. |
| Step 2 | Initialize matrix $\Delta$ of size $2$ x $M$, where each element is set at an initial value of 0.5. Initialize a null matrix $F$ to store the Pareto points. Set $i = 0$ and generate $I$ between $I_{min}$ and $I_{max}$. |
| Step 3 | Scale the $2M$ points in $P$ using eq. (1) and determine their fitness values. |
| Step 4 | do |
| Step 5 | for each objective $o \in M$, |
| Step 6 | if the second point of the $o^{th}$ pair of points in $P$ is fitter than the first point of the $o^{th}$ pair with respect to the $o^{th}$ objective fitness value, |
| Step 7 | Interchange the two points of the $o^{th}$ pair of points in $P$. Interchange the two elements of the $o^{th}$ column of the matrix $\Delta$, endif, endfor |
| Step 8 | Store the points in $P$ and their corresponding fitness values in archive $A$ and update $i = i+1$. |
| Step 9 | for each objective $o \in M$ and $j = 1$ to 2 |
| Step 10 | Execute the Splitting Stage for the $j^{th}$ point of the $o^{th}$ pair of points in $P$. The $j^{th}$ value in the $o^{th}$ column of the $\Delta$ matrix is utilized for this purpose. |
| Step 11 | if $i$ equals $I$ |
| Step 12 | Add all the current points in $P$ as well as in $F$ to the archive $A$. Calculate the crowding distance for each point in $A$ and obtain the front number corresponding to each point. Sort the points in the descending order of their crowding distance. |
| Step 13 | Replace all the points in $F$ with the points in the first front of archive $A$. If there is only a single point in the first front, add all points in the second front to the matrix $F$ as well. If the number of points in $F$ exceeds the maximum limit $P_{max}$, store the first $P_{max}$ points in $F$, reject all other points. |
| Step 14 | if the number of points in $F$ is lesser than $M+1$ |
| Step 15 | for each objective $o \in M$, |
| Step 16 | Replace both the $o^{th}$ pair of points in the matrix $P$ with a random point from the matrix $F$, endfor |
| Step 17 | else |
| Step 18 | for each objective $o \in M$, |
| Step 19 | Replace the first point of the $o^{th}$ pair of points in $P$ with the $o^{th}$ point in $F$, endfor |
| Step 20 | Replace all the even numbered points in matrix P with the $(M+1)^{th}$ point in matrix $F$, endif |

Step 21   Update all the terms in the matrix Δ using eq. (4).
          Clear the archive A, generate a new I between $I_{min}$ and
          $I_{max}$, set i = 0, enddo
Step 22   while termination criteria is not satisfied.
Step 23   Execute Step 12 and Step 13, matrix F is the ob-
          tained Pareto Front.

Each objective in the problem is assigned two points, allowing the algorithm to focus on each individual objective to a larger extent, hence giving it superior capabilities in exploring the corners of the Pareto fronts. Two search radii are assigned to each individual objective such that every point has a corresponding search radius $\delta$. The points are stored as rows in a matrix $P$ (of size $2M \times D$) where the points for the first objective occupy the first two rows. The points for the second objective occupy the next two rows and so on. The search radii for each of the points in $P$ are stored in a matrix $\Delta$ (of size $2 \times M$). Relationship between search the search radii $\Delta$ and the points $P$ is given below

$$\Delta_{p,o} \equiv P(2(o-1)+p), \ \forall o = 1, 2 \dots M, \ p = 1, 2$$

The $2M$ points in $P$ are randomly generated within the bounds 0 to 1 and their fitness values are evaluated after scaling using Eq. (1). Subsequently, all the search radii in $\Delta$ are initialized to 0.5 and a null matrix $F$ is defined for points which are not dominated. The Archive Stage is to be executed after $I$ iterations, hence $I$ is set to a random integer value between $I_{min}$ and $I_{max}$. The structure of the matrices $P$ and $\Delta$ are provided below

$$P = \begin{bmatrix} P_1 \\ P_2 \\ P_3 \\ P_4 \\ \vdots \\ \vdots \\ P_{2M-3} \\ P_{2M-2} \\ P_{2M-1} \\ P_{2M} \end{bmatrix} = \begin{bmatrix} x_1^1 & x_1^2 & \cdots & \cdots & x_1^D \\ x_2^1 & x_2^2 & \cdots & \cdots & x_2^D \\ x_3^1 & x_3^2 & \cdots & \cdots & x_3^D \\ x_4^1 & x_4^2 & \cdots & \cdots & x_4^D \\ \vdots & \vdots & & & \vdots \\ \vdots & \vdots & & & \vdots \\ x_{2M-3}^1 & x_{2M-3}^2 & \cdots & \cdots & x_{2M-3}^D \\ x_{2M-2}^1 & x_{2M-2}^2 & \cdots & \cdots & x_{2M-2}^D \\ x_{2M-1}^1 & x_{2M-1}^2 & \cdots & \cdots & x_{2M-1}^D \\ x_{2M}^1 & x_{2M}^2 & \cdots & \cdots & x_{2M}^D \end{bmatrix}$$

$$\Delta = \begin{bmatrix} \Delta_{1,1} & \Delta_{1,2} & \cdots & \Delta_{1,M-1} & \Delta_{1,M} \\ \Delta_{2,1} & \Delta_{2,2} & \cdots & \Delta_{2,M-2} & \Delta_{2,M} \end{bmatrix}$$

$\Delta_{1,1}$ is the radii of $P_1$                     $\Delta_{2,1}$ is the radii of $P_2$

$\Delta_{1,2}$ is the radii of $P_3$                     $\Delta_{2,2}$ is the radii of $P_4$

$\Delta_{1,M-1}$ is the radii of $P_{2M-3}$     $\Delta_{2,M-1}$ is the radii of $P_{2M-2}$

$\Delta_{1,M}$ is the radii of $P_{2M-1}$         $\Delta_{2,M}$ is the radii of $P_{2M}$

At the start of each iteration, the points in $P$ are rearranged according to Steps 4 to 7 in Algorithm 1, subsequent to which they are stored in an archive $A$ and undergo the Splitting Stage. It should be noted that the archive $A$ is separate from the archive which stores the Pareto optimal points (archive $F$).

The **Splitting Stage** is responsible for generating new points and updating the matrix $P$ and is executed for each point in $P$. The pseudo-code for this stage is separately presented in Algorithm 2. It requires three input arguments, (i) a single point (say $Q$), (ii) its search radius ($\delta$), and (iii) the objective number corresponding to this point (say objective $o$). The new points are generated by either Step 4 in Algorithm 2 (One-way splitting) or Steps 6 and 7 in Algorithm 2 (D-way) splitting for which details are available in [7]. Both these methods involve generating $2D$ points corresponding to the point $Q$. Thus, $2D$ copies of the point $Q$ are stored in a matrix $S$ for subsequent modification. The equations for one-way splitting and D-way splitting are provided in Eq. (2) and Eq. (3), respectively.

$$\left. \begin{array}{l} S_j^j = S^j + r_j\delta \\ S_{D+j}^j = S^j - r_{D+j}\delta \end{array} \right\} \forall\, j = 1, 2, 3 \ldots D \tag{2}$$

$$\left. \begin{array}{ll} S_k^j = S^j + r_k\left(\delta/\sqrt{2}\right) & \text{if } B_k^j = 1, \\ S_k^j = S^j - r_k\left(\delta/\sqrt{2}\right) & \text{otherwise} \end{array} \right\} \forall\, k = 1, 2, 3 \ldots 2D \text{ and } j = 1, 2, 3 \ldots D \tag{3}$$

where the subscript denotes the point number, the superscript denotes the variable number and $r$ is a random number between 0 and 1 which is uniquely generated for each variable and point in matrix $S$. In this work, the matrix $B$ is generated by randomly generating $2D$ unique integers between 0 and $2^D - 1$ and subsequently converting them to binary strings of length $D$.

At this stage, any violation of the variable bounds in the matrix $S$ is rectified by replacing it with a random number between 0 and 1. The points of matrix $S$ are scaled to its original bounds using Eq. (1) and evaluated for fitness. If there exists a single point $S$ such that it has the best fitness with respect to the objective $o$ in comparison with all other points in $S$, then this point is returned to the main algorithm along with its fitness values. If the best fitness of the objective $o$ corresponds to multiple points in $S$, then we determine the non-dominance fronts for these points (i.e., the points with the best fitness on objective $o$). The first front signifies better quality points, hence one of these points are then randomly selected to replace the point $Q$, and is returned to the main algorithm along with its fitness values.

**Algorithm 2**. Splitting Stage

| | |
|---|---|
| Step 1 | Generate random number $R$ $(0 \le R \le 1)$. |
| Step 2 | Generate $2D$ copies of $Q$ and store it in $S$. |
| Step 3 | if $R < 0.5$ |
| Step 4 | Modify each point in $S$ using eq. (2). |
| Step 5 | else |
| Step 6 | Generate a random binary matrix $B$ of size $(2D$ x $D)$ such that every row is unique. |
| Step 7 | Modify each point in $S$ using eq. (3), endif |
| Step 8 | Scale every point in $S$ according to eq. (1) and evaluate fitness. |
| Step 9 | Replace each variable in $S$ which is lesser than 0 or greater than 1 with a unique randomly generated number. |
| Step 10 | Updated $Q$ = fittest point in $S$ with respect to the objective $q$. If this corresponds to multiple points, sort all the points with the best objective $q$ and randomly pick a point from the first front. |
| Step 11 | Return the updated $Q$ along with its fitness values. |

It should be noted that the Splitting Stage invariably updates every point in $P$ despite the fact that the updated point could be inferior to the point used to generate it. This does not negatively affect the algorithm's performance as $P$ has been stored in the archive $A$ prior to being updated at the Splitting Stage. The Archive Stage is initiated every $I$ iteration. The iterations terminate if the Archive Stage is not initiated, in which case we check for satisfaction of the termination criteria.

The strategy proposed by Deb et al. [4] for classifying points into non-dominance fronts based on objective value is utilized in the Archive Stage. The points in $P$ and $F$ are stacked in $A$. The crowding distance is calculated corresponding to all the points in $A$ following which the points are assigned non-dominance fronts. The points in the first front are arranged based on their crowding distance (descending order) and stored in $F$.

If the number of points in F is more than the upper limit on the number of non-dominated ($P_{max}$), then the first $P_{max}$ points in $F$ are reported. F-YYPO utilizes $M + 1$ points in $F$ to replace the $2M$ points in $P$. The first $M$ points of $F$ would have crowding distances equal to infinity (the corner points for each objective). Each of these $M$ points would be utilized to update the first point in the pair of points corresponding to every objective. The $(M + 1)$th point in $F$ would be the least crowded point in $F$ (excluding the corner points) and hence is utilized to update every second point in the every pair of points. However, if the number of points in $F$ is lesser than $M + 1$, then a random point in $F$ is utilized by each pair of points to replace both the points in the pair. It should be noted that a random point from $F$ is selected corresponding to each pair of points.

The last step in the Archive Stage involves updating the values of $\Delta$ and $I$. The search radii corresponding to the first point in every pair of points, i.e., the first row of $\Delta$ is reduced. This contracts the hyper-volume of the search space explored by these points, hence can aid in exploitation. Similarly, the search radii in the second row of $\Delta$ are increased to expand the hyper-volume explored by the corresponding points. The search radii are updated in the following manner:

$$\left. \begin{array}{l} \Delta(1, o) = \Delta(1, o) - (\Delta(1, o)/\alpha) \\ \Delta(2, o) = \min(\Delta(2, o) + (\Delta(2, o)/\alpha), 0.75) \end{array} \right\} \forall \, o = 1, 2, 3 \ldots M \qquad (4)$$

$\Delta$ is limited to 0.75 similar to the original YYPO as a very high value of $\Delta$ could lead to ineffective exploration of the search space due to an increase in tendency to violate the variable bounds. A new interval of iterations $I$ is generated between its bounds $I_{min}$ and $I_{max}$ and the archive $A$ is emptied, thus concluding the archive stage. Subsequent to the satisfaction of the termination criteria, the points in $P$, $A$, and $F$ are combined and are sorted for the non-dominated points and are the solution determined by F-YYPO.

## 3 Performance Evaluation on Benchmark Functions

A suite of ten unconstrained benchmark problems for multi-objective optimization was proposed as part of the CEC2009 [11] and is utilized to compare the performance of F-YYPO with three other multi-objective optimization algorithms. Problems UF1 to UF7 have two objectives while problems UF8 to UF10 have three objectives, and all the ten problems have 30 decision variables. For the sake of brevity, additional details on these problems are not provided here as they are available in the literature [11]. For a fair comparison, 30 runs of each problem were considered for every algorithm thereby leading to a total of 1200 instances (4 algorithms × 10 problems × 30 runs). The termination criteria were fixed at 300,000 functional evaluations, and the maximum number of non-dominated points for measuring GD [1] and IGD [11] was limited to 100 for two objective problems and 150 for the three objective problems. These settings correspond to the evaluation criteria specified in the CEC2009 report [11]. The *twister* random number generator inbuilt in MATLAB 2015a was utilized to fix the random seeds (seed numbers 1 through 30) for all the algorithms. The algorithm-specific parameters are provided below

- MODE-RMO [12]: population size = 100 for UF1 to UF7, 150 for UF8 to UF10, differential weight = 0.5, and crossover probability = 0.3
- MOGWO [5]: population size = 100, grid inflation parameter = 0.1, leader selection pressure parameter = 4, extra repository member selection pressure = 2, and number of grids per dimension = 10

- NSGA-II [4]: population size = 100 for UF1 to UF7, 150 for UF8–UF10, binary tournament selection, simulated binary crossover, polynomial mutation, crossover probability = 0.9, and mutation probability = 0.1
- F-YYPO: $I_{min}$ = 1, $I_{max}$ = 3 and $\alpha$ = 20.

## 3.1 Results on Benchmark Functions and Discussions

The non-dominated fronts obtained by F-YYPO on the instances corresponding to the minimum IGD for each problem is presented in Fig. 1. Considering the difficulty of the benchmark problems, the convergence as well as spread obtained by F-YYPO is reasonably good. In particular, the problems with discontinuous Pareto fronts (UF5 and UF6) were challenging while UF1, UF2, UF4, and UF7 were effectively solved. The mean GD and IGD values of the various algorithms are provided as supplementary information (goo.gl/aqvtVK).

Multiple comparison tests give the overall ranks obtained by each algorithm, thus providing a better idea of their performances in relation with each other. The data sets provided to these tests are the mean IGD and GD obtained by each of these algorithms on each of the ten problems over 30 runs. The results of the Friedman, Friedman aligned, and Quade tests are provided in Table 1, where the lowest rank corresponds to the best performance. F-YYPO consistently attains the best rank among all the algorithms on all the tests considering the IGD as well as GD measure. Additionally, the corresponding p-values depict that the test results are statistically significant. Thus, it can be concluded that F-YYPO outperforms the competing algorithms on the CEC2009 benchmark problems.

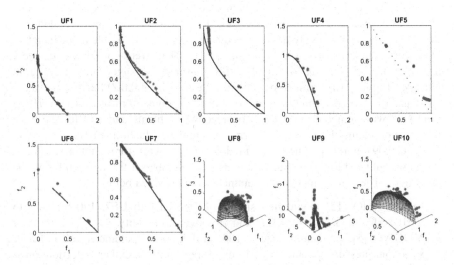

**Fig. 1** Plots of the non-dominated front (red stars) and the true Pareto fronts (blue dots) for each of the CEC2009 problems

**Table 1** Ranks obtained by all the algorithms based on the three nonparametric multiple comparison tests with respect to IGD and GD

| Algorithm | IGD | | | GD | | |
|---|---|---|---|---|---|---|
| | Friedman | Friedman aligned | Quade | Friedman | Friedman aligned | Quade |
| MODE-RMO | 3.90 | 33.60 | 3.87 | 3.80 | 33.40 | 3.76 |
| MOGWO | 2.30 | 16.40 | 2.29 | 2.00 | 13.90 | 2.09 |
| NSGA-II | 2.20 | 20.50 | 2.47 | 2.90 | 22.70 | 2.98 |
| F-YYPO | **1.60** | **11.50** | **1.36** | **1.30** | **12.00** | **1.16** |
| Test statistic | 17.40 | 15.94 | 9.22 | 21.24 | 18.06 | 13.19 |
| p-value | 2.77E-4 | 5.51E-4 | 1.65E-4 | 4.49E-5 | 2.03E-4 | 1.03E-5 |

## 3.2 Computational Time

The average time required by each algorithm on the CEC2009 problems is provided in Table 2. NSGA-II and MODE-RMO are observed to require time in the range of hundreds of seconds for each problem, while MOGWO requires time in the range of thousands of seconds on many problems. On the other hand, F-YYPO solves each problem with an average time in the range of a few seconds, which is 100 to 1000 times lesser than that required by the other algorithms. Additionally, F-YYPO requires similar computational time for three objective problems and two objective problems in the test set, while that is not valid for the other algorithms.

**Table 2** Average time required (in seconds) by all the algorithms on the CEC2009 problems

| Problem | MODE-RMO | MOGWO | NSGA-II | F-YYPO |
|---|---|---|---|---|
| UF1 | 325.6844 | 525.0368 | 236.9102 | 1.9142 |
| UF2 | 393.9782 | 1256.1989 | 228.5968 | 2.3180 |
| UF3 | 305.6660 | 806.1215 | 250.5302 | 3.2313 |
| UF4 | 626.7231 | 1286.4817 | 218.8316 | 2.0987 |
| UF5 | 303.3847 | 264.9186 | 271.5787 | 2.1717 |
| UF6 | 295.6485 | 303.0673 | 267.6559 | 2.1771 |
| UF7 | 314.3120 | 1028.4775 | 236.7026 | 1.9625 |
| UF8 | 525.8066 | 3291.0357 | 306.4153 | 2.0019 |
| UF9 | 527.5100 | 2181.5383 | 303.8686 | 1.9974 |
| UF10 | 497.1538 | 2131.5870 | 314.0272 | 2.2165 |

## 4 Conclusion

F-YYPO is a meta-heuristic for solving optimization problems with multiple objectives. Based on the number of objectives and decision variables in the problem, the algorithm can utilize an appropriate number of search points, thus adapting to the problem dimension. F-YYPO is formulated with an emphasis on the quality of results as well as computational speed. Its performance is compared with that of recent as well as popular algorithms such as MODE-RMO, MOGWO, and NSGA-II on the set of 10 multi-objective problems proposed as part of CEC2009. Two metrics (GD and IGD) and three nonparametric statistical tests (Friedman, Friedman Aligned, and Quade test) were employed to quantify and compare the quality of the obtained Pareto. Based on these metrics and tests, F-YYPO performs significantly better than its competitors. In addition, F-YYPO requires much lesser time as compared to its competitors to solve the CEC2009 problems and is able to provide results in a matter of a few seconds. This suggests that the algorithm can greatly reduce the computational time required to solve real problems. Future work could be to utilize this algorithm to solve various real-world problems and compare its performance with the other algorithms.

## Appendix 1

- MODE-RMO: http://www.mathworks.com/matlabcentral/fileexchange/49582-multi-objective-differential-evolution-with-ranking-based-mutation-operator
- MOGWO: http://www.alimirjalili.com/GWO.html
- NSGA-II: http://www.mathworks.com/matlabcentral/fileexchange/10429-nsga-ii--a-multi-objective-optimization-algorithm.

## References

1. Deb, K., *Multi-objective optimization using evolutionary algorithms*. Vol. 16. 2001: John Wiley & Sons.
2. Deep, K., et al., *A real coded genetic algorithm for solving integer and mixed integer optimization problems*. Applied Mathematics and Computation, 2009. **212**(2): p. 505–518.
3. Ramteke, M. and R. Srinivasan, *Large-Scale Refinery Crude Oil Scheduling by Integrating Graph Representation and Genetic Algorithm*. Industrial & Engineering Chemistry Research, 2012. **51**(14): p. 5256–5272.
4. Deb, K., et al., *A fast and elitist multiobjective genetic algorithm: NSGA-II*. IEEE Transactions on Evolutionary Computation, 2002. **6**(2): p. 182–197.
5. Mirjalili, S., et al., *Multi-objective grey wolf optimizer: A novel algorithm for multi-criterion optimization*. Expert Systems with Applications, 2016. **47**: p. 106–119.
6. Marler, R.T. and J.S. Arora, *Survey of multi-objective optimization methods for engineering*. Structural and multidisciplinary optimization, 2004. **26**(6): p. 369–395.

7. Punnathanam, V. and P. Kotecha, *Yin-Yang-pair Optimization: A novel lightweight optimization algorithm.* Engineering Applications of Artificial Intelligence, 2016. **54**: p. 62–79.

8. Punnathanam, V. and P. Kotecha, *Multi-objective optimization of Stirling engine systems using Front-based Yin-Yang-Pair Optimization.* Energy Conversion and Management, 2017. **133**: p. 332–348.

9. Punnathanam, V. and P. Kotecha. *Optimization of Multi-objective Dynamic Optimization Problems with Front-based Yin-Yang-Pair Optimization.* in *International Conference on Smart Innovations in Communications and Computational Sciences.* 2017. Moga, Punjab.

10. Derrac, J., et al., *A practical tutorial on the use of nonparametric statistical tests as a methodology for comparing evolutionary and swarm intelligence algorithms.* Swarm and Evolutionary Computation, 2011. **1**(1): p. 3–18.

11. Zhang, Q., et al., *Multiobjective optimization test instances for the CEC 2009 special session and competition.* University of Essex, Colchester, UK and Nanyang technological University, Singapore, special session on performance assessment of multi-objective optimization algorithms, technical report, 2008. **264**.

12. Chen, X., W. Du, and F. Qian, *Multi-objective differential evolution with ranking-based mutation operator and its application in chemical process optimization.* Chemometrics and Intelligent Laboratory Systems, 2014. **136**: p. 85–96.

# Thermal Model Parameters Identification of Power Transformer Using Nature-Inspired Optimization Algorithms

**Anshu Mala⊙, Chandra Madhab Banerjee⊙, Arijit Baral⊙ and Sivaji Chakravorti⊙**

**Abstract** The thermal property of transformer insulation is influenced by the design and size of transformer concern. As non-interrupted power supply is required to avoid financial loss and to provide reliable service to the consumer, thermal modeling of such power supply equipment has become an important tool for condition monitoring of power transformer. Thermal model equations are used to determine the hot spot temperature as it represents thermal condition and heat dissipation of transformer. This paper proposes a methodology to identify the thermal model of oil-immersed power transformer insulation. In this paper, measured top-oil and bottom-oil temperatures of a power transformer are used to find parameter of simple thermoelectric analogous thermal model. Nature-inspired algorithms are used to parameterize the thermal model. Three algorithms, namely genetic algorithm (GA), particle swarm optimization (PSO), and simulated annealing (SA), are used for the purpose. The *cost functions* of these algorithms are based on thermal model equations that are reported for the prediction of top-oil and bottom-oil temperatures. In addition, this paper also presents a comparative analysis between these optimization techniques.

**Keywords** Thermal model · Genetic algorithm (GA) · Particle swarm optimization (PSO) · Simulated annealing (SA) · Top-oil temperature Bottom-oil temperature

---

A. Mala (✉) · C. M. Banerjee · A. Baral
IIT (ISM) Dhanbad, Dhanbad, Jharkhand, India
e-mail: anshumala7aug@gmail.com

C. M. Banerjee
e-mail: cmbanerjee90@gmail.com

A. Baral
e-mail: a_baral@ieee.org

S. Chakravorti
Jadavpur University, Kolkata, India
e-mail: s_chakrav@yahoo.com

© Springer Nature Singapore Pte Ltd. 2019
B. K. Panigrahi et al. (eds.), *Smart Innovations in Communication and Computational Sciences*, Advances in Intelligent Systems and Computing 669, https://doi.org/10.1007/978-981-10-8968-8_34

# 1 Introduction

Power transformer is expensive equipment, and its performance is significantly affected by the condition of its insulation. Aging of transformer insulation is influenced by its capability to transfer the internally generated heat to its surrounding. So, it is important to estimate the thermal model of transformer during normal and overloading conditions. Today, for economic reasons, there is a great focus on extending the life of transformer in service. In this regard, a better knowledge of transformer thermal characteristics is essential.

The conventional temperature calculation methods, which are reported in the IEC [1] and the IEEE [2] regulations for oil-immersed power transformer, are used to estimate the top-oil temperatures (TOTs), bottom-oil temperatures (BOTs), and hot spot temperatures (HSTs). But, it is still typical to calculate the transformer's internal temperatures using conventional methods. The above-mentioned conventional calculations are based on assumptions of several thermal conditions [3, 4]. In addition, these parameters depend on the thermal profile of a particular transformer, and all the information related to this is not always present or changed with time. It is known that under realistic loading conditions, conventional method provides limited result [3]. To address these issues, several investigations have been reported which use optimization algorithms for investigating temperature dynamics of transformer [4]. For proper condition monitoring of transformer, it is very important to develop an accurate thermal model [2, 5]. A simple thermoelectric analogous thermal model is used [4] in this paper. The circuit is based on the theory of basic law of circuit and heat exchange [6, 7].

In addition, this paper also reports the performance of three optimization algorithms, namely genetic algorithm, particle swarm optimization, and simulated annealing, for identification of thermal model parameters. This paper presents a comparative analysis among the three optimization algorithms on the basis of their *cost function* values. The *cost functions* of these optimization algorithms are based on actual top- and bottom-oil temperatures that are reported for a real-life transformer [3]. The parameters of the thermal model (identified by the optimization algorithms) are then used to estimate the top-oil and bottom-oil temperatures. The calculated values of top-oil and bottom-oil temperatures are compared with the real data reported in the literature. Calculated top-oil temperature and bottom-oil temperature are thus obtained by using three different optimization algorithms. It can further be used for estimation of the hot spot temperature [4] of power transformer which is used for condition monitoring purpose. So an accurate identification of thermal model will be helpful to estimate the thermal condition of transformer.

## 2 Simplified Thermoelectric Analogous Thermal Model

On comparing the basic laws of heat transfer and the Ohm's law, the fundamental equation of heat transfer is analogous to the equation of basic electric circuits [6, 7]. Due to analogy between thermal and electrical system, it is relatively easy to develop an equivalent heat circuit [8]. The basic components of equivalent thermal circuits are heat conductance, heat capacitance, and heat current source [4]. However, it should be kept in mind that the active heat transfer parts of transformer are its winding, dielectric oil, transformer container, and outer cooler [3].

### 2.1 Equivalent Thermal Circuit Model

Description of an equivalent thermal circuit consisting of five nodes for an oil-immersed power transformer has been reported in [8]. Due to presence of many temperature nodes, the number of thermal parameters required to be optimized for this case is computationally extensive in nature. In addition, all the nodes of temperatures involved in the calculation were not introduced in the *cost function*, due to unavailability of online measurements [3]. It is understood that reduction in thermal parameters has the potential to increase the speed of estimation process. A simplified equivalent thermal circuit of oil-immersed power transformer is shown in Fig. 1 [4],

where

$Q_{all}$ Sum of iron losses, copper losses, and stray losses,
$G_1$ Thermal conductance between core and winding to oil,
$G_2$ Thermal conductance of coolers from outlet to inlet,
$G_3$ Thermal conductance from cooler to environment,
$C_i$ Corresponding thermal capacitance of each part of $G_i$,
$\Phi_1$ Rise in average oil temperature,
$\Phi_2$ Rise in top-oil temperature at the outlet,
$\Phi_3$ Rise in bottom-oil temperature at the inlet.

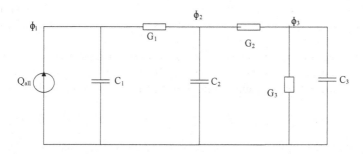

**Fig. 1** Equivalent thermal circuit of thermal model

## 2.2 Mathematical Formulation

Circuit shown in Fig. 1 works on the principle of Kirchhoff's law [3]. It is understood that the model shown in Fig. 1 can be successfully used to predict the value of top-oil and bottom-oil temperatures. Keeping in mind the structure of the thermal model shown in Fig. 1, it is obvious that the solution of $\Phi_2$ and $\Phi_3$ is represented in the form of non-homogeneous first-order differential equations. It can be observed that the equations related to parameters shown in Fig. 1 can be expressed in the form of Eq. (1).

$$\begin{bmatrix} Q_{all} \\ 0 \\ 0 \end{bmatrix} = \begin{bmatrix} C_1 & 0 & 0 \\ 0 & C_2 & 0 \\ 0 & 0 & C_3 \end{bmatrix} \begin{bmatrix} \frac{d\phi_1}{dt} \\ \frac{d\phi_2}{dt} \\ \frac{d\phi_3}{dt} \end{bmatrix} + \begin{bmatrix} G_1 & -G_1 & 0 \\ -G_1 & G_1+G_2 & -G_2 \\ 0 & -G_2 & G_2+G_3 \end{bmatrix} \begin{bmatrix} \phi_1 \\ \phi_2 \\ \phi_3 \end{bmatrix} \quad (1)$$

# 3 Estimation of Parameters for Thermal Model

## 3.1 Heat Generation

It is understood that the heat generation sources in core and winding are magnetic and electric losses respectively [9]. Broadly speaking, losses happening in a transformer can be subdivided into three categories: copper loss (which is due to $I^2R$), stray loss (occurs in the winding), and core loss (which is equal to the sum of the eddy current loss and hysteresis loss). The copper loss is directly influenced by the square of load current [5], and it is given below in Eq. (2).

$$P_{cu,o} = P_{cu,r} \left(\frac{I_o}{I_r}\right)^2 \left(\frac{235+\phi_o}{235+\phi_r}\right) \quad (2)$$

where $P_{Cu,O}$ is resistive loss due to operated load current, $P_{Cu,r}$ is resistive loss due to rated load, $I_o$ is the operated load current, $I_r$ is the rated load current, $\Phi_o$ is the operated winding temperature, and $\Phi_r$ is the winding temperature at rated losses. In the present paper, data corresponding to a real-life power transformer that is reported in the literature is used. Table 1 shows the details of power transformer [3] used for analysis.

It is understood that the total thermal loss of power transformer is the sum of core losses, copper losses, and stray losses. The value of total loss of power generated, $Q_{all}$ (according to Table 1), is calculated using Eq. (3) as below:

**Table 1** Rating of observed power transformer [3]

| Name plate rating | 180.0 MVA |
|---|---|
| $U_{HV}$ (kV)/$U_{LV}$ (kV) | 275.0/66.0 kV |
| Core losses | 128.3 kW |
| Copper losses (at rated load) | 848.0 kW |
| Stray losses | 113.0 kW |
| Mass of windings and core | 100.0 tons |
| Mass of oil | 80.0 tons |
| TOT (at half-rated load) | 41.3 °C |
| Cooling type | OFAF |
| Factory/year | Hack bridge/61 |

$$Q_{all} = 128.3\,kW + 848.0\,kW + 113.0\,kW = 1089.3\,kW \qquad (3)$$

It is reported in [3] that the thermal capacitance of oil ($C_{oil}$) and copper ($C_{Fe\&cu}$) for the transformer (details given in Table 1) is approximately equal to $144.0 \times 10^3$ kJ/K and $50.0 \times 10^3$ kJ/K, respectively. It is worth mentioning here that the thermal capacitance of the whole oil-immersed power transformer is the sum of thermal capacitance of oil and thermal capacitance of winding and core which is approximately equal to $195.0 \times 10^3$ kJ/K.

## 4 Identification of Thermal Parameters

Various offline experiments need to be performed for getting the values of thermal parameters [10]. It is practically difficult to perform a heat run test on long in-service power transformer by keeping it shut down. Furthermore, such method does not provide better results as parameters are varied with time [3]. To address this issue, evolutionary computation algorithms are used in the present paper. Three different search methods are used to identify the thermal parameters of the model shown in Fig. 1. Brief introduction of three optimization algorithms used in this paper is given below. Parameters in Fig. 1 are affected by several aging sensitive factors (like oil viscosity, conductive aging by-product, concentration of fural compound, dissolved gas, oil-moisture content). Each of the above-mentioned factors (and their interactions) influences TOT and BOT. Therefore, identification of Fig. 1 parameters can be considered to be analogous to finding the shortest path within a connected graph with the factors representing different nodes. Keeping in mind the connectivity and interactions of the above-mentioned factors in influencing the thermal model parameters, the present problem can be considered to a NP-hard problem.

## 4.1 Genetic Algorithm

Genetic algorithm (GA) is a population-based randomized search technique that works on the concept of natural selection and genetics [11]. GA is reported to have the potential to quickly figure out the global maximum or minimum values. It is different from the traditional optimization methods as initial estimation is not required in GAs, and it does not need to differentiate the cost function [11].

While using GAs, two major points should be remembered: First one is the conversion of a problem domain into bit pattern (genes), and other is the development of an effective *cost function* [12]. In order to obtain the *cost value*, the *cost function* needs to be defined on the basis of physical model of problem. In the case of binary coded GA, the strings which have higher *cost function* values have higher probability of selection to the next generation.

## 4.2 Particle Swarm Optimization

Particle swarm optimization (PSO) is another population-based technique inspired by the motion and intelligence of swarms. It is inspired from swarming habits of animals like fish and birds [13]. In this algorithm each individual, called particle, moves around the N-dimensional search space in search of the best solution [14]. In PSO, particles move randomly in search space in search of their best value. Each particle modifies its movement corresponding to its own motion and the movement of its neighbor. The following three steps of PSO algorithm are repeated until the optimum solution is obtained [14].

- Estimation of fitness of each particle;
- Updating of individual and global bests;
- Updating of position and velocity of each particle.

The *cost function* used here is similar to that used in the case of genetic algorithm. After each iteration, solution is obtained by evaluating the *cost function* corresponding to each particle to determine their relative fitness [15]. Thereafter, the position and velocity of each particle are updated depending on the following parameters:

- The best value achieved by each particle so far, called '$P_{best}$';
- Finest value obtained by any particle in population, called '$G_{best}$.'

## 4.3 Simulated Annealing (SA)

It is a local search-based method which comes from the name of physical process 'annealing' [16]. In annealing process, alloys of certain metal, crystal, or glass are tempered by heating above its melting point and then cooling it slowly until it solidifies into perfect crystalline form. The optimization algorithm inspired by this process is called simulated annealing. It is a probabilistic single-solution-based optimization technique [17]. SA is a general purpose algorithm used to find the global minimum solution of a *cost function* which may have many local minima. Simulated annealing comprises of two stochastic processes [16]: First one is generation of solution, and other is for acceptance of solution. The most important characteristics of SA over other methods are its ability for escaping from local minima [17]. It is very powerful tool to deal with nonlinear problems or functions which have multiple local minima. However, it is computation intensive in nature.

## 5 Implementation Procedure

$\Phi_2$ and $\Phi_3$ in Fig. 1 represent the top-oil temperature and bottom-oil temperature, respectively. The equation of top-oil temperature and bottom-oil temperature (derived from Eq. (1)) is shown as Eq. (4).

$$\phi_2 = \left[ \frac{G_2 + G_3}{2G_2 + G_3} \right] \left[ \left( \frac{Q_{all} - \exp^{\frac{-(G_1)t}{c_1}} - \exp^{\frac{-(G_1+G_2)t}{c_2}}}{G_2} \right) + \left( \frac{\exp^{\frac{-(G_2+G_3)}{c_3}t}}{G_2 + G_3} \right) \right]$$

$$\phi_3 = \left[ \frac{G_2}{2G_2 + G_3} \right] \left[ \left( \frac{Q_{all} - \exp^{\frac{-(G_1)t}{c_1}} - \exp^{\frac{-(G_1+G_2)t}{c_2}}}{G_2} \right) + \left( \frac{\exp^{\frac{-(G_2+G_3)t}{c_3}}}{G_2 + G_3} \right) \right] - \left( \frac{\exp^{\frac{-(G_2+G_3)t}{c_3}}}{G_2 + G_3} \right)$$

(4)

In order to show the capability of the three discussed optimization algorithms in estimating $\Phi_2$ and $\Phi_3$, 30 numbers of data points (top-oil and bottom-oil temperatures at different time) are selected from Figs. 5.14 and 5.15 in [3]. Table 2 shows ten such top- and bottom-oil temperatures that are obtained from Figs. 5.14 and 5.15 in [3].

It should be mentioned here that the definition of *cost function* considered in the present paper for all three optimization algorithms is given by Eq. (5).

$$F = \sum_{i=1}^{30} (TOT_{cal} - TOT_{mea})^2 + (BOT_{cal} - BOT_{mea})^2$$

(5)

The *cost function* shown by Eq. 5 is based on the total error between the measured values and predicted values of TOT and BOT. Here, Euclidean distance is

**Table 2** Measured value of TOT and BOT (in °C)

| Time (h) | Measured TOT (°C) | Measured BOT (°C) |
|----------|-------------------|-------------------|
| 0.0 | 30.0 | 29.0 |
| 9.6 | 32.0 | 17.0 |
| 14.4 | 32.0 | 16.0 |
| 19.2 | 37.0 | 30.0 |
| 36.0 | 34.0 | 16.0 |
| 43.2 | 37.0 | 18.0 |
| 57.6 | 34.0 | 16.0 |
| 67.2 | 38.0 | 34.0 |
| 79.2 | 23.0 | 16.0 |
| 84.0 | 38.0 | 19.0 |

used to calculate *cost function* as it is useful for continuous data analysis, and it reflects absolute distance between two points.

## 6 Results and Discussion

In this paper, six variables $G_1$, $G_2$, $G_3$, $C_1$, $C_2$, and $C_3$ of Eqs. 4 and 5 are identified using optimization algorithms. Values of above-mentioned thermal conductances and capacitances are obtained by developing MATLAB script. Once the values of these six parameters are identified, they are used to find the value of top-oil temperature and bottom-oil temperatures by using Eq. (4). These values are compared with the measured value of top-oil and bottom-oil temperatures (that are presented above in Table 2). A comparative analysis between measured and calculated temperature of TOT and BOT is shown below in Table 3 and Table 4, respectively.

**Table 3** Measured value and predicted value of TOT (°C) using three considered algorithms

| Measured value of TOT (°C) | Predicted value of TOT (°C) using GA | Predicted value of TOT (°C) using PSO | Predicted value of TOT (°C) using SA |
|----------------------------|--------------------------------------|---------------------------------------|--------------------------------------|
| 30.0 | 29.9977 | 30.00 | 29.9426 |
| 32.0 | 32.0492 | 32.00 | 31.9212 |
| 32.0 | 31.9767 | 32.00 | 31.9679 |
| 37.0 | 37.0525 | 37.00 | 36.9666 |
| 34.0 | 33.9767 | 34.00 | 34.0342 |
| 37.0 | 36.8915 | 37.00 | 37.0612 |
| 34.0 | 33.8843 | 34.00 | 33.9370 |
| 38.0 | 37.9630 | 38.00 | 37.9620 |
| 23.0 | 22.9568 | 23.00 | 23.0303 |
| 38.0 | 38.0004 | 38.00 | 38.0064 |

**Table 4** Measured value and predicted value of BOT (°C) using three considered algorithms

| Measured value of BOT (°C) | Predicted value of BOT (°C) using GA | Predicted value of BOT (°C) using PSO | Predicted value of BOT (°C) using SA |
|---|---|---|---|
| 29.0 | 28.9952 | 29.00 | 29.1268 |
| 17.0 | 17.2080 | 17.00 | 17.1097 |
| 16.0 | 15.9974 | 16.00 | 15.9914 |
| 30.0 | 30.0420 | 30.00 | 30.0246 |
| 16.0 | 15.9823 | 16.00 | 15.9292 |
| 18.0 | 17.8843 | 18.00 | 17.9837 |
| 16.0 | 16.8029 | 16.00 | 15.6969 |
| 34.0 | 33.9576 | 34.00 | 34.0313 |
| 16.0 | 15.9551 | 16.00 | 15.9454 |
| 19.0 | 18.9801 | 19.00 | 18.6034 |

A comparative analysis based on the optimized value of cost function is also shown in Table 5 for all three optimization techniques. Table 5 presents the performance of the three optimization algorithms in predicting top-oil and bottom-oil temperatures (shown in Table 2).

It can be observed from Table 5 that PSO gives minimum error among the three algorithms considered. In order to improve readability, the value of obtained parameters $G_1$, $G_2$, $G_3$, $C_1$, $C_2$, and $C_3$ at t = 14.4 h predicted by the three methods are presented in Table 6. Furthermore, the population size for PSO is suitably selected so that PSO does not get stuck to local minima during the course of optimization. Table 7 on the other hand shows the error between predicted and measured value of bottom- and top-oil temperatures at t = 14.4 h.

It is shown in Table 3 through Table 7 that the performance of PSO in estimating top-oil and bottom-oil temperatures is significantly better than that obtained by GA and SA. In order to improve the readability, the variation of *cost function* obtained for the PSO algorithm for data corresponding to t = 14.4 h is shown in Fig. 2. It can be observed from Fig. 2 that the *cost function* of PSO is minimized

**Table 5** Compared results between three different optimization techniques

| Cost function (F) | | | |
|---|---|---|---|
| Time (h) | GA | PSO | SA |
| 0.0 | 0.00002 | 0.00 | 0.01948 |
| 9.6 | 0.03346 | 0.00 | 0.01823 |
| 14.4 | 0.00055 | 0.00 | 0.00110 |
| 19.2 | 0.00451 | 0.00 | 0.00171 |
| 36.0 | 0.00085 | 0.00 | 0.00618 |
| 43.2 | 0.02517 | 0.00 | 0.00401 |
| 57.6 | 0.65800 | 0.00 | 0.09582 |
| 67.2 | 0.00316 | 0.00 | 0.00242 |
| 79.2 | 0.00388 | 0.00 | 0.00389 |
| 84.0 | 0.00039 | 0.00 | 0.15730 |

**Table 6** Obtained parameters of the model

| Parameters | PSO | GA | SA |
|---|---|---|---|
| $G_1$ | 24.6372 | 6.7467 | 30.4154 |
| $G_2$ | 22.6937 | 22.6815 | 22.6745 |
| $G_3$ | 22.6937 | 22.7072 | 22.6958 |
| $C_1$ | 131.5596 | 131.4610 | 177.1405 |
| $C_2$ | 18.6601 | 2.3260 | 13.7145 |
| $C_3$ | 44.7802 | 61.2129 | 4.1449 |

**Table 7** Error of predicted BOT and predicted TOT

| Optimization techniques | $\Delta$BOT | $\Delta$TOT |
|---|---|---|
| PSO | 0.00 | 0.00 |
| GA | 0.0026 | 0.0233 |
| SA | 0.0086 | 0.0321 |

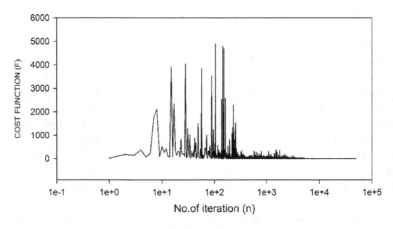

**Fig. 2** Convergence of PSO for data obtained t = 14.4 h

after approximately 8000 no. of iterations. Classic version of the PSO reported by [13] is used in the present paper. At present, the authors are engaged in studying the performance of other versions of PSO (including SA-aided PSO and hybrid PSO). The results obtained will be reported in future communications.

# 7 Conclusion

The present paper presents the performance of three different optimization algorithms in identifying the thermal model of a real-life power transformer. In the paper, a simplified thermoelectric analogous model of oil-immersed power transformer has been taken for analysis [4]. Thirty numbers of measured top- and bottom-oil temperatures are used to test the performance of the optimization algorithms. GA, PSO, and SA are used to identify and optimize the thermal parameters (thermal capacitance and conductance). It is observed from the above results that the predicted value of TOT and BOT obtained by using the PSO algorithm is more approaching to the measured valued of TOT and BOT. The comparative analysis based on the value of the *cost function* shows that the results PSO provide less error than other two optimization technique, GA and SA. So the related analysis presented in the paper shows that the performance of PSO is superior compared to the result obtained by GA and SA.

# References

1. International Electrotechnical Commission (1991) IEC60354: Loading guide for oil-immersed power transformers. International Electrotechnical Commission Standard, Geneva, Switzerland
2. Transformers committee of the IEEE Power Engineering Society (1991) IEEE Guide for loading mineral oil-immersed transformers. IEEE Standard, C 57.91–1995
3. Tang WH, Wu QH (2011) Condition monitoring and assessment of power transformers using computational intelligence. Springer Science & Business Media. https://doi.org/10.1007/978-0-85729-052-6_4
4. Tang WH, Wu QH, Richardson ZJ (2004) A simplified transformer thermal model based on thermal-electric analogy. IEEE transactions on power delivery 19(3), 1112–1119. https://doi.org/10.1109/tpwrd.2003.822926
5. Radakovic Z, Feser K (2003) A new method for the calculation of the hot-spot temperature in power transformers with ONAN cooling. In: IEEE Transactions on Power Delivery, vol. 18, no. 4, pp. 1284–1292. https://doi.org/10.1109/tpwrd.2003.817740
6. Fourier (1992) Analytical theory of heat (translated by Freeman, A), Stechert, New York
7. Simonson JR (1981) Engineering heat transfer. Springer, The city University, London. https://doi.org/10.1007/978-1-349-19351-6
8. Tang WH, Wu QH, Richardson ZJ (2002) Equivalent heat circuit based power transformer thermal model. In IEE Proceedings on Electric Power Applications 149, no. 2, pp. 87–92. https://doi.org/10.1049/ip-epa:20020290
9. Popescu MC, Bulucea CA and Perescu LI (2009) Improved Transformer Thermal Models. In WSEAS Transactions on Heat and Mass Transfer 4(4): 87–97
10. Tenbohlen, Stefan, Coenen S, Djamali M, Müller A, Samimi M H, Siegel M (2016) Diagnostic measurements for power transformers. Energies 9, no. 5, 347. https://doi.org/10.3390/en9050347
11. Goldberg D (1989) Genetic algorithms in search, optimization, and machine learning. Addison-Wesley publishing company, Inc., USA. https://doi.org/10.1023/a:1022602019183
12. Holland J (1975) Adaptation in natural and artificial systems. An introductory analysis with application to biology, control, and artificial intelligence, Ann Arbor, MI: University of Michigan Press, USA
13. Kennedy J, Eberhart RC (1995) Particle swarm optimization. In Proc. IEEE int'l conf. on neural networks Vol. IV, pp. 1942–1948. IEEE service center, Piscataway, NJ
14. Blondin J (2009) Particle swarm optimization: A tutorial. Available from: http://cs.armstrong.edu/saad/csci8100/psotutorial.pdf

15. Ali M, Taghikhani (2012) Power transformer top oil temperature estimation with GA and PSO methods Energy and Power Engineering. https://doi.org/10.1109/61.891504
16. Du KL, Swamy MN (2016) Simulated Annealing. In Search and Optimization by Metaheuristics. Springer International Publishing, pp. 29–36. https://doi.org/10.1007/978-3-319-41192-7_2
17. Sindekar AS, Agrawal AR, Pande VN (2013) Comparison of Some Optimization Techniques for Efficiency Optimization of Induction Motor, International Journal of Engineering Science and Technology 5.6: 1303. 10.1.1.348.4648

# Hierarchical Modeling Approaches for Generating Author Blueprints

**G. S. Mahalakshmi, G. Muthu Selvi and S. Sendhilkumar**

**Abstract** Bibliometrics has developed into a prominent research field that provides way for evaluating and bench-marking research performance. However, the performance of researchers is graded on par with one another within a scientific publication. Though theories do exist in the literature for provision of reportedly higher credits for the first author, reality is very trying due to the extent of possible manipulations. Mining author contributions would bring in semantic perspective to a certain extent. But this mining has to happen with the foundation of author profiles. Generation of individual author profiles or author blueprints would create provisions for analyzing the extent of author's contribution to a publication. This paper proposes the idea of generating author blueprints from author's publication histories across domains using hierarchical Dirichlet processes (HDP).

**Keywords** Document Topic Model · Author-Topic Model · Author blueprint
Latent Dirichlet allocation · Hierarchical Dirichlet process

## 1 Introduction

Bibliometrics is the process of analyzing the research, researcher, and the research articles in a quantitative manner. Authors of a scientific paper should be limited to those individuals based on their attribution and contribution. Attribution is a kind of

G. S. Mahalakshmi · G. Muthu Selvi (✉)
Department of Computer Science and Engineering, Anna University,
Chennai 600025, Tamil Nadu, India
e-mail: gmuthuselvi16@gmail.com; gmselvi2012@gmail.com

G. S. Mahalakshmi
e-mail: gsmaha@annauniv.edu

S. Sendhilkumar
Department of Information Science and Technology, Anna University,
Chennai, Tamil Nadu, India
e-mail: ssk_pdy@yahoo.co.in

© Springer Nature Singapore Pte Ltd. 2019
B. K. Panigrahi et al. (eds.), *Smart Innovations in Communication and Computational Sciences*, Advances in Intelligent Systems and Computing 669, https://doi.org/10.1007/978-981-10-8968-8_35

classification problem which deals with the determination of an author of an anonymous document or text with the additional features called stylometry [5, 18]. Author contribution [2, 10] deals with identifying the role of an author in a research paper [6]. However, analyzing the researcher's performance [7, 4] in a multi-author paper is a difficult task. Every author has their own style and idea which reflects during article writing. In order to derive contribution statistics, we need to tap on the topics contributed by every author, and these topics might overlap with that of co-author's topic contributions. Topic modeling approaches shall be applied for topic contribution analysis. These topics shall be bundled to form the Author Profile Models (APMs) (or) author blueprints.

## 2   Related Work

Generation of APMs depends on the availability of author publication history. Rexha A et al. [14] analyzed the Research Papers Dataset, a free database created by the US National Library which houses full-text articles. Seroussi Y et al. [16] have done the experimentation with five research article datasets like Judgment, PAN 11, IMDb62, IMDb1M, and Blog. Song M et al. [17] analyzed the author's publishing history dataset which is collected from DBLP: Computer Science Bibliography.

Modeling the entire research article with topic models is the fundamental step for APM generation and author contribution mining. It brings out the most frequently and most likely used words in the documents. Topic models work by defining a probabilistic representation of the latent factors of corpora called topics. Girgis MR et al. [3] discuss the authorship attribution using different Topic Modeling Algorithms. Pratanwanich et al. [13] proposed Latent Dirichlet Allocation (LDA) implementation using the Collapsed Gibbs Sampling Method. Rubin TN et al. [15] model Flat-LDA, Prior-LDA, and Dependency-LDA, but the performance rapidly drops off as the total number of labels and the number of labels per document increases.

LDA cannot be used to model the authors. Author-Topic Model (ATM) is proposed with the aim of identifying topic word distribution of anonymous documents and authors. Author-Topic Model identifies the topics of authors based on the publishing history dataset to predict author's interest [16]. However, ATMs shall not be used for modeling the research articles. In DADT, words are generated from two disjoint sets of topics: Document Topics (DT) and Author Topics (AT). Disjoint Author-Document-Topic Model (DADT) is a hybrid model which uses both document topics and author topics [16]. DADT yields a document topic distribution for each document, and an author-topic distribution for each author. Author identification derives the Author topics from a large collection of unstructured texts [22]. Author identification is also approached using sequential minimal optimization [7]. Authorship profiling is the problem of determining the characteristics of a set of authors based on the text they produce [1]. "Authorship Index" is an alternative to measure an author's contribution to the literature [9].

# 3 Generation of Author Blueprint via HDP

In topic-based generative models [12], a document is described by a particular topic proportion, where a topic is defined as a distribution over words. LDA, a mixed-membership topic model, was introduced whereas the authorship model works based on the interests of author and word distributions instead of the latent topics. The hierarchical Dirichlet process (HDP) is a Bayesian approach that works in a nonparametric manner. It uses a Dirichlet process which involves parameters with assigned distributions initially. As the processes evolve, the distributions recurse naturally to form distributions within themselves thereby introducing new parameters, and this continues until all the topics are examined [11, 19–21].

In this section, we compare LDA and HDP on the research articles to derive the Document Topics (DT) and Author Topics (AT). HDP provides best topic distribution over LDA because it derives an unlimited number of topics as well as the relation between topics based on the content of the document, but LDA derives only specified number of topics [8].

## 3.1 Hierarchical Dirichlet Process Topic Model

In this section, we attempt to explain the parameters of HDP and the process behind generation of topics. Let us assume $|D|$ documents, $|K|$ topics, and $|V|$ vocabulary.

1. For $k = 1, \ldots, K$ topics: $\phi^{(k)} \sim Dirichlet(\beta)$

   $\beta$ is a vector of length $|W|$ which is used to affect how word distribution of topic $k$, $\phi^{(k)}$, is generated. This step is essentially saying that, since we know you have $k$ topics, you should expect to have $k$ different word distributions.
2. Now we have $k$ word distributions. Next we want to generate documents. For each document $d \in D: \theta_d \sim Dirichlet(\alpha)$.

We first need to determine its topics. Actually, document $d$ has a topic distribution $\theta_d$. Even though a document has a strong single theme, it more or less covers many different topics. For each word $\omega_i \in d$:

(i) $$z_i \sim Discrete(\theta_d)$$

(ii) $$\omega_i \sim Discrete\left(\phi^{(z_i)}\right)$$

Before each word in the document $d$ is written, the word should be assigned to one single topic $z_i$ at first place. Such topic assignment for each word $\omega_i$ follows the topic distribution $\theta_d$ you got in (a). After we know which topic this word belongs to, we finally determine this word from the word distribution $\phi^{(z_i)}$. $\alpha$ and $\beta$ are called hyperparameters.

LDA provides Dirichlet distribution, word distribution for a specific topic, or topic distribution for a specific document. It is from the drawn word distribution that words are finally generated but not the Dirichlet distribution [11].

$\alpha, \beta$ and $\vec{\omega}$—known initially, and $\theta, \phi$ and $\vec{z}$—unknown initially, where $\vec{\omega}$ is a vector of word counts of the documents, $\theta$ is a topic distributions of the documents, $\phi$ is a word distributions of topics, and $\vec{z}$ is a vector of length $|W|$ representing each word's topic. So ideally, with $p(\vec{\omega}|\phi_z)$ as conditional probability, we want to know:

$$p(\theta, \phi, \vec{z}|\vec{\omega}, \alpha, \beta) = \frac{p(\theta, \phi, \vec{z}, \vec{\omega}|\alpha, \beta)}{p(\vec{\omega}|\alpha, \beta)} = \frac{p(\phi|\beta)p(\theta|\alpha)p(\vec{z}|\theta)p(\vec{\omega}|\phi_z)}{\iiint p(\phi|\beta)p(\theta|\alpha)p(\vec{z}|\theta)p(\vec{\omega}|\phi_z)d\theta d\phi d\vec{z}} \quad (1)$$

## 3.2 Generation of Author Blueprints

In this section, we propose methods for generation of author blueprints. For this, we utilize the author's publishing history retrieved from DBLP. If the author has produced at least one single-author publication (authored one and only by oneself) in respective domain, we assume the topic distribution out of that research article. If there are no single-author research articles, we proceed to retrieve the particular author's topics by analyzing the document's topics over author topics including the co-authors as well. Therefore, this is a recursive process if the co-author does not possess single-author history.

Let us consider there is a research article $P_0$ authored by two authors $(R_0, R_1)$ (refer Table 1). Let us assume that the single-author history of research articles is available only for $R_0$. In other words, the single-author history of $R_0$ is also known as $ATM(R_0)$. ATM is aka Author-Topic Model or author blueprint. This is obtained by aggregating the Document Topic Model (DTM) of every single-author publications of $R_0$. DTM is obtained by generating topic model for the respective research article. Therefore, if $R_0$ has authored $S$ research articles as sole author, then,

$$ATM(R_0) = DTM(s_0) + \cdots + DTM(s_{n-1}) \quad (2)$$

However, aggregation of topics of sole author publications means normalizing (or averaging) the respective probabilities. Let us consider an author has published articles on topic 'gene.' When this topic appearing in one article with probability 0.45 is aggregated with the occurrences of 'gene' from other two articles written by

**Table 1** Methodologies for generation of author blueprints

| Article # | Authors | Single-author history (availability) | Method |
|---|---|---|---|
| $P_0$ | $R_0, R_1$ | $R_0$ | $ATM(R_1) = DTM(P_0) - ATM(R_0)$ |
| $P_0$ $P_1$ | $R_0, R_1$ $R_0, R_1, R_2$ | $R_0, R_1$ | $ATM(R_2) = DTM(P_1) - DTM(P_0)$ $ATM(R_2) = DTM(P_1) - (ATM(R_0) \cup ATM(R_1))$ |
| $P_1$ $P_2$ | $R_0, R_1, R_2$ $R_2, R_3, R_4$ | Not available | $ATM(R_2) = DTM(P_1) \cap DTM(P_2)$ |
| $P_0$ $P_3$ | $R_0, R_1$ $R_0, R_1, R_2$ | $R_0$ | $ATM(R_1) = DTM(P_0) - ATM(R_0)$ $ATM(R_2) = DTM(P_3) - (ATM(R_0) \cup ATM(R_1))$ |

the same author with topic probability 0.7, 0.89, respectively, the aggregated topic probability of 'gene' is 0.68. This indicates that the author has proficiency of 68% related to 'gene.' However, this is not true if we consider the periodicity of article publication. If the first article is published earlier and had been followed by second and third articles, the latest research article on 'gene' has the proficiency 89%. But, by aggregating the topic probabilities of author, we have reduced the author proficiency on the same topic to 68% which is absurd. This misinterpretation if allowed to persist might cascade into deciding reduced contribution of the author on research topic 'gene.' However, for discussion on ATM, we have assumed the aggregation of topic probabilities.

$$ATM(R_1) = DTM(P_0) - ATM(R_0) \tag{3}$$

The above process (Eq. 3) is referred to as 'ATM by Subtraction.' Suppose if there are three authors $R_0, R_1, R_2$ who have authored research article $P_1$; and a research article $P_0$ authored by two authors $(R_0, R_1)$ also exists. The single-author history of $R_0, R_1$ exists; and the objective is to find the ATM of researcher $R_2$. This could be approached by two variations: 1. utilizing DTMs and 2. utilizing both DTM and ATM. In the first approach, the DTM of both the articles $P_0, P_1$ is to be obtained separately and later, the DTM of two author research articles $P_0$ has to be subtracted from the DTM of $P_1$.

$$ATM(R_2) = DTM(P_1) - DTM(P_0) \tag{4}$$

This is under the assumption that the overlap boundary of research topics falls only over the zone of popular authors $R_0, R_1$ and concepts otherwise new, if any, only would be assigned to $R_2$. This is very similar to the second variation: aggregating the blueprints of $R_0, R_1$ and then subtracting it from the DTM of $P_1$. This procedure is known as 'ATM by Nested Subtraction.'

$$ATM(R_2) = DTM(P_1) - (ATM(R_0) \cup ATM(R_1)) \tag{5}$$

Even if no single-author history is available, we can still find the ATM of the author. Consider articles $P_1$ authored by $R_0, R_1, R_2$ and $P_2$ authored by $R_2, R_3, R_4$. The intersection of DTMs of both the articles would deliver the common topics in both the articles. Since the author $R_2$ is common in both the articles, we conclude that the common topics are of the author $R_2$. This is called 'ATM by Intersection.'

$$ATM(R_2) = DTM(P_1) \cap DTM(P_2) \tag{6}$$

Consider articles $P_0$ authored by $R_0, R_1$ and $P_3$ authored by $R_0, R_1, R_2$. If only one single-author history is available, i.e., $ATM(R_0)$, then first the ATM of $R_1$ shall be formed by Eq. (3). Following this, the ATM of $R_2$ is found using Eq. 7.

$$ATM(R_2) = DTM(P_3) - (ATM(R_0) \cup ATM(R_1)) \tag{7}$$

## 4 Results and Discussion

The objective of generating author blueprints is to facilitate author contribution mining. For this, we have assumed the articles of author 'Catherine Blake' from Biomedical Informatics Domain. The author has five articles as sole author and 15 articles as co-author (refer Tables 2 and 5). Table 3 gives the top ten words of top five topics under LDA for research articles in Table 2. Table 4 lists the topics obtained under HDP.

**Table 2** Publications of 'Catherine Blake' as sole author

| S. no. | Single-author publications |
|---|---|
| 1 | Blake, Catherine. "A Technique to Resolve Contradictory Answers." New Directions in Question Answering. 2003 |
| 2 | Blake, Catherine. "Information synthesis: A new approach to explore secondary information in scientific literature." Proceedings of the 5th ACM/IEEE-CS joint conference on Digital libraries. ACM, 2005 |
| 3 | Blake, Catherine. "A comparison of document, sentence, and term event spaces." Proceedings of the 21st International Conference on Computational Linguistics and the 44th annual meeting of the Association for Computational Linguistics. Association for Computational Linguistics, 2006 |
| 4 | Blake, Catherine. "Beyond genes, proteins, and abstracts: Identifying scientific claims from full-text biomedical articles." Journal of biomedical informatics 43.2 (2010): 173–189 |
| 5 | Catherine Blake. "Text mining." ARIST 45(1): 121–155 (2011) |

**Table 3** Topic words for 'Catherine Blake' sole author—LDA

| Topic 0 | | Topic 1 | | Topic 2 | | Topic 3 | | Topic 4 | |
|---|---|---|---|---|---|---|---|---|---|
| Information | 0.156 | Claim | 1.181 | Text | 1.240 | Information | 1.277 | Term | 1.171 |
| Cancer | 0.117 | Claims | 1.178 | Literature | 1.175 | Text | 1.240 | Idf | 1.091 |
| Breast | 0.078 | Change | 1.165 | Extraction | 1.083 | Mining | 1.215 | Document | 1.045 |
| Analysis | 0.078 | Collection | 1.164 | Journal | 1.071 | Proceedings | 1.203 | Documents | 1.036 |
| Articles | 0.039 | Development | 1.157 | Based | 1.066 | Conference | 1.188 | Terms | 1.035 |
| Alcohol | 0.039 | Quality | 1.133 | Document | 0.481 | System | 1.181 | Language | 1.030 |
| Study | 0.034 | Grammar | 1.125 | Provide | 0.429 | Knowledge | 1.142 | Sentence | 1.002 |
| Metis | 0.033 | Framework | 1.122 | Words | 0.416 | Methods | 1.037 | Corpus | 0.511 |
| Synthesis | 0.032 | Precision | 1.095 | Terms | 0.412 | Association | 1.014 | Frequency | 0.505 |
| Article | 0.031 | Sentence | 0.487 | Relationship | 0.404 | Discovery | 0.434 | Spaces | 0.496 |

**Table 4** Topic words for 'Catherine Blake' sole author—HDP

| Topic 0 | | Topic 1 | | Topic 2 | | Topic 3 | | Topic 4 | |
|---|---|---|---|---|---|---|---|---|---|
| People | 0.993 | Relationship | 0.891 | Constrained | 0.892 | Report | 0.893 | Multiple | 0.921 |
| Cancer | 0.986 | People | 0.855 | Problem | 0.859 | Sources | 0.790 | Hierarchy | 0.919 |
| Explores | 0.865 | Worldwide | 0.821 | Biostatistics | 0.813 | Age | 0.768 | Contrast | 0.918 |
| Motivated | 0.854 | Information | 0.754 | Technique | 0.725 | Geographic | 0.658 | Banko | 0.893 |
| Interactions | 0.843 | Behavior | 0.742 | Resolve | 0.715 | Scientific | 0.644 | Full | 0.833 |
| Seventy | 0.733 | Generation | 0.613 | Contradictory | 0.703 | Breast | 0.598 | Considered | 0.810 |
| Epidemiology | 0.532 | Manual | 0.545 | Answers | 0.523 | Describe | 0.567 | Resolve | 0.754 |
| Prototype | 0.431 | Answers | 0.533 | Therine | 0.515 | Respect | 0.531 | Contradictory | 0.715 |
| Sector | 0.426 | Therine | 0.475 | Publication | 0.490 | Boulder | 0.522 | Answers | 0.705 |
| Variety | 0.413 | Blake | 0.458 | Technical | 0.470 | Gravano | 0.499 | Therine | 0.670 |

**Table 5** Publications of 'Catherine Blake' as co-author

| Article Id. | Multi-author publications |
|---|---|
| 1 | Blake, Catherine, and Wanda Pratt. "Multiple categorization of search results." Proceedings of the AMIA Symposium. American Medical Informatics Association, 2000 |
| 2 | Blake, Catherine, and Wanda Pratt. "Better rules, fewer features: a semantic approach to selecting features from text." Data mining, 2001. ICDM 2001, Proceedings IEEE international conference on. IEEE, 2001 |
| 3 | Blake, Catherine, Wanda Pratt, and Tammy Tengs. "Automated information extraction and analysis for information synthesis." Proceedings of the AMIA Symposium. American Medical Informatics Association, 2002 |
| 4 | Blake, Catherine, et al. "A study of annotations for a consumer health portal." Digital Libraries, 2005. JCDL'05. Proceedings of the 5th ACM/IEEE-CS Joint Conference on. IEEE, 2005 |
| 5 | Blake, Catherine, et al. "Cataloging on-line health information: a content analysis of the NC Health Info portal." AMIA Annual Symposium Proceedings. Vol. 2005. American Medical Informatics Association, 2005 |
| 6 | Blake, Catherine, and Meredith Rendall. "Scientific Discovery: A view from the trenches." International Conference on Discovery Science. Springer Berlin Heidelberg, 2006 |
| 7 | AbdelRahman, Samir, and Catherine Blake. "A rule-based human interpretation system for semantic textual similarity task." Proceedings of the First Joint Conference on Lexical and Computational Semantics-Volume 1: Proceedings of the main conference and the shared task, and Volume 2: Proceedings of the Sixth International Workshop on Semantic Evaluation. Association for Computational Linguistics, 2012 |
| 8 | Guo, Jinlong, Yujie Lu, Tatsunori Mori, and Catherine Blake. "Expert-guided contrastive opinion summarization for controversial issues." In Proceedings of the 24th International Conference on World Wide Web, pp. 1105–1110. ACM, 2015 |
| 9 | Blake, Catherine, et al. "The Role of Semantics in Recognizing Textual Entailment." TAC. 2010 |
| 10 | Blake, Catherine, and Wanda Pratt. "Collaborative information synthesis." Proceedings of the American Society for Information Science and Technology 39.1 (2002): 44–56 |
| 11 | Blake, Catherine, and Wanda Pratt. "Collaborative information synthesis I: A model of information behaviors of scientists in medicine and public health." Journal of the Association for Information Science and Technology 57.13 (2006): 1740–1749 |
| 12 | West, Suzanne L., Catherine Blake, Zhiwen Liu, J. Nikki McKoy, Maryann D. Oertel, and Timothy S. Carey. "Reflections on the use of electronic health record data for clinical research." Health Informatics Journal 15, no. 2 (2009): 108–121 |
| 13 | Lučić, Ana, and Catherine L. Blake. "A syntactic characterization of authorship style surrounding proper names." Digital Scholarship in the Humanities 30.1 (2015): 53–70 |
| 14 | Zheng, Wu, and Catherine Blake. "Using distant supervised learning to identify protein subcellular localizations from full-text scientific articles." Journal of biomedical informatics 57 (2015): 134–144 |
| 15 | Blake, Catherine, and Ana Lucic. "Automatic endpoint detection to support the systematic review process." Journal of biomedical informatics 56 (2015): 42–56 |

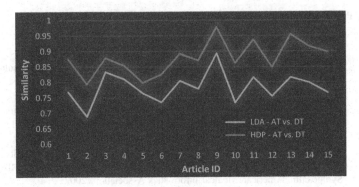

**Fig. 1** Comparison of author topics $\tau(R_t)$ with various document topics $\delta_k$

We examined the cosine similarity of author topics obtained via LDA with that of the author's co-authored publications (Table 5). Let us assume $\tau$ indicates ATM, and $\delta$ indicates DTM, then, cosine similarity is given by,

$$\cos(\tau(R_t), \delta_k) = \frac{\sum_{i=1}^{m} \sum_{j=1}^{n} \tau(R_{ti}) * \delta_{kj}}{\sqrt{\sum_{i=1}^{m} \tau(R_{ti})^2} * \sqrt{\sum_{j=1}^{n} (\delta_{kj})^2}} \tag{8}$$

where $R_t$ indicates the author t; $\delta_k$ indicates the DTM of $k$th article authored by $R_t \& k > 0; i, j$ indicates the topics of $\tau$ and $\delta$.

This is to indirectly convey that the author has contributed the 'indicated' percentage of contents to the research article among the co-authors. We repeated the experiments for HDP topics. It is evident (Fig. 1) that the enrichment available in HDP topics elevated the similarity value much better when compared to LDA.

HDP is unable to out-perform LDA when there is not enough publishing history available for authors. Challenges remain if the list of co-authors in an article is high in number, the hardly found single research article that supports the ATM generation might cease to exist or had been published too early in the timeline; and the co-author(s) do not have any sole publishing history. In such cases, the ATMs need to be generated for those new co-author(s) beforehand to be able to conclude the ATM of 'Catherine Blake.'

## 5 Conclusion

This paper proposes utilization of HDP for generation of author blueprints or Author-Topic Models. The ATM is generated using various statistical approaches. However, deriving the ATM of an author via one research article will not be equivalent to that of obtaining from another research article written by same author. In other words, the statistical approaches proposed involved topics obtained via topic modeling approaches and therefore, topics from research article and topics from author blueprint are considered on par with one another. This is not very desirable given the high quantum and periodicity of research articles generated every day. Further, we have not had any reservations over the ATMs derived using articles across the research timelines. If such reservations, say, articles of latest N years are to be considered, more interesting research issues might open up. A deep research into deciding upon the boundaries of author topics would facilitate the real blueprint of a given author.

**Acknowledgements** The authors would like to thank Department of Science and Technology for supporting this research by Inspire Fellowship No: DST/INSPIRE Fellowship/2013/505.

# References

1. Das, P. K. (2015), "Authorship Pattern and Research Collaboration of Journal of Informatics". *International Journal of Information Dissemination and Technology*, 5(1), 53.
2. Ge .Z and Sun .Y, "Author Identification using Sequential Minimal Optimization", In SoutheastCon, 33:4, pp. 1–2, 2016.
3. Girgis .MR, Aly .AA and Azzam .F.M.E, "Authorship Attribution with Topic Models", Computational Linguistics, 40:2, pp. 269–310, 2014.
4. G. MuthuSelvi, G.S. Mahalakshmi and S. Sendhilkumar, "An Investigation on Collaboration Behavior of highly Cited Authors in Journal of Informetrics (2007–2016)", Journal of Computational and Theoretical Nanoscience, 3803 (Accepted).
5. G. MuthuSelvi, G.S. Mahalakshmi and S. Sendhilkumar, "Author attribution using Stylometry for Multi-author Scientific Publications", Advances in Natural and Applied Sciences, 2016 June, 10(8): Pages: 42–47.
6. G.S. Mahalakshmi, G .MuthuSelvi and S. Sendhilkumar, "A Bibliometric Analysis of Journal of Informetrics—A Decade Study", International Conference on Recent Trends and Challenges in Computational Models (ICRTCCM'17), organized by Department of Computer Science and Engineering, University College of Engineering, Tindivanam, 2017.
7. G.S. Mahalakshmi, G. MuthuSelvi and S. Sendhilkumar, "Authorship Analysis of JOI Articles (2007–2016)", International Journal of Control Theory and Applications, 9(10), 2016, pp. 1–11 ISSN: 0974-5572.
8. G.S. Mahalakshmi, G. MuthuSelvi and S. Sendhilkumar, "Generation of Author Topic Models using LDA", International Conference on Intelligent Computing and Control Systems (ICICCS 2017) is being organized on 15–16, June 2017 by The Vaigai College Engineering (VCE), Madurai.
9. G.S. Mahalakshmi, G. MuthuSelvi, and S. Sendhilkumar, "Measuring Authorial Indices from the Eye of Co-Author(s)", International Conference on Smart Innovations in Communications and Computational Sciences (ICSICCS-2017) organizing by North West Group of Institutions, Moga, Punjab, India during 23–24 June 2017.
10. G.S. Mahalakshmi, G. MuthuSelvi and S. Sendhilkumar, "Measuring author contributions via LDA", 2nd International Conference on Advanced Computing and Intelligent Engineering, Central University of Rajasthan, Ajmer, India. (Under Review).
11. G.S. Mahalakshmi, G. MuthuSelvi and S. Sendhilkumar, "Gibbs Sampled Hierarchical Dirichlet Mixture Model based Approach for Clustering Scientific Articles", Pattern Recognition and Machine Intelligence (PREMI-2017), Kolkata, India, 2017 (Under Review).
12. Poulston .A, Stevenson .M and Bontcheva .K, "Topic Models and Ngram Language Models for Author Profiling", In Proceedings of CLEF, 15:3, pp. 113–117, 2015.

13. Pratanwanich .N and Lio .P, "Who Wrote This? Textual Modeling with Authorship Attribution in Big Data", In the Proceedings of the 2014 IEEE International Conference on Data Mining Workshop, 16:4, pp. 645–652, 2014.
14. Rexha .A, Klampfl .S, Kroll .M and Kern R. "Towards a more Fine Grained Analysis of Scientific Authorship: Predicting the Number of Authors using Stylometric Features", In the Proceedings of the 3rd Workshop on Bibliometric-enhanced Information Retrieval (BIR2016),1567, pp. 26–31, 2016.
15. Rubin .TN, Chambers .A, Smyth .P and Steyvers .M, "Statistical Topic Models for Multi-Label Document Classification", Machine learning, 88:(1–2), pp. 157–208, 2012.
16. Seroussi .Y, Bohnert .F and Zukerman .I, "Authorship Attribution with Author-aware Topic Models", In the Proceedings of the 50th Annual Meeting of the Association for Computational Linguistics, 2, pp. 264–269, 2012.
17. Song .M, Heo .GE and Kim .SY, "Analyzing Topic Evolution in Bioinformatics: Investigation of Dynamics of the Field with Conference Data in DBLP", Scientometrics, 101:1, pp. 397–428, 2014.
18. Steven H. H. Ding, Benjamin C. M. Fung, Farkhund Iqbal, William K. Cheung, "Learning Stylometric Representations for Authorship Analysis", arXiv preprint, arXiv:1606.01219, pp. 22–33, 2016.
19. Teh, Y. W.; Jordan, M. I.; Beal, M. J.; Blei, D. M. (2006). "Hierarchical Dirichlet Processes" (PDF). *Journal of the American Statistical Association.* 101: *pp.* 1566–1581.
20. Teh, Y. W., & Jordan, M. I. (2010). Hierarchical Bayesian nonparametric models with applications. Bayesian nonparametrics, 1.
21. Teh, Yee Whye, Michael I. Jordan, Matthew J. Beal, and David M. Blei. "Sharing Clusters among Related Groups: Hierarchical Dirichlet Processes." In NIPS, pp. 1385–1392. 2004.
22. Zhang .C, Wu .X, Niu .Z and Ding .W, "Authorship identification from unstructured texts", Knowledge-Based Systems, 66, pp. 99–111, 2014.

# Measuring Authorial Indices from the Eye of Co-author(s)

**G. S. Mahalakshmi, G. Muthu Selvi and S. Sendhilkumar**

**Abstract** The author's productivity and impact in a research paper mainly deal with the order of the positions in their publications. However, it is actually very difficult to measure the author's contribution on a particular article based on the author position. In this paper, we propose three authorial indices for measuring scientific authorships. Modified authorship index (MAI) is proposed as a measure for analysing the author's overall contribution towards a scientific publication. MAI accommodates the author position information within a scientific publication, thereby generating position-based contribution score for every publication of the respective author. Further, the proposed Author Contribution to Article Visibility ($ACAV_{MAI}$) score accommodates the role of citations in creating sufficient visibility for the respective publication. In addition, this paper also proposes consolidated authorship index ($AI_c$) for each author which is derived from the author's $ACAV_{MAI}$. The proposed indices are compared with the citation count, h-index and i10-index across the authors of "Journal of Informetrics", and the results are tabulated.

**Keywords** Author's productivity · Modified authorship index (MAI)
Consolidated authorship index ($AI_c$) · Citations · h-index · i10-index

G. S. Mahalakshmi · G. Muthu Selvi (✉)
Department of Computer Science and Engineering, Anna University,
Chennai 600025, Tamil Nadu, India
e-mail: gmuthuselvi16@gmail.com; gmselvi2012@gmail.com

G. S. Mahalakshmi
e-mail: gsmaha@annauniv.edu

S. Sendhilkumar
Department of Information Science and Technology, Anna University, Chennai,
Tamil Nadu, India
e-mail: ssk_pdy@yahoo.co.in

© Springer Nature Singapore Pte Ltd. 2019
B. K. Panigrahi et al. (eds.), *Smart Innovations in Communication
and Computational Sciences*, Advances in Intelligent Systems
and Computing 669, https://doi.org/10.1007/978-981-10-8968-8_36

423

# 1   Introduction

Research paper publication is an important aspect in assessing the author's reputation. A strict criterion has been laid down for determining authorship; it is difficult to exactly assess the relative credit each author gets for a particular paper towards the total work done. The scientific content in the research paper is quantified based on the quality of journal [7] in which the paper has been published and the number of citations received by the paper. The position of the authors in a research paper [33] serves as an important aspect in assessing the author's impact on a particular publication. Author-level metrics like citations, h-index [3, 20, 27], i10-index measure the author's impact on a research paper. According to the guidelines given by International Committee of Medical Journal Editors (ICJME), an author is generally considered to be someone who has made substantial intellectual contribution towards a scientific literature.

The author's intellectual output can be assessed by various indices which help us to explore related research communities [26, 29, 31]. Such indices would further help us to identify similar authors [32] which aid author contribution analysis [17, 34]. Measuring scientific research shall be dealt in two ways: statistical- [30] as well as semantic-based [12, 15, 16, 18, 19] approaches. In this paper, we propose authorial indices like modified authorship index (MAI) and consolidated authorship index ($AI_c$) for measuring research.

The visibility of an article indicates the number of citations received by the article. MAI and $AI_c$ have been included as a metric in the author-level metrics (ALMETRICS). The author-level metrics influence the visibility of the article and increases the quality of the paper [10, 11]. One approach to determine the pure impact of the author is to assume that the article visibility depends on the past publications made by the author. Basically, the first author's impact is more compared to the second and other authors. The $ACAV_{MAI}$ is calculated for each of the sample case taken which indicates the author's contribution on a particular article. This is influenced by the co-authors; i.e., the value of $ACAV_{MAI}$ differs for an author on a research article, based on the co-authors. The MAI only takes into account individual author's publications. The consolidated $ACAV_{MAI}$ includes individual author's publications plus the co-author's research strength.

# 2   Related Works

In recent days, publications of research papers have been widely increased. Measuring the author's contribution [4] to an article becomes a challenging task. CAV indicator [8] has been used to measure the author's contribution to the visibility of the article. Two different approaches have been used. The first approach deals with determining an indicator, and the second approach indicates weighted average method and union method. The number of citations has been considered as the main measurement for the author's productivity. The relationship between article's visibility and certain metrics such as number of citations, h-index and i10-index has been investigated [2]. Determining the actual contribution made by each author in a particular article is really very difficult. The authorship index has been proposed which calculates a particular author's overall contribution towards a scientific publication [22].

The revised ICMJE guidelines have been used to quantify author contributions and responsibilities. The authorship matrix [6] has been proposed which is fully integrated with ICMJE guidelines. The authorship matrix consists of element-allocation matrix and an authorship contribution/responsibility matrix which estimates individual contribution levels. Quantity measures and quality measures are defined for measuring the author's contribution in Wikipedia [1]. Number of edits, text only, edit only, text longevity, edit longevity, ten revisions and text longevity with penalty are indicated as contribution measures.

# 3  Measuring Modified Authorship Indices Based on Positions

The topic of deriving authorship index of publications is delicate and challenging. The authorial index such as MAI has been calculated for assessing the author's credit in a multi-authored publication. Table 1 indicates the cases of research papers for determining MAI. Totally, there are seven cases with two or three authors. The authors are listed down based on their order of the positions in their publications. The research papers are indicated by the article ID and they are cited in the case references. In Article $p_1$, totally there are two authors; the author Werner Marx serves as the first author, and the author Lutz Bornmann serves as the second author.

In Article $p_2$, totally there are three authors; the author Lutz Bornmann serves as the first author; the author Werner Marx serves as the second author, and the author Andreas Barth serves as the third author. MAI and $AI_c$ have been derived for the below-mentioned authors based on their order of positions in their publications.

**Table 1** Articles with author's positions

| Article ID | A1 | A2 | A3 |
|---|---|---|---|
| $p_1$ [35] | Werner Marx | Lutz Bornmann | NA |
| $p_2$ [25] | Lutz Bornmann | Werner Marx | Andreas Barth |
| $p_3$ [23] | Lutz Bornmann | Felix de Moya Anegon | NA |
| $p_4$ [28] | Robin Haunschild | Andreas Barth | Werner Marx |
| $p_5$ [24] | Lutz Bornmann | Felix de Moya Anegon | Rudiger Mutz |
| $p_6$ [9] | Felix de Moya Anegon | Victor Herrero Solana | NA |
| $p_7$ [5] | Carmen Galvez | Felix de Moya Anegon | Victor Herrero Solana |

In articles $p_2$, $p_3$ and $p_5$, Lutz Bornmann acts as the first author, so $ACAV_{MAI}$ for the author Lutz Bornmann is high when compared to the co-authors.

## 3.1 Modified Authorship Index (MAI)

Author's MAI has been calculated based on their position in their publications using the following formula derived and modified from [22]. It is a simple way to measure an author's contribution to the literature. Table 2 denotes each author's MAI based on their position in their publications. Let us assume the author $R_t$ (or $t$th author) has authored $P$ publications, where $P = \{p_0, \ldots, p_k\}$, then MAI of the author $R_t$ is given by

$$MAI(R_t) = \frac{P_{first} + 0.5 * P_{other}}{P} * 100 \qquad (1)$$

where

$P_{first} = \{p_0, \ldots, p_u\}$, where $u$ denotes the no. of publications in which the author is in first position

$P_{other} = \{p_0, \ldots, p_v\}$, where $v$ denotes the no. of publications in which the author is in other positions

$$u \leq k; \ v \leq k; \ P_{first} \cap P_{other} = \emptyset$$

Table 2 indicates the author's total publications as first author and other author with the estimated MAI. Werner Marx is the first author; the total number of publications made by Werner Marx is 48. Out of 48 publications, 15 research papers are published by Werner Marx as first author and remaining 33 publications

**Table 2** Modified authorship index

| Author index— $R_t$ | Author name | P | $P_{first}$ | $P_{other}$ | C | H | i10 | $MAI(R_t)$ |
|---|---|---|---|---|---|---|---|---|
| $R_0$ | Werner Marx | 48 | 15 | 33 | 1344 | 22 | 38 | 65.63 |
| $R_1$ | Lutz Bornmann | 239 | 160 | 79 | 8370 | 45 | 155 | 83.47 |
| $R_2$ | Andreas Barth | 18 | 10 | 8 | 5238 | 28 | 58 | 77.78 |
| $R_3$ | Felix de Moya Anegon | 99 | 10 | 89 | 6341 | 40 | 144 | 55.05 |
| $R_4$ | Robin Haunschild | 22 | 8 | 14 | 523 | 14 | 17 | 68.18 |
| $R_5$ | Rudiger Mutz | 33 | 4 | 29 | 2219 | 22 | 35 | 56.06 |
| $R_6$ | Victor Herrero Solana | 14 | 2 | 12 | 3024 | 26 | 57 | 57.14 |
| $R_7$ | Carmen Galvez | 8 | 7 | 1 | 456 | 11 | 11 | 93.75 |

are published as second and other author positions. The MAI for Werner Marx is 65.63. Lutz Bornmann is the second author; the total number of publications made by Lutz Bornmann is 239. Out of 239 publications, 160 research papers are published by Lutz Bornmann as first author and remaining 79 publications are published as second and other author positions. The MAI for Lutz Bornmann is 83.47. Lutz Bornmann's first author publications are more when compared to Werner Marx, so the MAI is more. The author Andreas Barth has published 18 papers; out of 18 publications, 10 publications are published as first author and 2 publications are published as second and other authors. Since the first author publications are more, MAI is high when compared to Werner Marx.

The author Carmen Galvez has totally eight publications. Out of eight publications, seven publications are published as first author and one publication is published as second and other author. So the MAI for Carmen Galvez is more when compared to all the other authors. Table 2 depicts the author-level metrics like author's citation, h-index, i10-index and MAI. Citation indicates how many times a research paper has been used by other papers. h-index assesses the research performance of an author by analysing the citations [20, 21]. i10-index indicates at least 10 citations received by the author for his/her publications. MAI is based on the positions of the author. The author who has published many papers as first author receives more MAI as indicated in Table 2. MAI has been compared with the other indices for assessing the author's impact. Table 3 shows the author's contribution to article's visibility (**ACAV**) score calculated based on their 'Alt-metrics' [8]. The author-level metrics for a researcher $R_t$ with respect to a given research article $p_k$ are calculated as follows:

$$p_k(ACAV_C(R_t)) = \frac{C(R_t)}{\sum_{i=0}^{n} C(p_k(A_n))} \tag{2}$$

$$p_k(ACAV_H(R_t)) = \frac{h(R_t)}{\sum_{i=0}^{n} h(p_k(A_n))} \tag{3}$$

$$p_k(ACAV_{i10}(R_t)) = \frac{i10(R_t)}{\sum_{i=0}^{n} i10(p_k(A_n))} \tag{4}$$

$$p_k(ACAV_{MAI}(R_t)) = \frac{MAI(R_t)}{\sum_{i=0}^{n} MAI(p_k(A_n))} \tag{5}$$

where $ACAV_C$, $ACAV_H$, $ACAV_{i10}$, $ACAV_{MAI}$ are ACAV scores from citations, h-index, i10-index and MAI perspectives; $A_i$ denotes $i$th author of research article $p_k$.

Case 1.1 indicates the research paper written by two authors such as Werner Marx and Lutz Bornmann. The author Werner Marx has totally 1344 citations; the $ACAV_C$ is calculated by dividing $1344/9714 = 0.14$. The h-index value is 22; the $ACAV_H$ is calculated by dividing $22/67 = 0.33$. The i10-index value is 38; the

**Table 3** Author's contribution to article's visibility

| Research article | Authors | C | h-index | i10-index | $MAI(R_r)$ | $ACAV_C(R_r)$ | $ACAV_H(R_r)$ | $ACAV_{i10}(R_r)$ | $ACAV_{MAI}(R_r)$ |
|---|---|---|---|---|---|---|---|---|---|
| $P_1$ Case 1.1 | $R_0$ | 1344 | 22 | 38 | 65.63 | 0.14 | 0.33 | 0.20 | 0.44 |
| | $R_1$ | 8370 | 45 | 155 | 83.47 | 0.86 | 0.67 | 0.80 | 0.56 |
| | Total | 9714 | 67 | 193 | 149.1 | 1.00 | 1.00 | 1.00 | 1.00 |
| $P_2$ Case 1.2 | $R_1$ | 8370 | 45 | 155 | 83.47 | 0.56 | 0.47 | 0.62 | 0.37 |
| | $R_0$ | 1344 | 22 | 38 | 65.63 | 0.09 | 0.23 | 0.15 | 0.29 |
| | $R_2$ | 5238 | 28 | 58 | 77.78 | 0.35 | 0.29 | 0.23 | 0.34 |
| | Total | 14952 | 95 | 251 | 226.88 | 1.00 | 1.00 | 1.00 | 1.00 |
| $P_3$ Case 2.1 | $R_1$ | 8370 | 45 | 155 | 83.47 | 0.57 | 0.53 | 0.52 | 0.60 |
| | $R_3$ | 6341 | 40 | 144 | 55.05 | 0.43 | 0.47 | 0.48 | 0.40 |
| | Total | 14711 | 85 | 299 | 138.52 | 1.00 | 1.00 | 1.00 | 1.00 |
| $P_4$ Case 2.2 | $R_4$ | 523 | 14 | 17 | 68.18 | 0.07 | 0.22 | 0.15 | 0.32 |
| | $R_2$ | 5238 | 28 | 58 | 77.78 | 0.74 | 0.44 | 0.51 | 0.37 |
| | $R_0$ | 1344 | 22 | 38 | 65.63 | 0.19 | 0.34 | 0.34 | 0.31 |
| | Total | 7105 | 64 | 113 | 211.59 | 1.00 | 1.00 | 1.00 | 1.00 |
| $P_5$ Case 2.3 | $R_1$ | 8370 | 45 | 155 | 83.47 | 0.49 | 0.42 | 0.46 | 0.43 |
| | $R_3$ | 6341 | 40 | 144 | 55.05 | 0.37 | 0.37 | 0.43 | 0.28 |
| | $R_5$ | 2219 | 22 | 35 | 56.06 | 0.13 | 0.21 | 0.10 | 0.29 |
| | Total | 16930 | 107 | 334 | 194.58 | 1.00 | 1.00 | 1.00 | 1.00 |
| $P_6$ Case 3.1 | $R_3$ | 6341 | 40 | 144 | 55.05 | 0.68 | 0.61 | 0.72 | 0.49 |
| | $R_6$ | 3024 | 26 | 57 | 57.14 | 0.32 | 0.39 | 0.28 | 0.51 |
| | Total | 9365 | 66 | 201 | 112.19 | 1.00 | 1.00 | 1.00 | 1.00 |
| $P_7$ Case 3.2 | $R_7$ | 456 | 11 | 11 | 93.75 | 0.05 | 0.14 | 0.05 | 0.46 |
| | $R_3$ | 6341 | 40 | 144 | 55.05 | 0.65 | 0.52 | 0.68 | 0.27 |
| | $R_6$ | 3024 | 26 | 57 | 57.14 | 0.31 | 0.34 | 0.27 | 0.28 |
| | Total | 9821 | 77 | 212 | 205.94 | 1.00 | 1.00 | 1.00 | 1.00 |

$ACAV_{i10}$ is calculated by dividing $38/193 = 0.20$. The MAI value is 65.63; the $ACAV_{MAI}$ is calculated by dividing $65.63/149.1 = 0.44$.

The author Lutz Bornmann has totally 8370 citations; the $ACAV_C$ is calculated by dividing $8370/9714 = 0.86$. The h-index value is 45; the $ACAV_H$ is calculated by dividing $45/67 = 0.67$. The i10-index value is 155; the $ACAV_{i10}$ is calculated by dividing $155/193 = 0.80$. The MAI value is 83.47; the $ACAV_{MAI}$ is calculated by dividing $83.47/149.1 = 0.56$. The MAI value is low when compared to other metrics because it mainly depends on the position of the author in the publications. Since the author Lutz Bornmann has only 160 publications as first author out of 239, the MAI value is comparatively less.

Consider the cases 1.2, 2.1 and 2.3 in which Lutz Bornmann is the first author. The value of $ACAV_{MAI}$ is more when compared to other co-authors. So the author's position in his/her publications plays a major role in determining the author's contribution. The first author in every publication receives a higher $ACAV_{MAI}$ value when compared to the author's in other positions. In case 1.1, the number of citations received by the author Lutz Bornmann is higher so the value of MAI and $ACAV_{MAI}$ is also high. In case 2.2, though the author Andreas Barth is the second author, he has received higher value in MAI and $ACAV_{MAI}$ since the number of citations received by the author is more when compared to the other co-authors.

In case 3.1, Felix is the first author but his MAI and $ACAV_{MAI}$ values are less because Felix has published 99 articles; out of 99, only in 10 articles the author serves as first author, whereas the author Victor has published totally 14 articles; out of 14, in 2 articles the author serves as first author so the author Victor's MAI and $ACAV_{MAI}$ values are more. In case 3.2, the author Carmen Galvez has highest number of first author publications so his MAI and $ACAV_{MAI}$ values are high. Table 3 clearly indicates that MAI and $ACAV_{MAI}$ mainly depend on two factors; one is first author publication, and the other one is the number of citations received by the author in his publications.

Table 4 indicates the estimated value of $ACAV$ for the sample papers taken for each author based on their positions. For instance, the author Andreas Barth has two publications, and the value of $ACAV_{MAI}$ differs in both the publication because of the influence of the co-authors.

Table 5 indicates the consolidated values of ACAV. The consolidated values are calculated based on the average value of $ACAV_{MAI}$ which is calculated for the sample papers taken. The research strength of each author is indicated by the estimated $ACAV_{MAI}$. In case 1.2 and 2.3 (in articles $p_3$, $p_5$—Table 1), the author Lutz Bornmann serves as the first author. The values of MAI and $ACAV_{MAI}$ vary because these values are influenced by the values of co-authors MAI and $ACAV_{MAI}$. Consider the sample papers $p_1$ and $p_2$ (Table 1) which have common authors, Lutz Bornmann and Werner Marx. The MAI and $ACAV_{MAI}$ values are different for these two papers because the authors have combined with another author Andreas Barth in $p_2$ (Table 1). It shows that MAI and $ACAV_{MAI}$ values differ when co-authors differ.

**Table 4** ACAV score for each author based on the varied positions in respective articles

| Author | Article ID | $ACAV_c$ | $ACAV_H$ | $ACAV_{i10}$ | $ACAV_{MAI}$ |
|--------|-----------|----------|----------|--------------|--------------|
| $R_1$ | $p_1$ | 0.86 | 0.67 | 0.80 | 0.56 |
|       | $p_2$ | 0.56 | 0.47 | 0.62 | 0.37 |
|       | $p_3$ | 0.57 | 0.53 | 0.52 | 0.60 |
|       | $p_5$ | 0.49 | 0.42 | 0.46 | 0.43 |
| $R_0$ | $p_1$ | 0.14 | 0.33 | 0.20 | 0.44 |
|       | $p_2$ | 0.09 | 0.23 | 0.15 | 0.29 |
|       | $p_4$ | 0.19 | 0.34 | 0.34 | 0.31 |
| $R_2$ | $p_2$ | 0.35 | 0.29 | 0.23 | 0.34 |
|       | $p_4$ | 0.74 | 0.44 | 0.51 | 0.37 |
| $R_3$ | $p_3$ | 0.43 | 0.47 | 0.48 | 0.40 |
|       | $p_5$ | 0.37 | 0.37 | 0.43 | 0.28 |
|       | $p_6$ | 0.68 | 0.61 | 0.72 | 0.49 |
|       | $p_7$ | 0.65 | 0.52 | 0.68 | 0.27 |
| $R_6$ | $p_6$ | 0.32 | 0.39 | 0.28 | 0.51 |
|       | $p_7$ | 0.31 | 0.34 | 0.27 | 0.28 |
| $R_7$ | $p_7$ | 0.05 | 0.14 | 0.05 | 0.46 |
| $R_4$ | $p_4$ | 0.07 | 0.22 | 0.15 | 0.32 |
| $R_5$ | $p_5$ | 0.13 | 0.21 | 0.10 | 0.29 |

**Table 5** Consolidated values of ACAV

| Author | $ACAV_c$ | $ACAV_H$ | $ACAV_{i10}$ | $ACAV_{MAI}$ |
|--------|----------|----------|--------------|--------------|
| $R_1$ | 0.62 | 0.52 | 0.60 | 0.49 |
| $R_0$ | 0.14 | 0.30 | 0.30 | 0.35 |
| $R_2$ | 0.54 | 0.36 | 0.37 | 0.35 |
| $R_3$ | 0.49 | 0.48 | 0.54 | 0.39 |
| $R_6$ | 0.32 | 0.36 | 0.27 | 0.40 |
| $R_7$ | 0.05 | 0.14 | 0.05 | 0.46 |
| $R_4$ | 0.07 | 0.22 | 0.15 | 0.32 |
| $R_5$ | 0.13 | 0.21 | 0.10 | 0.29 |

## 4   Conclusion

In this paper, we propose MAI and $ACAV_{MAI}$ as authorial metrics which depend on the position of author and the number of citations received by co-authors. These proposed indicators have been calculated for the authors of "Journal of Informetrics" [13, 14]. These indicators are mainly derived based on the publications as a first author. The main issue of this method is that if an author has published papers with the same co-authors, then the value of $ACAV_{MAI}$ is same, otherwise the value of $ACAV_{MAI}$ differs across different authors across different publications. In future, we plan to indices for author contribution for determining the exact contribution and impact of the author on a particular article.

# References

1. Adler, B. T., de Alfaro, L., Pye, I., & Raman, V. (2008, September). Measuring author contributions to the Wikipedia. *In the Proceedings of the 4th International Symposium on Wikis* (p. 15).
2. Ale Ebrahim .N, Salehi .H, Embi .M .A, Habibi .F, Gholizadeh .H and Motahar .S .M (2014). Visibility and citation impact. *International Education Studies, 7(4),* 120–125.
3. Bornmann, L., & Marx, W. (2012). HistCite analysis of papers constituting the h index research front. *Journal of Informetrics, 6(2),* 285–288.
4. Brand, A., Allen, L., Altman, M., Hlava, M., & Scott, J. (2015). Beyond authorship: attribution, contribution, collaboration, and credit. *Learned Publishing, 28(2),* 151–155.
5. Carmen Galvez, Félix de Moya Anegón, Victor HerreroSolana: Term conflation methods in information retrieval: Non-linguistic and linguistic approaches. Journal of Documentation 61 (4): 520–547 (2005).
6. Clement .T .P, "Authorship Matrix: A Rational Approach to Quantify Individual Contributions and Responsibilities in Multi-Author Scientific Articles", Journal of Science and Engineering Ethics, 20:2, pp. 345–361, 2014.
7. Deepika J. and Mahalakshmi G.S., Towards Knowledge based Impact Metrics for Open Source Research Publications, International Journal on Internet and Distributed Computing Systems, Vol. 2, No. 1, 2012, pp. 102–108.
8. Egghe .L, Guns .R and Rousseau .R (2013). Measuring Co-authors Contribution to an Articles Visibility, Journal of Scientometrics, 95(1), 55–67.
9. Félix de Moya Anegón, Victor HerreroSolana: Worldwide topology of the scientific subject profile: a macro approach on the country level. CoRR abs/1005.2223 (2010).
10. G. MuthuSelvi, G.S. Mahalakshmi and S. Sendhilkumar, "An Investigation on Collaboration Behavior of highly Cited Authors in Journal of Informetrics (2007–2016)", Journal of Computational and Theoretical Nanoscience, 3803.
11. G. MuthuSelvi, G.S. Mahalakshmi and S. Sendhilkumar, "Author attribution using Stylometry for Multi-author Scientific Publications", Advances in Natural and Applied Sciences, 2016 June, 10(8): Pages: 42–47.
12. G. S. Mahalakshmi, and S. Sendhilkumar: Optimizing Research Progress Trajectories with Semantic Power Graphs. Springer 2013 Lecture Notes in Computer Science ISBN 978-3-642-45061-7, PReMI 2013: 708–713.
13. G.S. Mahalakshmi, G. MuthuSelvi and S. Sendhilkumar, "A Bibliometric Analysis of Journal of Informetrics – A Decade Study", International Conference on Recent Trends and Challenges in Computational Models (ICRTCCM'17), organized by Department of Computer Science and Engineering, University College of Engineering, Tindivanam, 2017.
14. G.S. Mahalakshmi, G. MuthuSelvi and S. Sendhilkumar, "Authorship Analysis of JOI Articles (2007–2016)", International Journal of Control Theory and Applications, 9(10), 2016, pp. 1–11 ISSN: 0974-5572.
15. G.S. Mahalakshmi, G. MuthuSelvi and S. Sendhilkumar, "Generation of Author Topic Models using LDA", International Conference on Intelligent Computing and Control Systems (ICICCS 2017) is being organized on 15–16, June 2017 by The Vaigai College Engineering (VCE), Madurai.

16. G.S. Mahalakshmi, G. MuthuSelvi and S. Sendhilkumar, "Hierarchical Modeling Approaches for Generating Author Blueprints", International Conference on Smart Innovations in Communications and Computational Sciences (ICSICCS-2017) organizing by North West Group of Institutions, Moga, Punjab, India during 23–24 June 2017.

17. G.S. Mahalakshmi, G. MuthuSelvi and S. Sendhilkumar, "Measuring author contributions via LDA", 2nd International Conference on Advanced Computing and Intelligent Engineering, Central University of Rajasthan, Ajmer, India. (Under Review).

18. G.S. Mahalakshmi, G. MuthuSelvi, and S. Sendhilkumar, "Gibbs Sampled Hierarchical Dirichlet Mixture Model based Approach for Clustering Scientific Articles", Pattern Recognition & Machine Intelligence (PREMI-2017) Kolkata 2017., (Under Review).

19. G.S. Mahalakshmi, S. Sendhilkumar, and S. Dilip Sam, Refining Research Citations through Context Analysis, Lecture notes on Intelligent Informatics, Springer-Verlag, Proceedings of the International Symposium on Intelligent Informatics ISI'12, Chennai, India, pp. 65–71, August 2012.

20. Hirsch, J. E. (2005). An index to quantify an individual's scientific research output. Proceedings of the National academy of Sciences of the United States of America, 16569–16572.

21. Hirsch, J. E. (2010). An index to quantify an individual's scientific research output that takes into account the effect of multiple coauthorship. Scientometrics, 85(3), 741–754.

22. Kaushik, R. (2013). The "Authorship Index"-a simple way to measure an author's contribution to literature. International Journal of Research in Medical Sciences, 1(1), 1–3.

23. Lutz Bornmann, Félix de Moya Anegón: What proportion of excellent papers makes an institution one of the best worldwide? Specifying thresholds for the interpretation of the results of the SCImago Institutions Ranking and the Leiden Ranking. JASIST 65(4): 732–736 (2014).

24. Lutz Bornmann, Felix Moya Anegón, and RüdigerMutz. "Do universities or research institutions with a specific subject profile have an advantage or a disadvantage in institutional rankings?" Journal of the American Society for Information Science and Technology 64, no. 11 (2013): 2310–2316.

25. Lutz Bornmann, Werner Marx, Andreas Barth: The Normalization of Citation Counts Based on Classification Systems. Publications 1(2): 78–86 (2013).

26. Mahalakshmi G.S., Dilip Sam S. and Sendhilkumar S., Establishing Knowledge Networks via Analysis of Research Abstracts, Special Issue of Journal of Universal Computer Science (JUCS) on "Advances on Social Network Applications", Vol. 18, No. 8 (2012), pp. 993–1021.

27. McCarty, C., Jawitz, J. W., Hopkins, A., & Goldman, A. (2013). Predicting author h-index using characteristics of the co-author network. Scientometrics, 96(2), 467–483.

28. Robin Haunschild, Andreas Barth, Werner Marx: Evolution of DFT studies in view of a scientometric perspective. J. Cheminformatics 8(1): 52:1–52:12 (2016).

29. Sendhilkumar S., Dilipsam and Mahalakshmi G.S. Enhancement of Co-authorship networks with content similarity information, International Conference on Advances in Computing, Communications and Informatics, ICACCI, 2012, ACM digital library, pp. 1225–1228.

30. Siva R, Mahalakshmi G.S. and S. Sendhilkumar, 1-hop Greedy cite order plagiarism detection, International Journal of Control theory and applications, Vol. 10 (8), pp. 585–588.

31. Vasantha Kumar, S. Sendhilkumar, and G.S. Mahalakshmi A Power-graph based approach to Detection of Research Communities from Co-Authorship Networks, Journal of Computational Theoretical Nanoscience, 2017. (to appear).

32. Vasantha Kumar, S. Sendhilkumar, and G.S. Mahalakshmi, Author similarity identification using citation context and proximity, International Journal of Control theory and applications, 2017 (to appear).

33. Walters, G. D. (2016). Adding authorship order to the quantity and quality dimensions of scholarly productivity: evidence from group-and individual-level analyses. Scientometrics, 106(2), 769–785.
34. Warrender .J .M. (2016) "A Simple Framework for Evaluating Authorial Contributions for Scientific Publications", Journal of Science and engineering ethics, 22(5), 1419–1430.
35. Werner Marx, Lutz Bornmann: Change of perspective: bibliometrics from the point of view of cited references - a literature overview on approaches to the evaluation of cited references in bibliometrics. Scientometrics 109(2): 1397–1415 (2016).

# Evaluation of Teaching Learning Based Optimization with Focused Learning on Expensive Optimization Problems (CEC2017)

Remya Kommadath and Prakash Kotecha

**Abstract** Teaching learning based optimization (TLBO) simulates the transfer of knowledge in a classroom environment for solving various optimization problems. In the current work, we propose a variant of TLBO which incorporates a focused learning strategy and evaluates its performance on bound constrained single objective computationally expensive problems provided for CEC2017. The proposed variant of TLBO uses the functional evaluations effectively to handle expensive optimization problems and has lower computational complexity.

**Keywords** Teaching learning based optimization · Focused learning
Computationally expensive optimization problems · CEC 2017

## 1 Introduction

Most of the real-life optimization problems are multimodal, discrete, noisy, or even multi-objective in nature which requires the use of evolutionary techniques. Several new evolutionary techniques such as Teaching Learning Based Optimization (TLBO) [1], Yin-Yang-Pair Optimization [2], Virus Colony Search [3], Simultaneous Heat Transfer Search [4] have been recently proposed. Unlike many other algorithms, TLBO requires only the size of the population and the termination criteria [1] as user-defined parameters. It has found applications in diverse fields of engineering such as heat exchanger design, economic load dispatch problem, machining processes, fault diagnostics, power systems, scheduling problems, and controller parameter tuning. Moreover, the multi-objective variant of TLBO [5] has been utilized to solve various multi-objective optimization problems. Despite several criticisms in terms of estimation of functional evaluations per iterations and

R. Kommadath · P. Kotecha (✉)
Department of Chemical Engineering, Indian Institute of Technology Guwahati,
Guwahati 781039, Assam, India
e-mail: pkotecha@iitg.ernet.in

© Springer Nature Singapore Pte Ltd. 2019
B. K. Panigrahi et al. (eds.), *Smart Innovations in Communication and Computational Sciences*, Advances in Intelligent Systems and Computing 669, https://doi.org/10.1007/978-981-10-8968-8_37

discrepancies in the description of the algorithm and its implementation [6, 7], TLBO has been widely used.

In TLBO, a solution undergoes two different stages, namely the teacher phase and the learner phase. Some recent variants of TLBO include TLBO-Global crossover [8], modified TLBO [9], and single phase multi-group TLBO [10]. As part of the Congress on Evolutionary Computation, several competitions are conducted to benchmark novel or modified evolutionary algorithms by providing test suites on real-parameter bound constrained problems, computationally expensive single objective problems, constrained single objective problems, constrained and unconstrained multi-objective optimization problems. The different categories of problems include unimodal and multimodal functions. Additionally, it also has hybrid and composition functions. This helps to obtain a comprehensive performance evaluation of an algorithm to solve different categories of problems.

Many of the engineering design problems are computationally expensive, and it may require a few seconds to days for a single evaluation of the fitness function. Such problems can be challenging for evolutionary techniques as they require repeated evaluation of the fitness function to determine the optimal solution. Hence, it is necessary to tune evolutionary techniques to handle such computationally expensive optimization problems. The special session on real-parameter single objective optimization, CEC 2017 has provided a platform to benchmark such modified algorithms on expensive optimization problems [11]. In this work, a variant of TLBO is proposed and its performance is evaluated using the computationally expensive problems of CEC 2017.

The article is structured as follows: In Sect. 2, a description of TLBO and its proposed variant is provided. The details of the benchmark functions and the various experimental settings are given in Sect. 3. Subsequently, the results are discussed in Sect. 4 and we conclude the work by summarizing the overall performance of the proposed variant.

## 2 Algorithm Description

### 2.1 Teaching Learning Based Optimization

Teaching and learning is integral in the life of an individual to acquire knowledge. TLBO [1] is a population-based meta-heuristic which borrows the concept of knowledge transfer in a classroom through teaching and learning process. It was proposed in the year 2011 and has found wide acceptance in solving a variety of single and multi-objective optimization problems, constrained and unconstrained problems, continuous and discrete problems. Each solution in a population is termed as a student, and the design parameters are equivalent to the various subjects offered to the students. The student with the best fitness value will act as the teacher for the rest of the class. The algorithm consists of two phases, namely the teacher phase and the learner phase, and every student undergoes both the phases till the stopping criteria are satisfied.

*Teacher Phase*: In this phase, the teacher delivers knowledge to the students and has been reported to improve the mean result of the whole class. Every student strives to gain knowledge from the teacher, and this process is modeled as in Eq. (1). The improvement in the result of the student undergoing the teacher phase depends on the position of the solution acting as teacher as well as the mean of the learners in the class.

$$X^{new}_{(i,j)} = X_{(i,j)} + r_{(i,j)} \left( X_{(best,j)} - T_F X_{(mean,j)} \right) \quad \forall i = 1, \ldots, N_P, \ \forall j = 1, \ldots, D \quad (1)$$

Here $N_P$ indicates the size of the population and $D$ indicates the total number of design variables. $X_{(i,j)}$ is the student undergoing the teacher phase, $X_{(best,j)}$ is the teacher of the class, and $X_{(mean,j)}$ is the mean of the current class. $r_{(i,j)}$ is a random number in the range of [0, 1] whose value is different for each decision variable. $T_F$ is termed as the teaching factor and its value is determined as

$$T_F = round(1 + rand(0, 1)) \quad (2)$$

If the objective function value of this new potential student is better than the original student, the new student replaces the original student, else it is discarded. It should be noted that the class is updated after each member undergoes the teaching phase.

*Learner Phase*: After the completion of the teacher phase, the student undergoes a learning process through interacting with a partner student which is selected randomly from the rest of the class. This is mathematically modeled as shown in Eq. (3).

$$\left. \begin{array}{ll} X^{new}_{(i,j)} = X_{(i,j)} + r_{(i,j)} \left( X_{(i,j)} - X_{(rand,j)} \right); & if \ f\left(X_{(i,j)}\right) < f\left(X_{(rand,j)}\right) \\ X^{new}_{(i,j)} = X_{(i,j)} + r_{(i,j)} \left( X_{(rand,j)} - X_{(i,j)} \right); & if \ f\left(X_{(i,j)}\right) \geq f\left(X_{(rand,j)}\right) \end{array} \right\} \forall i = 1, \ldots, N_P; \ \forall j = 1, \ldots, D$$

$$(3)$$

In the above equation, $X_{(i,j)}$ is the student undergoing the learner phase and $X_{(rand,j)}$ is the partner selected randomly for updating the current student. $r_{(i,j)}$ denotes a random vector (of size $1 \times D$), with the values in the range [0, 1]. The new potential student is accepted if it provides a better fitness than the student undergoing the learner phase. Every student undergoes the teacher phase followed by the learner phase, and this is repeated till the satisfaction of the termination criteria.

## 2.2 Teaching Learning Based Optimization with Focused Learning

In TLBO, the result of a student is improved either through the teacher phase or the learner phase or both. Even though a student might have improved the result in the teacher phase, it still has to undergo the learner phase. However, it is not guaranteed that the student will improve its result through the learner phase too. The competition on computationally expensive problems requires the algorithm to reach the optima by utilizing very few function evaluations. Hence for the purpose of proper utilization of the function evaluations, it might be beneficial that the learner phase is performed only if the teacher phase is a failure. In TLBO-FL, all the students make an attempt to improve their knowledge through the teacher phase and if they succeed, the modified student replaces the original one. However, if the student fails to utilize the teacher phase for improving its fitness, it will undergo the learner phase.

In the teacher phase, all the students (except the one acting as the teacher) utilize the Eq. (1) to generate new solutions. However, the teacher of the class is modified by randomly changing the value of $D-1$ decision variables, in order to exploit the search space. The learner phase of basic TLBO is improved by incorporating focused learning in which an attempt to obtain an improved result is performed through interacting with a student better than the current student as well as with a student which has an inferior fitness than the current student. At any stage, before the evaluation of fitness for a solution, if the solution violates any of the bound constraints, it is bounded by fixing it to the nearest lower or upper bound. The various steps involved in the implementation of TLBO-FL are given in the pseudo-code. It has been given for the termination criterion of maximum number of iterations but can be easily modified to accommodate other criteria.

---

Pseudo-code for teaching learning based optimization with focused learning

*Inputs*: Objective function (*f*), lower (*lb*) and upper (*ub*) bounds of decision variables, population size ($N_P$), the number of decision variables (*D*), and the number of iterations/generations.
*Outputs*: Optimal solution and its corresponding fitness value.

---

*Step 1*. Initialize $N_P$ number of students uniformly using the bounds of the decision variables. Evaluate the fitness of these students.

*Step 2*. Implement Step 3 to Step 12 till the specified number of generations.

*Step 3*. Implement Step 4 to Step 12 for all the students in the class.

*Step 4*. Assign the best student as teacher $X_{best}$.

*Step 5*. Sort the population with respect to the ascending order of their fitness value.

*Step 6*. Determine the mean of the class.

*Step 7.* **If** the current student is teacher, go to Step 8 **else** go to Step 10

*Step 8.* Choose a random number $k$ between 1 and $D$. Determine the new solution

$$
\begin{aligned}
X_{(best,j)}^{new} &= X_{(best,j)} + r_j X_{(best,j)} \quad \forall j = 1, \ldots, D, j \neq k \\
X_{(best,j)}^{new} &= X_{(best,j)} \qquad\qquad\quad\ if\ j = k
\end{aligned}
\tag{4}
$$

*Step 9.* Bound the solution, if required. Determine the fitness value of the newly generated potential teacher. **If** a better fitness value is obtained for the new solution than the solution representing the teacher ($X_{best}$), then replace $X_{best}$ with the new solution and go to Step 3, **else** the new solution is rejected and go to Step 12.

*Step 10.* Determine a new solution in the teacher phase using Eq. (1).

*Step 11.* Bound the solution (as explained earlier), if required. Determine the fitness value of the newly generated student. **If** the new potential student provides better fitness value, it replaces the current student $X_i$, and go to Step 3, **else** it is discarded and go to Step 12.

*Step 12.* Determine a partner $X_b$ with better fitness and a partner $X_w$ with inferior fitness value than the current student $X_i$ randomly.

*Step 13.* Determine a new potential student based on the current student as shown below

$$
X_{(i,j)}^{new} = X_{(i,j)} + r_{(i,j)} \cdot \left( X_{(b,j)} - X_{(i,j)} \right) + r_{(i,j)} \cdot \left( X_{(i,j)} - X_{(w,j)} \right) \quad \forall j = 1, \ldots, D
\tag{5}
$$

Bound the new solution, if required. Evaluate the fitness of the new solution. This new potential student replaces the current student if it has a better fitness value than the current solution, else discard. Go to Step 3.

---

In teacher phase, the best solution as well as the mean of the class is updated after updating the position of the student. In a single iteration of TLBO, each solution will require two functional evaluations, in the absence of duplicate removal. However, it is not true in the case of TLBO-FL. The learner phase of a student is performed if and only if the student fails to determine a better solution in the teacher phase. This leads to the possibility for fewer fitness evaluations per iteration of TLBO-FL compared to the basic TLBO. Hence, TLBO-FL performs more number of iterations than basic TLBO (without duplicate removal) for a specified number of evaluations (of the fitness function) and population size. The greedy selection performed in both TLBO as well as TLBO-FL guarantees a monotonic convergence of these algorithms.

## 3   Experimental Settings

The performance of TLBO-FL is demonstrated on the test suite comprising of 15 scalable functions that are used for the competition, as part of the CEC 2017 [11], on computationally expensive problems. The test suite constitutes 2 unimodal functions, 7 multimodal functions, 3 hybrid functions, and 3 composition functions. As per the requirement for the competitions, the functions are to be tested for two different dimensions ($D = 10$ and $D = 30$). The termination criterion for the algorithm is set as a maximum of 50D (=MaxFEs) function evaluations, where $D$ is the dimension of the function. All the 15 functions are minimization functions having a search domain in $[-100, 100]^D$ and their global optima are known. The details of the functions are provided in Table 1.

The codes of the test suite are available for several platforms. In this work, we have utilized the MATLAB codes of the test suite. Most of the meta-heuristic techniques are stochastic in nature and in view of this, all the 15 functions with the individual dimensions are to be solved for 20 times. The score of the algorithm is determined as follows [11].

$$
\begin{aligned}
Total\ Score = &\sum_{i=1}^{15} (mean(f_a))_i|_{D=10} + \sum_{i=1}^{15} (mean(f_a))_i|_{D=30} \\
+ &\sum_{i=1}^{15} (median(f_a))_i|_{D=10} + \sum_{i=1}^{15} (median(f_a))_i|_{D-30}
\end{aligned}
\tag{6}
$$

**Table 1**  Summary of the expensive numerical test functions [11]

| Tag | Functions | Optima |
|-----|-----------|--------|
| F1 | Rotated Bent Cigar Function | 100 |
| F2 | Rotated Discus Function | 200 |
| F3 | Shifted and Rotated Weierstrass Function | 300 |
| F4 | Shifted and Rotated Schwefel's Function | 400 |
| F5 | Shifted and Rotated Katsuura Function | 500 |
| F6 | Shifted and Rotated HappyCat Function | 600 |
| F7 | Shifted and Rotated HGBat Function | 700 |
| F8 | Shifted and Rotated Expanded Griewank's plus Rosenbrock's Function | 800 |
| F9 | Shifted & Rotated Expanded Scaffer's F6 Function | 900 |
| F10 | Hybrid Function 1 (Three functions) | 1000 |
| F11 | Hybrid Function 2 (Four functions) | 1100 |
| F12 | Hybrid Function 3 (Five functions) | 1200 |
| F13 | Composition Function 1 (Five functions) | 1300 |
| F14 | Composition Function 2 (Three functions) | 1400 |
| F15 | Composition Function 3 (Five functions) | 1500 |

where $f_a = 0.5 \times (f_{MaxFEs} + f_{0.5MaxFEs})$ which corresponds to the mean of the best objective function value at 100% and 50% of MaxFEs for each run of a single problem. Hence, it is clear from the above equation that the scoring mechanism provides equal preference to the mean as well as median value of both the dimensions of all the functions.

The best objective function value obtained till 1% of MaxFEs, 2% of MaxFEs, ..., 9% of MaxFEs, 10% of MaxFEs, 20% of MaxFEs, ..., 100% of MaxFEs, and their corresponding error values are determined for creating an information matrix for both the dimensions of all problems in all 20 runs. The best, worst, mean, median, and standard deviation of best objective value for each of the dimension constitute the "statistics results" which provide a better idea about the performance of the algorithm on different problems over various runs. Both the information matrix and the statistical results can be generated as specified in the technical report [11] and can be utilized for the performance analysis of the proposed variant.

Since TLBO-FL does not possess any additional parameters except for the population size and termination criterion, the possibilities for any specific encoding or parameter tuning does not arise. TLBO-FL has been implemented in MATLAB, and its performance was evaluated using a personal computer with an Intel i7 3.40 GHz processor and 16 GB RAM. The "*twister*" algorithm of MATLAB 2016a is utilized for realizing the various independent runs which is used to analyze the performance of TLBO-FL. The seed for the "*twister*" algorithm is varied from 1 to 20 for accomplishing 20 independent runs and helps in the reproducibility of the reported results.

# 4 Results and Discussion

In this section, we provide the performance of TLBO-FL on the two different dimensions of the 15 computationally expensive problems. The resulting analysis includes the performance evaluation of the proposed variant of TLBO in terms of its capability to determine the optimum solution and its time complexity.

## 4.1 Error Values Obtained

The error value is the difference between the obtained results and the global optima. The statistical analysis of error values obtained at MaxFE on 10D and 30D of all problems are reported in Table 2 and Table 3 respectively. For the convergence analysis of the proposed variant, we have considered 19 different fractions of function evaluations (as specified in the report [11]). The best error value till the specified fraction of function evaluations is determined for each run. In view of multiple runs, the mean of the error in best fitness values are plotted with respect to

**Table 2** Error values obtained at MaxFEs of 10D problems

| F | Best | Worst | Median | Mean | SD |
|------|----------|----------|----------|----------|----------|
| F1 | 2.32E+08 | 2.14E+09 | 8.41E+08 | 9.03E+08 | 7.76E+08 |
| F2 | 2.50E+04 | 8.51E+04 | 4.64E+04 | 4.98E+04 | 1.11E+05 |
| F3 | 5.64E+00 | 1.20E+01 | 8.30E+00 | 8.62E+00 | 1.18E+00 |
| F4 | 1.47E+03 | 2.34E+03 | 2.00E+03 | 2.00E+03 | 2.06E+02 |
| F5 | 1.39E+00 | 4.01E+00 | 2.55E+00 | 2.68E+00 | 7.43E-01 |
| F6 | 5.87E-01 | 2.29E+00 | 1.32E+00 | 1.35E+00 | 5.79E-01 |
| F7 | 8.48E-01 | 1.88E+01 | 5.51E+00 | 6.81E+00 | 6.70E+00 |
| F8 | 8.62E+00 | 2.30E+02 | 1.76E+01 | 3.39E+01 | 1.80E+03 |
| F9 | 3.57E+00 | 4.29E+00 | 4.03E+00 | 3.96E+00 | 1.90E-01 |
| F10 | 1.34E+05 | 3.20E+06 | 6.77E+05 | 7.94E+05 | 5.23E+05 |
| F11 | 6.32E+00 | 2.42E+01 | 1.07E+01 | 1.16E+01 | 4.85E+00 |
| F12 | 6.29E+01 | 4.10E+02 | 2.77E+02 | 2.64E+02 | 7.83E+01 |
| F13 | 3.24E+02 | 4.11E+02 | 3.43E+02 | 3.51E+02 | 3.50E+01 |
| F14 | 2.03E+02 | 2.15E+02 | 2.09E+02 | 2.09E+02 | 4.55E+00 |
| F15 | 6.03E+01 | 5.42E+02 | 3.70E+02 | 3.03E+02 | 1.26E+02 |

**Table 3** Error values obtained at MaxFEs of 30D problems

| Tag | Best | Worst | Median | Mean | SD |
|------|----------|----------|----------|----------|----------|
| F1 | 5.15E+09 | 1.58E+10 | 9.80E+09 | 1.06E+10 | 5.08E+09 |
| F2 | 6.06E+04 | 1.28E+05 | 1.01E+05 | 9.91E+04 | 1.18E+04 |
| F3 | 2.37E+01 | 3.72E+01 | 3.05E+01 | 3.02E+01 | 1.94E+00 |
| F4 | 6.99E+03 | 8.17E+03 | 7.79E+03 | 7.72E+03 | 4.86E+02 |
| F5 | 2.10E+00 | 4.99E+00 | 4.31E+00 | 4.28E+00 | 5.39E-01 |
| F6 | 5.36E-01 | 3.45E+00 | 2.38E+00 | 2.12E+00 | 4.10E-01 |
| F7 | 1.32E+01 | 3.55E+01 | 2.25E+01 | 2.40E+01 | 9.67E+00 |
| F8 | 4.54E+03 | 2.90E+05 | 5.66E+04 | 8.12E+04 | 3.43E+05 |
| F9 | 1.30E+01 | 1.40E+01 | 1.38E+01 | 1.37E+01 | 2.10E-01 |
| F10 | 9.75E+05 | 3.92E+07 | 1.07E+07 | 1.37E+07 | 1.09E+07 |
| F11 | 2.39E+01 | 1.40E+02 | 5.04E+01 | 5.77E+01 | 3.85E+01 |
| F12 | 7.58E+02 | 1.48E+03 | 1.12E+03 | 1.16E+03 | 3.11E+02 |
| F13 | 3.93E+02 | 5.48E+02 | 4.75E+02 | 4.80E+02 | 7.09E+01 |
| F14 | 2.38E+02 | 3.09E+02 | 2.66E+02 | 2.68E+02 | 3.08E+01 |
| F15 | 5.62E+02 | 1.23E+03 | 1.07E+03 | 1.05E+03 | 7.29E+01 |

the fractions of maximum allowed function evaluations (MaxFEs). The convergence plots corresponding to both dimensions of F1 to F15 functions are depicted in Fig. 1.

A few observations that can be inferred from the tables and the figure are given below.

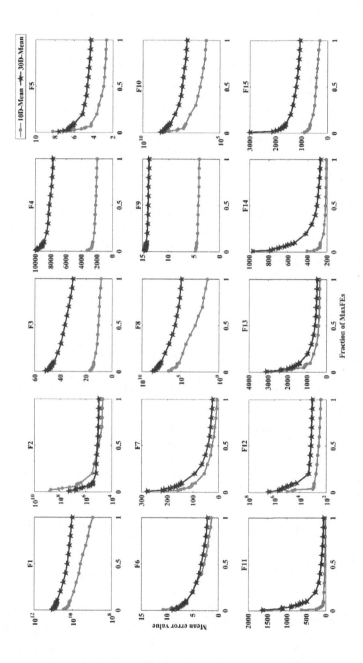

**Fig. 1** Convergence curves for mean error of 10D and 30D function

- It is evident from the tables that the mean error of all functions increases as the dimension of the problem increases.
- The performance of TLBO-FL on unimodal functions (F1-F2) is not satisfactory.
- In multimodal functions, TLBO-FL determines a solution nearby global optima on F5 and F6 and their magnitude of the error values in both the dimensions are similar.
- It is observed that the mean error obtained for F11 has a similar magnitude in both dimensions whereas the performance on the other two hybrid functions is not satisfactory.
- The mean and median error values of the composition functions F13 and F14 have an error value of similar magnitude which is not true in the case of function F15.

In view of the large variation in error values, the plots of few functions such as F1, F2, F8, F10, and F12 use log axis. In all the functions except F2, F5, and F10, it is observed that the error values obtained in every specified function evaluations of 30D functions are inferior to the 10D functions. The total score for TLBO-FL determined as per Eq. (6) is 3.26E+10.

## 4.2 Algorithm Complexity

The procedure for determining the time complexity of the algorithm is briefly described here and additional details can be obtained from the report [11]. The computational time complexity of the proposed variant is calculated and the results are provided in Table 4.

Step 1: Determine the value of $T_0$ by evaluating a set of basic mathematical operations given in the technical report of CEC 2017 [11].

Step 2: Determine the meantime ($T_1$) required for the optimization of a specific function over 20 runs for both the dimensions. The termination criteria for this step are the maximum number of permissible function evaluations which is equal to 50D.

Step 3: The algorithm complexity is determined using the formula $T_1/T_0$.

Table 4  Time complexity of TLBO-FL

| Tag | $T_1$ (s) | $T_1/T_0$ | Tag | $T_1$ (s) | $T_1/T_0$ | Tag | $T_1$ (s) | $T_1/T_0$ |
|-----|-----------|-----------|-----|-----------|-----------|-----|-----------|-----------|
| F1 | 0.02 | 0.14 | F6 | 0.02 | 0.12 | F11 | 0.03 | 0.22 |
| F2 | 0.02 | 0.11 | F7 | 0.02 | 0.12 | F12 | 0.02 | 0.15 |
| F3 | 0.09 | 0.6 | F8 | 0.02 | 0.12 | F13 | 0.03 | 0.19 |
| F4 | 0.02 | 0.12 | F9 | 0.02 | 0.11 | F14 | 0.03 | 0.18 |
| F5 | 0.05 | 0.37 | F10 | 0.02 | 0.12 | F15 | 0.1 | 0.69 |

## 5 Conclusion

A variant of TLBO by modifying the learning strategy is proposed in this work in which the learner phase is encountered only if there is no improvement in the teacher phase. The performance of the TLBO-FL is evaluated by the computationally expensive test functions which are proposed as a part of one of the competitions in CEC2017. On analyzing the results, it has been noticed that TLBO-FL is able to satisfactorily solve problems in all the categories of the test suite except for unimodal functions. A hybrid approach for solving the unimodal functions could be a possible future aspect of this work.

## References

1. R. V. Rao, V. J. Savsani, and D. P. Vakharia, "Teaching–learning-based optimization: A novel method for constrained mechanical design optimization problems," *Computer-Aided Design,* vol. 43, pp. 303–315, 2011.
2. V. Punnathanam and P. Kotecha, "Yin-Yang-pair Optimization: A novel lightweight optimization algorithm, *Engineering Applications of Artificial Intelligence,* vol. 54, pp. 62–79, 2016.
3. M. D. Li, H. Zhao, X. W. Weng, and T. Han, "A novel nature-inspired algorithm for optimization: Virus colony search," *Advances in Engineering Software,* vol. 92, pp. 65–88, 2016.
4. D. Maharana and P. Kotecha, "Simultaneous Heat Transfer Search for single objective real-parameter numerical optimization problem," in *2016 IEEE Region 10 Conference (TENCON),* 2016, pp. 2138–2141.
5. F. Zou, L. Wang, X. Hei, D. Chen, and B. Wang, "Multi-objective optimization using teaching-learning-based optimization algorithm," *Engineering Applications of Artificial Intelligence,* vol. 26, pp. 1291–1300, 2013.
6. M. Črepinšek, S.-H. Liu, and L. Mernik, "A note on teaching–learning-based optimization algorithm," *Information Sciences,* vol. 212, pp. 79–93, 2012.
7. S. Chinta, R. Kommadath, and P. Kotecha, "A note on multi-objective improved teaching–learning based optimization algorithm (MO-ITLBO)," *Information Sciences,* vol. 373, pp. 337–350, 2016.
8. H.-b. Ouyang, L.-q. Gao, X.-y. Kong, D.-x. Zou, and S. Li, "Teaching-learning based optimization with global crossover for global optimization problems," *Applied Mathematics and Computation,* vol. 265, pp. 533–556, 2015.
9. S. C. Satapathy and A. Naik, "Modified Teaching–Learning-Based Optimization algorithm for global numerical optimization—A comparative study," *Swarm and Evolutionary Computation,* vol. 16, pp. 28–37, 2014.
10. R. Kommadath, C. Sivadurgaprasad, and P. Kotecha, "Single phase multi-group teaching learning algorithm for computationally expensive numerical optimization (CEC 2016)," in *2016 IEEE Congress on Evolutionary Computation (CEC),* 2016, pp. 2989–2995.
11. Q. Chen, B. Liu, Q. Zhang, J. Liang, P. Suganthan, and B. Qu, "Problem Definitions and Evaluation Criteria for CEC 2015 Special Session on Bound Constrained Single-Objective Computationally Expensive Numerical Optimization," Technical Report, Computational Intelligence Laboratory, Zhengzhou University, Zhengzhou, China and Technical Report, Nanyang Technological University 2014.

# Optimization of Stirling Engine Systems using Single Phase Multi-group Teaching Learning Based Optimization and Genetic Algorithm

Remya Kommadath and Prakash Kotecha

**Abstract** In this work, Single Phase Multi-Group Teaching Learning Based Optimization (SPMGTLO) is used to optimize various objectives of three different models of Stirling heat engines, viz. (i) finite-time thermodynamic model, (ii) Stirling engine thermal model with associated irreversibility, and (iii) model based on polytropic finite-speed-based thermodynamics. The performance of SPMGTLO on solving these problems is compared with the inbuilt genetic algorithm (GA) function of MATLAB. It is observed that SPMGTLO performs better than GA in all the eight cases arising from the three models.

**Keywords** Stirling engine systems · Single phase multi-group teaching learning based optimization · Genetic algorithm · Teaching learning based optimization

## 1 Introduction

Solar energy is one among the abundant renewable energy resources which can be harnessed using various technologies such as photovoltaics, concentrating solar power (CSP) systems, solar heating, and cooling systems. The solar dish Stirling engine is categorized as CSP and uses reflective materials to concentrate sunlight for generating thermal energy which is effectively used to produce electricity. The Stirling engines are well known for its higher efficiency in converting thermal energy into mechanical work. It is an external combustion engine in which a compressible working fluid is employed to absorb heat from the heat source and

R. Kommadath · P. Kotecha (✉)
Department of Chemical Engineering, Indian Institute of Technology Guwahati,
Guwahati 781039, India
e-mail: pkotecha@iitg.ernet.in

© Springer Nature Singapore Pte Ltd. 2019
B. K. Panigrahi et al. (eds.), *Smart Innovations in Communication and Computational Sciences*, Advances in Intelligent Systems and Computing 669, https://doi.org/10.1007/978-981-10-8968-8_38

release to the heat sink [1]. Additional details on Stirling engines can be obtained from the literature [2].

The metaheuristic nature of evolutionary techniques has contributed to its widespread use in solving diverse optimization problems. Evolutionary techniques have been extensively used in optimizing various models of Stirling engines by considering single [3] as well as multiple conflicting objectives [4, 5]. Teaching learning based optimization (TLBO) is one of the several nature-inspired algorithms that has been recently proposed and has shown impressive results. In this work, a variant of TLBO named as Single Phase Multi-Group Teaching Learning Based Optimization (SPMGTLO) algorithm [6] is employed to optimize various design parameters of three widely used Stirling engine models to attain better performance. In particular, there are eight cases which arise from considering various alternative objectives in the three models.

This work is organized as follows: In Sect. 2, a brief description of SPMGTLO is provided, whereas the three different models of Stirling engines have been presented in Sect. 3. This is followed by the discussion of the results in Sect. 4. The work is concluded by summarizing the major results and by presenting possible future work.

## 2 Single Phase Multi-group Teaching Learning Based Optimization Algorithm

TLBO [7] is population-based metaheuristic technique that has been motivated from the knowledge transfer through the teaching and the learning process in a classroom. SPMGTLO [6] is a recently proposed variant of TLBO that has lower computational complexity. In this variant, the entire population is divided into multiple groups and each solution is updated using either the teacher phase or the learner phase. The selection of one of these phases is determined randomly by providing equal probability for both the phases. Each solution in the population is termed as student, and the best solution among each group is termed as the teacher for that specific group.

*Teacher phase*: Similar to TLBO, each student tries to update its position with respect to the teacher and the mean of the group. In addition to this, a tutorial learning is included, in which a partner student is incorporated to aid the update of a particular student. This is mathematically modeled as below

$$
\begin{aligned}
X_i^n &= X_i + r(X_{best} - T_F \bar{X}_m) + r'(X_i - X_p) \quad \text{if } f(X_i) < f(X_p) \\
X_i^n &= X_i + r(X_{best} - T_F \bar{X}_m) + r'(X_p - X_i) \quad \text{if } f(X_p) \leq f(X_i)
\end{aligned}
\tag{1}
$$

Here $X_i$ denotes the student undergoing teacher phase and $X_i^n$ denotes the new position of $X_i$. The notation $X_{best}$ denotes the position of teacher and $\bar{X}_m$ is the mean position of the group. The partner student with which $X_i$ interacts is represented as $X_p$. $r$ and $r'$ indicates two random vectors (of size 1 x number of variables) whose values lies in between 0 and 1. $T_F$ is the teaching factor and is determined as the ratio of objective function value of the student $X_i$ to the objective function value of the teacher. If the teacher has a fitness value of zero, then the value of $T_F$ is assumed to be one. The new solution is evaluated, and its fitness value is compared with the student undergoing the teacher phase. The solution which is better in fitness value is retained in the population for the next iteration, and the other one is discarded.

*Learner Phase*: The position update of a student in the learner phase is performed with the help of a partner student and through self-motivated learning. In self-motivated learning, students try to improve their results by utilizing the knowledge of their teacher. The position updating procedure is modeled as

$$X_i^n = X_i + r(X_i - X_p) + r'(X_{best} - E_g X_i) \ \ if \ f(X_i) < f(X_p)$$
$$X_i^n = X_i + r(X_p - X_i) + r'(X_{best} - E_g X_i) \ \ if \ f(X_p) \leq f(X_i)$$

(2)

where $E_g$ is the exploration factor whose value is either 1 or 2. The fitness of the newly generated solution is evaluated, and it replaces the current student $(X_i)$ undergoing the learner phase if it provides better fitness and is otherwise discarded.

It should be noted that in both the phases, before the evaluation of the fitness, the newly generated solutions are bounded to their nearest terminal bound if any of their decision variables exceeds their corresponding lower or upper bounds. For a single student, SPMGTLO performs only one function evaluation, whereas in TLBO (without the removal of duplicates) it performs two (once in the teacher phase and another one in the learner phase) functional evaluations. So it should be noted that, for a fixed number of maximum function evaluations and population size, SPMGTLO would execute more iterations compared to TLBO. The user-defined parameter in SPMGTLO is the number of groups which is additional to the usual parameters such as the population size and the stopping criteria. A detailed pseudo-code is available in the literature [6].

# 3 Problem Description

In this section, we restrict ourselves with presenting the details that are required from an optimization perspective for the three Stirling engines. Detailed information on various aspects of the respective models can be obtained from the literature [1, 4, 8, 9]. The values of parameters used in the three models are given in Table 1. Two of the three models contain three objectives each, whereas the other model has only two objectives.

**Table 1** Values of the various parameters required in the three models

| Case A | | Case B | | Case C | |
|---|---|---|---|---|---|
| $I$ | 1000 | $m_g$ | 0.001135 | $f$ | 41.7 |
| $C$ | 1300 | $\rho_R$ | 8030 | $V_p$ | $113.1 \times 10^{-6}$ |
| $\varepsilon$ | 0.9 | $N_r$ | 8 | $V_d$ | $120.8 \times 10^{-6}$ |
| $T_o$ | 288 | $b$ | $0.0688 \times 10^{-3}$ | $N_R$ | 8 |
| $\sigma_o$ | $5.67 \times 10^{-8}$ | $d$ | $0.04 \times 10^{-3}$ | $d$ | $40 \times 10^{-6}$ |
| $R$ | 4.3 | $v$ | $3.249 \times 10^{-7}$ | $n'_{ex}$ | 1.77 |
| $T_{HI}$ | 1300 | $\lambda$ | 1.2 | $n'_{cm}$ | 1.56 |
| $T_{LI}$ | 290 | $C_r$ | 502.48 | $m_g$ | $1.13 \times 10^{-3}$ |
| $K_o$ | 2.5 | $C_p$ | 5193 | $\rho_R$ | 8030 |
| $n$ | 1 | $C_v$ | 3115.6 | $\varepsilon_R$ | 0.9 |
| $C_v$ | 15 | $P_r$ | 0.71 | $y$ | 0.72 |
| $\xi$ | $2 \times 10^{-10}$ | $F$ | 0.556 | $N$ | 308 |
| $1/M_1 + 1/M_2$ | $2 \times 10^{-5}$ | $\Gamma$ | 1.667 | $C_p$ | 5193 |
| $h$ | 20 | $\varepsilon_R$ | 0.9 | $C_v$ | 3115.6 |
| $\lambda$ | 2 | $y$ | 0.72 | $C_r$ | 502.48 |
| $\eta_o$ | 0.9 | | | $v$ | $0.3249 \times 10^{-6}$ |
| | | | | $P_r$ | 0.71 |
| | | | | $b$ | $0.0688 \times 10^{-3}$ |
| | | | | $V_{dead}$ | $193.15 \times 10^{-6}$ |

## 3.1  Case A: Finite-Time Thermodynamic Model of Solar Dish—Stirling Engine [1]

$$Max \quad P = \frac{W}{t} = \frac{Q_H - Q_L}{t_{cycle}} \tag{3}$$

$$Max \quad \eta_m = \eta_t \left( \eta_0 - \frac{1}{IC} \left( h(T_{H_{ave}} - T_0) + \varepsilon \sigma_0 \left( T_{H_{ave}}^4 - T_0^4 \right) \right) \right) \tag{4}$$

$$Min \quad \sigma = \frac{1}{t_{cycle}} \left( \frac{Q_L}{T_{L_{ave}}} - \frac{Q_H}{T_{H_{ave}}} \right) \tag{5}$$

$$Q_h = nRT_h ln(\lambda) + nC_v(1 - \varepsilon_R)(T_h - T_c)$$
$$Q_c = nRT_c ln(\lambda) + nC_v(1 - \varepsilon_R)(T_h - T_c) \tag{6}$$

$$Q_H = Q_h + Q_0; \quad Q_L = Q_c + Q_0$$
$$Q_0 = \frac{K_0}{2} [(2 - \varepsilon_H)T_{H_1} - (2 - \varepsilon_L)T_{L_1} + (\varepsilon_H T_h - \varepsilon_L T_c)] t_{cycle} \tag{7}$$

$$t_{cycle} = \frac{nRT_h \ln\lambda + nC_v(1-\varepsilon_R)(T_h-T_c)}{C_H\varepsilon_H(T_{H_1}-T_h) + \xi C_H\varepsilon_H(T_{H_1}^4-T_h^4)} + \frac{nRT_c \ln\lambda + nC_v(1-\varepsilon_R)(T_h-T_c)}{C_L\varepsilon_L(T_c-T_{L_1})} + \left(\frac{1}{M_1}+\frac{1}{M_2}\right)(T_h-T_c)$$

(8)

$$T_{H_{ave}} = \frac{T_{H_1}+T_{H_2}}{2}, \quad T_{H_2} = (1-\varepsilon_H)T_{H_1} + \varepsilon_H T_h$$

$$T_{L_{ave}} = \frac{T_{L_1}+T_{L_2}}{2}, \quad T_{L_2} = (1-\varepsilon_L)T_{L_1} + \varepsilon_L T_c$$

(9)

$$\eta_t = \frac{nR(T_h-T_c)\ln\lambda}{\left(nRT_h\ln\lambda + nC_v(1-\varepsilon_R)(T_h-T_c) + \frac{K_0}{2}[(2-\varepsilon_L)T_{L_1}+\varepsilon_H T_h-\varepsilon_L T_c]t_{cycle}\right)}$$

(10)

$$0.4 \leq \varepsilon_R \leq 0.9; \quad 300 \leq C_L \leq 1800; \quad 300 \leq C_H \leq 1800; \quad 0.4 \leq \varepsilon_L \leq 0.8;$$
$$0.4 \leq \varepsilon_H \leq 0.8; \quad 400 \leq T_c \leq 510; \quad 800 \leq T_h \leq 1000;$$

## 3.2 Case B: Stirling Engine with Associated Irreversibility [8]

$$Min \quad \Delta p_{tot} = \Delta p_R + \Delta p_f + \Delta p_p$$

(11)

$$Max \quad \eta = \frac{\eta_{II,\Delta p}\eta_{cr}}{1+\left(\frac{X}{(\gamma-1)\ln\lambda}\right)\eta_{cr}}$$

(12)

$$Max \quad P = \eta Q_H$$

(13)

$$\tau = \frac{T_h}{T_c}; \quad T_h = T_H - \Delta T_H; \quad T_c = \Delta T_C + T_C$$

(14)

$$h = \frac{0.395\left(\frac{4p_m}{RT_c}\right)\left(\frac{sN}{30}\right)^{0.424}c_p v^{0.576}}{(1+\tau)\left[1-\frac{\pi}{4\left(\frac{b}{d}+1\right)}\right]D_R^{0.576}\,Pr^{0.667}}$$

(15)

$$\Delta p_R = \frac{15}{\gamma}\left(\frac{p_m}{2R(\tau+1)(T_C+\Delta T_C)}\right)\left(\frac{s^2N^2}{900}\right)\left(\frac{D_{cr}^2}{N_r D_R^2}\right)N_R$$

(16)

$$\Delta p_f = \left(\frac{(0.94+0.0015sN)10^5}{3\mu'}\right)\left(1-\frac{1}{\lambda}\right)$$

(17)

$$\Delta p_p = \left(\frac{sN}{60}\right)\left(\frac{4p_m}{(1+\lambda)(1+\tau)}\right)\left(\frac{\lambda \ln \lambda}{\lambda - 1}\right)\left(\frac{1}{\sqrt{T_C + \Delta T_C}}\right)$$
$$\left(1 + \sqrt{\frac{T_H - \Delta T_H}{T_C + \Delta T_C}}\right)\left(\sqrt{\frac{\gamma}{R}}\right) \tag{18}$$

$$\eta_{II,\Delta p} = 1 - \left(\frac{3\mu' \Delta p_{tot}(1+\lambda)(1+\tau)}{4p_m\left(\eta_{cr}\eta_{II(1-\varepsilon_r)}\right)\left(\frac{T_H - \Delta T_H}{T_C + \Delta T_C}\right)\ln \lambda}\right) \tag{19}$$

$$\eta_{II(1-\varepsilon_r)} = \frac{1}{1 + \left(\frac{1-\varepsilon_R}{(\gamma-1)\ln \lambda}\right)\eta_{cr}} \tag{20}$$

$$\eta_{cr} = 1 - \frac{T_C + \Delta T_C}{T_H - \Delta T_H} \tag{21}$$

$$X = y\left(\frac{1 + 2M' + \exp(-B)}{2(1+M')}\right) + (1-y)\left(\frac{M' + \exp(-B)}{1+M'}\right) \tag{22}$$

$$\mu' = 1 - \left(\frac{1}{3\lambda}\right); \quad M' = \frac{m_g c_v}{m_R c_R}; \quad B = (1+M')\left(\frac{hA_R}{m_g C_v}\right)\left(\frac{30}{N}\right) \tag{23}$$

$$A_R = \frac{\pi^2 D_R^2 L}{4(b+d)}; \quad m_R = \frac{\pi^2 D_R^2 L d\rho_R}{16(b+d)}; \quad R = C_p - C_v \tag{24}$$

$$Q_H = \left(R(T_H - \Delta T_H)\left(1 - \frac{\Delta p_v(\lambda+1)(\tau+1)}{4p_m} - \frac{b\Delta p_R}{2p_m} - \frac{f\Delta p_f}{p_m}\right)\ln \lambda + c_v XR(T_H - \Delta T_H - T_C - \Delta T_C)\right)\frac{Nm_g}{60} \tag{25}$$

$1200 \le N \le 3000; \quad 0.69 \times 10^6 \le p_m \le 6.89 \times 10^6; \quad 0.06 \le s \le 0.1;$
$250 \le N_r \le 400; 0.05 \le D_{cr} \le 0.14; \quad 0.02 \le D_R \le 0.06; \quad 0.006 \le L \le 0.073;$
$800 \le T_H \le 1300; \quad 288 \le T_C \le 360; \quad 64.2 \le \Delta T_H \le 237.6; \quad 5 \le \Delta T_C \le 25$

### 3.3 Case C: Polytropic Finite-Speed Thermodynamic (PFST) Thermal Model for Stirling Engine [9]

$$Min \quad \Delta p_{tot} = \Delta p_{th} + \Delta p_f + \Delta p_p \tag{26}$$

$$Max \quad P = f W_{net} = f \eta_{II,\Delta P} W_{poly} \tag{27}$$

$$\Delta p_{th} = N_r \frac{15}{\gamma} \left( \frac{D_c^2}{N_R D_R^2} \right)^2 \left[ \frac{p_m}{2R(1 + (1/\tau))(T_c + \Delta T_C)} \left( \frac{s^2 N^2}{900} \right) \right] \tag{28}$$

$$\Delta p_p = \frac{sN}{60} \sqrt{\frac{\gamma}{R}} \left( \frac{4p_m}{(1 + (1/\tau))(1 + r)} \right) \left( \frac{r \ln r}{r - 1} \right)$$
$$\left( \frac{1}{\sqrt{T_c + \Delta T_C}} \right) \left( 1 + \sqrt{\frac{T_h - \Delta T_H}{T_c + \Delta T_C}} \right) \tag{29}$$

$$\Delta p_f = \frac{(0.94 + (0.045 sN/30)) \times 10^5}{3\left(1 - \frac{1}{3r}\right)} \left( 1 - \frac{1}{r} \right) \tag{30}$$

$$r = \frac{V_{max} + V_{dead}}{V_{min} + V_{dead}} = \frac{(1 + k')(1 + K')}{1 + K'(1 + k')}; \quad K' = \frac{V_{dead}}{V_d}; \quad k' = \frac{V_p}{V_d} \tag{31}$$

$$\eta_{cr} = 1 - \frac{T_C + \Delta T_C}{T_H - \Delta T_H}; \quad \mu' = 1 - \left( \frac{1}{3\lambda} \right); \quad \tau = \frac{T_h}{T_c}; \tag{32}$$

$$W_{poly} = m_g R T_h \left[ \frac{\left( r^{(1 - n'_{ex})} - 1 \right)}{1 - n'_{ex}} + \frac{\tau \left( r^{(n'_{cm} - 1)} - 1 \right)}{1 - n'_{cm}} \right] \tag{33}$$

$$\eta_{II,R} = \frac{1}{1 + \frac{X \eta_{cr}}{(\gamma - 1) \ln \lambda}} \tag{34}$$

$X$ can be calculated using Eqs. (15) and (22–24)

$$\eta_{II,\Delta P} = 1 - \left[ \frac{3\mu'(1 + r)(1 + (1/\tau))\Delta p_{tot}}{4 p_m \eta_{cr} \eta_{II,R} \left( \frac{T_h - 2\Delta T_H}{T_c + 2\Delta T_C} \right) \ln r} \right] \tag{35}$$

$1200 \leq N \leq 3000; \quad 0.69 \times 10^6 \leq p_m \leq 6.89 \times 10^6; \quad 0.06 \leq s \leq 0.1; \quad 0.05 \leq D_c \leq 0.14;$
$0.02 \leq D_R \leq 0.06; \quad 288 \leq T_C \leq 360; \quad 800 \leq T_H \leq 1300; \quad 800 \leq T_H \leq 1300;$
$0.006 \leq L \leq 0.073; \quad 5 \leq \Delta T_C \leq 25; \quad 64.2 \leq \Delta T_H \leq 237.6;$

# 4 Experimental Settings

SPMGTLO is employed to optimize the three different models of solar dish Stirling engine problems. The population size and the maximum allowed functional evaluations are set to 100 and 50,100 respectively. In SPMGTLO, the whole population is divided into four equal sized groups. The results obtained by SPMGTLO is compared with the results obtained by employing the inbuilt MATLAB 2016a *ga* function (with default settings except for the size of the population, the number of generations and the parameters related to termination). For a fair comparison, the number of generations for GA is set to be 500 with a population size of 100 and these settings ensure that only the specified number of functional evaluations is permitted. On considering the stochastic nature of the algorithm, both the algorithms are executed for 10 independent runs of each case which are realized by varying the seeds from 1 to 10 of the '*twister*' algorithm of MATLAB 2016a. Thus, the results are reported on the basis of 1600 instances (8 objectives $x$ 10 runs $x$ 2 algorithms).

## 5 Results and Discussion

The different objectives of the three considered cases of solar dish Stirling engine are optimized using SPMGTLO algorithm. In the statistical analysis, the best, mean, and standard deviation of the obtained results by both SPMGTLO and GA are reported. The performance of SPMGTLO regarding its converging ability to the optima is also analyzed in comparison with GA.

### 5.1 Statistical Analysis

The statistical values for each objective function by considering all the runs are provided in Table 2. It is noted that SPMGTLO is able to relatively outperform GA in terms of the mean values for seven objective functions. It is observed that GA has determined an identical best fitness value as SPMGTLO in all the cases except in maximization of efficiency in PFST thermal model of Stirling engine. The mean value as well as the best values obtained by SPMGTLO in all the runs is almost equal except in the minimization of pressure drop of Stirling engine problem with associated irreversibility (Case B). It is observed that the standard deviation

**Table 2** Statistical values for all the three cases

|  | Objective function | Best | | Mean | | SD | |
|---|---|---|---|---|---|---|---|
|  |  | SPMGTLO | GA | SPMGTLO | GA | SPMGTLO | GA |
| Case A | Max P | 7.03E+04 | 7.03E+04 | 7.03E+04 | 7.01E+04 | 1.51E-11 | 4.46E+02 |
|  | Max $\eta_m$ | 3.68E-01 | 3.68E-01 | 3.68E-01 | 3.67E-01 | 0.00E+00 | 1.14E-03 |
|  | Min $\sigma$ | 2.19E+01 | 2.19E+01 | 2.19E+01 | 2.21E+01 | 7.35E-15 | 2.32E-01 |
| Case B | Max P | 1.77E+04 | 1.77E+04 | 1.77E+04 | 1.77E+04 | 3.51E-12 | 2.00E+02 |
|  | Max $\eta$ | 1.88E-01 | 1.88E-01 | 1.88E-01 | 1.87E-01 | 1.20E-16 | 2.10E-03 |
|  | Min. $\Delta p$ | 9.64E+03 | 9.64E+03 | 9.64E+03 | 9.65E+03 | 3.92E+00 | 2.01E+01 |
| Case C | Max $W_{net}$ | 1.49E+04 | 1.49E+04 | 1.49E+04 | 1.48E+04 | 5.65E-12 | 1.06E+02 |
|  | Max $\eta$ | 6.32E-01 | 6.31E-01 | 6.32E-01 | 6.27E-01 | 1.15E-16 | 3.02E-03 |

obtained by SPMGTLO is almost equal to zero except in the minimization of pressure drop in Stirling engine problem with associated irreversibility.

## 5.2 Convergence Analysis

For the convergence plot, the mean of the best objective function value obtained till a particular fraction of maximum allowed function evaluation (MaxFE) of all runs are determined and it is plotted against the fraction of MaxFE. Figures 1, 2, and 3 shows the mean value of the objective functions obtained at various fractions of MaxFE for the three models. On analyzing the convergence plot for Case A, it is observed that SPMGTLO has determined an optimum solution by utilizing less than 0.1MaxFE for two of the objective functions and less than 0.01MaxFE for another objective function. However, for all the three objective functions, GA has determined an inferior solution and has also shown comparatively slow convergence with respect to SPMGTLO. For maximization of power and maximization of efficiency, it is observed that GA has not converged to a final solution even after completing the considered termination criterion. This can be inferred to imply that GA is unable to determine a solution within the MaxFE for this work. In the minimization of entropy of Case A, SPMGTLO has provided a better solution as well as a faster convergence to the optimum solution in comparison to GA.

In Case B, SPMGTLO is able to determine a considerably better solution within very few functional evaluations as compared to GA. Moreover, in this case also GA is not able to converge to a final solution even after reaching the termination

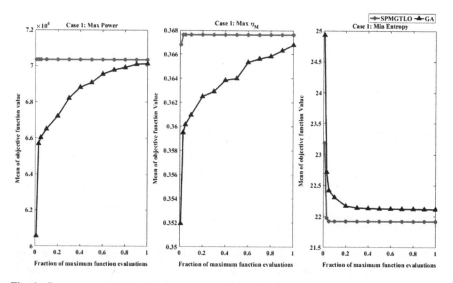

**Fig. 1** Convergence plot for Case A

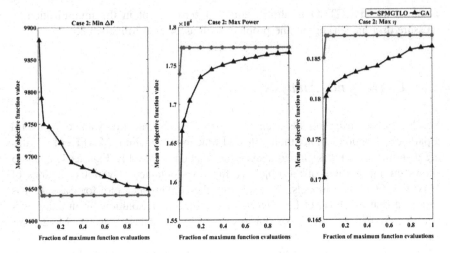

**Fig. 2** Convergence plot for Case B

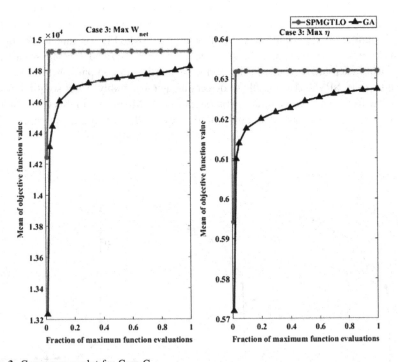

**Fig. 3** Convergence plot for Case C

criterion. For Case C, SPMGTLO requires only 0.05MaxFE functional evaluations for determining the optimum in both the objectives, whereas GA was not able to converge to a final solution. It should be noted that in all the eight objective functions considered in this work, SPMGTLO was able to obtain a better solution at all the fractions of function evaluations than the inbuilt GA function of MATLAB. It is also observed that the solution determined by SPMGTLO in the initial fractions of MaxFE is much better than the solution determined by GA. The poor performance of the widely used *ga* function in MATLAB, with the default settings and absence of guidelines on the selection of various operators and their parameters, observed in this work is consistent with an earlier finding in the literature [10]. Analysis on the sensitivity of population size can provide additional insights.

# 6 Conclusion

Solar energy is an abundant renewable resource, and the thermal energy can be converted to a useful energy form using Stirling engines. For the proper utilization of solar energy, the various parameters of the heat engine need to be systematically optimized in the presence of several nonlinearities. In this work, SPMGTLO is employed for optimizing the design parameters of three different thermodynamic models of a Stirling engine giving rise to eight unique optimization problems. It was observed that SPMGTLO was able to outperform the inbuilt MATLAB *ga* function by determining a better solution and also exhibited a faster convergence. In view of its supremacy, SPMGTLO can be used and tested to optimize various other problems arising in diverse fields. Moreover, the extension of SPMTLO to handle multiple objectives can also be a possible future work.

# References

1. M. H. Ahmadi, A. H. Mohammadi, S. Dehghani, and M. A. Barranco-Jiménez, "Multi-objective thermodynamic-based optimization of output power of Solar Dish-Stirling engine by implementing an evolutionary algorithm," *Energy Conversion and Management*, vol. 75, pp. 438–445, 2013.
2. B. Kongtragool and S. Wongwises, "A review of solar-powered Stirling engines and low temperature differential Stirling engines," *Renewable and Sustainable Energy Reviews*, vol. 7, pp. 131–154, 2003.
3. L. Yaqi, H. Yaling, and W. Weiwei,"Optimization of solar-powered Stirling heat engine with finite-time thermodynamics," *Renewable Energy*, vol.36, pp. 421–427, 2011.
4. V. Punnathanam and P. Kotecha, "Effective multi-objective optimization of Stirling engine systems," *Applied Thermal Engineering*, vol. 108, pp. 261–276, 2016.
5. V. Punnathanam and P. Kotecha, "Multi-objective optimization of Stirling engine systems using Front-based Yin-Yang-Pair Optimization," *Energy Conversion and Management*, vol. 133, pp. 332–348, 2017.
6. R. Kommadath, C. Sivadurgaprasad, and P. Kotecha, "Single phase multi-group teaching learning algorithm for single objective real-parameter numerical optimization (CEC2016)," in *2016 IEEE Congress on Evolutionary Computation*, pp. 1165–1172.

7. R. V. Rao, V. J. Savsani, and D. P. Vakharia, "Teaching–learning-based optimization: A novel method for constrained mechanical design optimization problems," *Computer-Aided Design*, vol. 43, pp. 303–315, 2011.
8. M. H. Ahmadi, H. Hosseinzade, H. Sayyaadi, A. H. Mohammadi, and F. Kimiaghalam, "Application of the multi-objective optimization method for designing a powered Stirling heat engine: Design with maximized power, thermal efficiency and minimized pressure loss," *Renewable Energy*, vol. 60, pp. 313–322, 2013.
9. H. Hosseinzade, H. Sayyaadi, and M. Babaelahi, "A new closed-form analytical thermal model for simulating Stirling engines based on polytropic-finite speed thermodynamics," *Energy Conversion and Management*, vol. 90, pp. 395–408, 2015.
10. V. Punnathanam, C. Sivadurgaprasad, and P. Kotecha, "On the performance of MATLAB's inbuilt genetic algorithm on single and multi-objective unconstrained optimization problems," in *2016 International Conference on Electrical, Electronics, and Optimization Techniques (ICEEOT)*, 2016, pp. 3976–3981.

# Performance Evaluation of Grey Wolf Optimizer and Symbiotic Organisms Search for Multi-level Production Planning with Adaptive Penalty

Sandeep Singh Chauhan and Prakash Kotecha

**Abstract** Production planning is a combinatorial optimization problem and involves a large number of semi-continuous variables and complex constraints. In this work, we propose an efficient strategy to handle the domain hole constraints and demonstrate its supremacy over the hard penalty approach used in the literature. Additionally, we employ the proposed strategy with two recently developed meta-heuristics algorithms, viz. Grey Wolf Optimizer and Symbiotic Organisms Search algorithm, to evaluate their performance for solving the production planning problem arising in the petrochemical industry.

**Keywords** Production planning · Meta-heuristics · Petrochemical industry Grey Wolf Optimizer · Symbiotic Organisms Search algorithm

## 1 Introduction

The global market of petrochemicals is worth billions of dollars, and the petrochemical sector plays a significant role in the economic development of a country as many of its products serve as raw materials for the manufacturing sector. Petrochemical industries are complex in nature and need to account multiple conflicting factors in order to be sustainable in an increasingly competitive environment. This necessitates the systematic optimization to determine the best set of decisions. An important problem that arises in petrochemical and other manufacturing industries is the determination of the products that need to be produced along with the process that needs to be used to produce the selected products. A mixed-integer linear programming has been proposed in the literature for the determination of the optimal production plan of the Saudi Arabian petrochemical industry [1]. This work has been further extended using mathematical programming [2] as well as com-

S. S. Chauhan · P. Kotecha (✉)
Department of Chemical Engineering, Indian Institute of Technology,
Guwahati 781039, Assam, India
e-mail: pkotecha@iitg.ernet.in

© Springer Nature Singapore Pte Ltd. 2019
B. K. Panigrahi et al. (eds.), *Smart Innovations in Communication and Computational Sciences*, Advances in Intelligent Systems and Computing 669, https://doi.org/10.1007/978-981-10-8968-8_39

459

putational intelligence (CI) techniques [3, 4] to incorporate production from more than a single level. In this work, we provide an efficient strategy to handle the domain hole constraints which arise from the semi-continuous variables and also evaluate the performance of two recently proposed meta-heuristic algorithms [5, 6].

## 2  Problem Statement

In the production planning problem, a set of processes $(S)$ are available for the production of a set of products $(U)$. Some of these products can be produced by multiple processes, but none of the processes produces more than one product. An instance of the production planning involving 54 processes and 24 products is given in the literature and has also been provided as supplementary information [7]. It can be observed that the product U1 can be produced by process S1, S2 and S3 whereas product U4 and U5 can be produced only by process S8 and S9, respectively. The production and investment cost of each of the process is specified at $K$ capacity levels (three levels: low, medium and high for the instance in the supplementary information [7]). The production and investment cost for any production capacity in between the specified capacity levels varies in accordance with a piecewise linear function [4]. It should be noted that it is not mandatory to employ all the processes and the production, if any, from the processes is to be within the minimum and maximum capacity levels. The selling price of product from each process and the amount of raw material required in a process for the production of a unit quantity of the product is also specified. The design of optimal production plan requires the determination of the amount of product that has to be produced from each of the processes so as to maximize the profit subject to the satisfaction of all the constraints. In certain cases, a unique process requirement is to be satisfied wherein a product is not to be produced from more than one process.

In the multi-level production strategy [4], a total of $(K - 1)S$ continuous decision variables are employed where $S$ is the total number of processes. Each decision variable corresponds to the amount of production between two consecutive capacity levels from a particular process. Thus, $x_k^s$ denotes the production from a process $s$ between the consecutive capacity levels of $k$ and $k + 1$. It is to be noted that if $x_k^s > 0$, then it has to be greater than or equal to $L_k^s$. However, it can also be equal to zero. Thus, $x_k^s$ is a semi-continuous variable as it has a domain hole (i.e. it cannot take a value greater than zero but less than $L_k^s$), but most computational intelligence-based optimization techniques are designed for continuous variables. This discontinuity has been handled in the literature using a hard penalty approach wherein a penalty of $10^5$ was assigned for every violation of the domain hole. In this work, we propose an adaptive approach, represented in Eq. 1, wherein the amount of penalty is not constant but is dependent on the amount of violation from either bounds.

$$P_k^{domain,\, s} = \begin{cases} 0 & x_k^s = 0 \\ 0 & L_k^s \leq x_k^s \leq L_{k+1}^s \quad \forall s = 1, \ldots, S, \ \forall k = 1, 2, \ldots, K-1 \\ \min\left( (d^s - x_k^s)^2, (L_k^s - x_k^s)^2 \right) & otherwise \end{cases}$$

$$(1)$$

The above equation is generic, and a value of zero is to be assigned for $d^s$ in this production planning problem to account for the possibility of zero production from a process. This adaptive penalty approach helps the CI-based optimization technique to effectively determine the optimal solution.

The total amount of production from each process $s$ from all the levels can be determined as $X^s = \sum_{k=1}^{K-1} x_k^s \ \forall s = 1, \ldots, S$.

If the amount of raw material of type $t$ required for producing a unit quantity of the product in process $s$ is denoted by $r_t^s$ and if $R_t$ indicates the total amount of raw material of type $t$ that is available, then the penalty for the violation of the raw material constraint is determined using Eq. 2. It should be noted that $P_t^{rawmaterial}$ would be zero if the amount of raw material (of type $t$) required for implementing a production plan is lower than the available amount.

$$P_t^{rawmaterial} = \left( \min\left( \left( R_t - \sum_{s=1}^{S} r_{st} X^s \right), 0 \right) \right)^2 \quad \forall t = 1, 2, \ldots, T \qquad (2)$$

If $V_k^s$ denotes the investment cost of capacity level $k$ of process $s$, then the investment cost $(I^s)$ required for process $s$ can be determined using the binary parameter $y_k^s$ as shown in Eqs. 3 and 4.

$$y_k^s = \begin{cases} 1 & L_k^s \leq x_k^s \leq L_{k+1}^s \\ 0 & otherwise \end{cases} \forall s = 1, 2, \ldots, S, \ \forall k = 1, 2, \ldots, K-1 \qquad (3)$$

$$I^s = \sum_{k=1}^{K-1} y_k^s \left( V_k^s + \left( \frac{V_{k+1}^s - V_k^s}{L_{k+1}^s - L_k^s} \right)(x_k^s - L_k^s) \right) \quad \forall s = 1, 2, .., S \qquad (4)$$

The total investment $I$ that is required for the entire production plan can be determined as $I = \sum_{s=1}^{S} I^s$. If the investment cost of a production plan exceeds the available budget $B$, the penalty for violation is determined as shown in Eq. 5.

$$P^{Budget} = (\min((B-I), 0))^2 \qquad (5)$$

The production cost of process $s$ can be determined using Eq. 6.

$$C^s = \sum_{k=1}^{K-1} y_k^s \left( Q_k^s + \left( \frac{Q_{k+1}^s - Q_k^s}{L_{k+1}^s - L_k^s} \right)(x_k^s - L_k^s) \right) \quad \forall s = 1, 2, \ldots, S \qquad (6)$$

where $Q_k^s$ denotes the production cost of process $s$ of capacity level $k$. The total number of processes that produce a particular product can be determined using Eq. 7.

$$Y^s = \begin{cases} 1 & X^s > 0 \\ 0 & otherwise \end{cases} \forall s = 1, 2, \ldots, S \tag{7}$$

If $S_u$ denotes the set of processes that can produce the product $u$ and if $U$ denotes the set of all products, then the number of processes utilized to produce the product $u$ is given by

$$n_u = \sum_{s=1}^{S_u} Y^s \; \forall u = 1, 2, \ldots, U \tag{8}$$

The penalty for violation of the unique process is determined using Eq. 9.

$$P_u^{uni} = \begin{cases} 1000^{n_u} & n_u > 1 \\ 0 & otherwise \end{cases} \forall u = 1, 2, \ldots, U \tag{9}$$

The objective function is the maximization of the profit, which is given by the difference between the selling price ($E^s$) of the product from process $s$ and the production cost of process $s$.

$$Profit = \sum_{s=1}^{S} (E^s X^s - C^s) \tag{10}$$

The fitness function for the production planning problem with the incorporation of the constraints is given by

$$\text{Min } f = -Profit + \lambda \left( \left( \sum_{s=1}^{S} \sum_{k=1}^{k-1} P_k^{domain, s} \right) + \left( \sum_{t=1}^{T} P_t^{rawmaterial} \right) + \left( P^{Budget} \right) + \left( \sum_{u=1}^{U} P_u^{uni} \right) \right) \tag{11}$$

where $\lambda$ is a penalty factor that ensures preference to feasible solutions. Thus, the production planning problem can be modelled with the help of only continuous variables despite its combinatorial nature. We have used the Grey Wolf Optimizer (GWO) [5] and Symbiotic Organisms Search (SOS) [6] to demonstrate the benefits of the proposed domain hole handling strategy. For the sake of brevity, a detailed explanation of these two algorithms is not provided and can be obtained from the literature.

# 3  Results and Discussion

In this section, we demonstrate the performance of the proposed domain hole handling strategy and evaluate the performance of GWO and SOS algorithm to solve the combinatorial production planning problem reported in the literature. The case study of the Saudi Arabian petrochemical industry involves 54 chemical processes, categorized into four different classes of products such as propylene, ethylene, synthetic gas and aromatic derivatives, and can produce 24 different petrochemical products. The investment and production costs of the 54 processes are available at three different production levels (low, medium and high). Each process requires a principal raw material, and there are three raw materials required for the 54 processes, of which two (propylene and ethylene) are available in limited quantities.

The case study consists of eight different cases that have been classified based on the availability of resources and the unique process requirement. Case 1–Case 4 includes the unique process constraint (i.e. an identical product cannot be produced by more than one process). The investment budget for Case 1 and Case 2 is $1000 \times 10^6$ and is $2000 \times 10^6$ for Case 3 and Case 4. The amount of raw material (propylene and ethylene) available for Case 1 and Case 4 is $500 \times 10^3$ tons/year, whereas $1000 \times 10^3$ tons/year are available for Case 3 and Case 4. The resource availability for Case 5 to Case 8 is identical to Case 1 to Case 4, respectively, but it does not enforce the unique process constraint. It can be noted that there are 108 decision variables (in view of three production capacity levels and 54 processes) whereas there are 135 constraints for Case 1–Case 4 as they include 24 unique process constraints (corresponding to 24 products) and 111 constraints for Case 5–Case 8.

In view of the stochastic nature of the algorithms, both the algorithms are executed for 51 independent runs which are realized by varying the seed of the random number generator ('v5normal' in MATLAB2016a from 1 to 51) thereby giving rise to a total of 816 instances (2 Algorithms $\times$ 2 domain hole handling technique $\times$ 8 Cases $\times$ 51 runs). The population size was set to 100, and the termination criteria for both algorithms are 50,000 functional evaluations. Since the problem of maximization of profit has been converted into minimization of the fitness function given by Eq. (11), it can be realized that a feasible solution would have a fitness value of zero or lower than zero.

The statistical results of the 51 runs are consolidated for each case in Table 1. In these tables, the term 'Best', 'Worst', 'Mean' and 'St. dev.' corresponds to the best, worst, mean and standard deviation of the optimal solutions obtained in the 51 runs. It can be observed from Table 1 that neither of the algorithms were able to identify even a feasible solution with the hard penalty approach. On the contrary, it can be observed from the 'Best' values in Table 1 that both the algorithms are able to determine feasible solutions with the proposed adaptive penalty approach. In view of the supremacy of the adaptive penalty approach, the rest of the discussion is given only for the adaptive penalty approach. It can be observed, from the 'Worst' values being negative, that SOS was able to determine a feasible solution in all the 51 runs (for all the eight cases) whereas GWO was not able to determine feasible solutions in many of the runs. The best value, worst value and mean determined using SOS in all the cases are better than those determined using GWO.

From Table 1, it can be observed that the profit in Case 1–Case 4 is lower than Case 5–Case 8, respectively, due to the presence of unique process constraint.

**Table 1** Statistical analysis of the results from the GWO and SOS algorithms

| Case | | Hard penalty approach | | Adaptive penalty approach | |
|------|------|------|------|------|------|
| | | GWO | SOS | GWO | SOS |
| Case 1 | Best | 8.6E+21 | 8.3E+21 | −335.4 | **−543.2** |
| | Worst | 9.0E+21 | 8.6E+21 | 1.7E+03 | −100.7 |
| | Mean | 8.8E+21 | 8.5E+21 | −106.7 | −336.1 |
| | St. dev. | 9.3E+19 | 7.3E+19 | 289.4 | 105.0 |
| Case 2 | Best | 8.4E+21 | 8.3E+21 | −462.2 | **−542.6** |
| | Worst | 9.0E+21 | 8.6E+21 | 10.4 | −98.5 |
| | Mean | 8.8E+21 | 8.5E++21 | −211.0 | −341.4 |
| | St. dev. | 1.1E+20 | 6.9E+19 | 104.8 | 105.8 |
| Case 3 | Best | 7.8E+21 | 7.6E+21 | −533.8 | **−717.0** |
| | Worst | 8.4E+21 | 7.9E+21 | 8.4E+05 | −370.1 |
| | Mean | 8.1E+21 | 7.7E+21 | 1.9E+04 | −515.1 |
| | St. dev. | 1.4E+20 | 7.4E+19 | 1.2E+05 | 86.8 |
| Case 4 | Best | 7.7E+21 | 7.5E+21 | −582.9 | **−873.7** |
| | Worst | 8.3E+21 | 7.9E+21 | 6.2E+05 | −392.9 |
| | Mean | 8.0E+21 | 7.7E+21 | 2.4E+04 | −611.8 |
| | St. dev. | 1.4E+20 | 9.1E+19 | 1.1E+05 | 118.3 |
| Case 5 | Best | 8.1E+21 | 8.0E+21 | −412.0 | **−560.2** |
| | Worst | 8.7E+21 | 8.2E+21 | 1.1E+05 | −155.4 |
| | Mean | 8.4E+21 | 8.1E+21 | 2.1E+03 | −344.8 |
| | St. dev. | 1.2E+20 | 4.2E+19 | 1.5E+04 | 98.1 |
| Case 6 | Best | 8.2E+21 | 8.0E+21 | −425.5 | **−637.8** |
| | Worst | 8.6E+21 | 8.2E+21 | 461.6 | −93.4 |
| | Mean | 8.3E+21 | 8.0E+21 | −206.7 | −374.3 |
| | St. dev. | 8.9E+19 | 4.6E+19 | 157.9 | 126.0 |
| Case 7 | Best | 7.3E+21 | 7.0E+21 | −371.6 | **−771.5** |
| | Worst | 8.0E+21 | 7.2E+21 | 2.5E+05 | −319.3 |
| | Mean | 7.6E+21 | 7.1E+21 | 1.1E+04 | −562.2 |
| | St. dev. | 1.4E+20 | 5.2E+19 | 3.8E+04 | 122.6 |
| Case 8 | Best | 7.1E+21 | 6.8E+21 | −498.1 | **−1127.2** |
| | Worst | 7.7E+21 | 7.1E+21 | 1.9E+06 | −335.4 |
| | Mean | 7.4E+21 | 6.9E+21 | 4.9E+04 | −668.4 |
| | St. dev. | 1.2E+20 | 5.4E+19 | 2.6E+05 | 158.8 |

However, the profit for Case 2 is lower than Case 1 despite the availability of raw material being twice the quantity. Moreover, the solutions determined by SOS are inferior to those determined using the mathematical programming approach [2]. The objective function determined by both the algorithms in the 51 runs is shown in Fig. 1. It should be noted that the values of the infeasible runs are not depicted in this figure and that SOS is able to determine feasible solutions in all the 408 instances whereas GWO was not able to determine a single feasible solution in 143 instances out of 408 instances.

The convergence curve corresponding to the best run for each of the two algorithms is given in Fig. 2. For better clarity, the figure only shows the

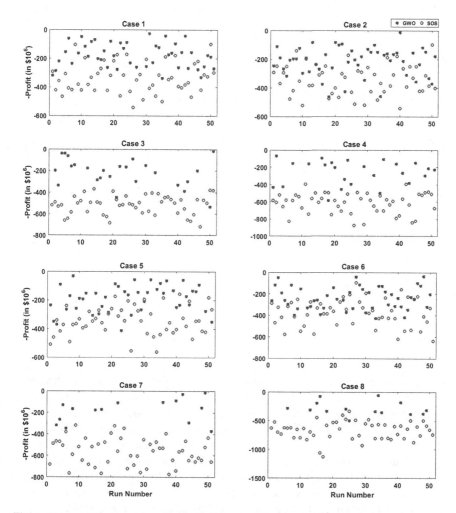

**Fig. 1** Values of the fitness function in the 51 runs of the two algorithms

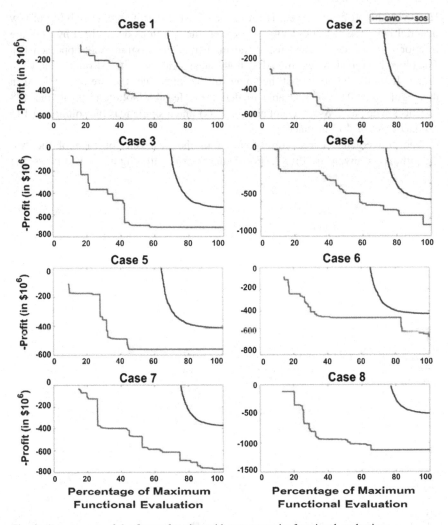

**Fig. 2** Convergence of the fitness function with respect to the functional evaluations

convergence curve from the functional evaluation in which the algorithm discovers the first feasible solution. In all the eight cases, it can be observed that SOS is able to determine a feasible solution in fewer functional evaluations than GWO. After the discovery of feasible solutions, the convergence is exponential in GWO whereas it is gradual in SOS. In all the cases, the solution determined by SOS is significantly better than the solution determined by GWO. The raw material and the investment cost required for the best production plan for all the eight cases are shown in Fig. 3, and it can be observed that SOS is able to better employ the limited resources which aids in the increase of profit. The production plan determined using SOS for each of the eight cases is given in Table 2.

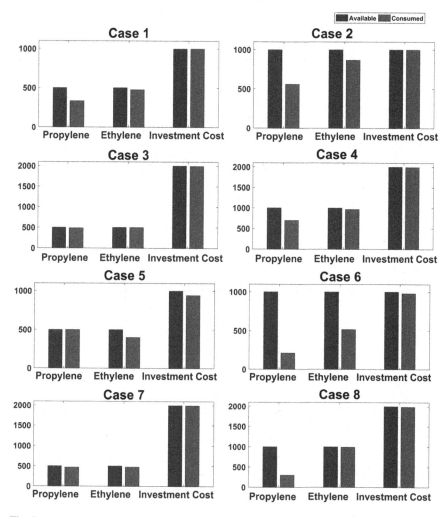

**Fig. 3** Resource utilization for the best production plan determined using SOS

It can be observed that each product, from Case 1 to Case 4, is produced by only one process that shows the satisfaction of the unique process requirement. In all the eight cases, the product U21 is produced and the number of products produced in Case 1–Case 4 is never lower than the number of products, than their corresponding cases, in Case 5–Case 8. It can also be observed that in Case 5–Case 8, many of the products are produced by more than one process. For example, in Case 5 and Case 7, the product U1 is produced by more than one process, whereas in all the four cases, the product U21 is produced using at least two processes. This leads to utilization of the resources to produce larger quantities of profitable products and leads to an increase in overall profit as shown in Table 1.

**Table 2** Optimal production plan determined using SOS

Unique process constraint

| | Product | Process | $x_1^s$ | $x_2^s$ | $X^s$ |
|---|---|---|---|---|---|
| Case 1 | U2 | S5 | 95.0 | 0 | 0 |
| | U3 | S7 | 0 | 134.0 | 134.0 |
| | U5 | S9 | 0 | 160.0 | 160.0 |
| | U16 | S34 | 0 | 146.8 | 146.8 |
| | U19 | S42 | 22.4 | 0 | 22.4 |
| | U21 | S48 | 450.0 | 680.0 | 1130.0 |
| Case 2 | U1 | S2 | 0 | 300.0 | 300.0 |
| | U2 | S4 | 0 | 290.0 | 290.0 |
| | U14 | S28 | 0 | 270.0 | 270.0 |
| | U15 | S33 | 0 | 490.0 | 490.0 |
| | U21 | S49 | 450.0 | 0 | 450.0 |
| | U24 | S54 | 197.3 | 0 | 197.3 |
| Case 3 | U1 | S2 | 150.0 | 0 | 150.0 |
| | U6 | S11 | 0 | 360.0 | 360.0 |
| | U9 | S20 | 0 | 300.0 | 300.0 |
| | U14 | S30 | 112.4 | 0 | 112.4 |
| | U15 | S31 | 200.0 | 0 | 200.0 |
| | U17 | S37 | 360.0 | 550.0 | 910.0 |
| | U19 | S41 | 24.8 | 0 | 24.8 |
| | U20 | S45 | 567.1 | 0 | 567.1 |
| | U21 | S47 | 0 | 680.0 | 680.0 |

Without unique process constraint

| | Product | Process | $x_1^s$ | $x_2^s$ | $X^s$ |
|---|---|---|---|---|---|
| Case 5 | U1 | S2 | 0 | 218.0 | 528.0 |
| | | S3 | 0 | 310.0 | |
| | U14 | S30 | 0 | 135.0 | 135.0 |
| | U17 | S36 | 365.0 | 0 | 365.0 |
| | U21 | S46 | 450.0 | 0 | 900.0 |
| | | S47 | 450.0 | 0 | |
| | U22 | S51 | 0 | 50.0 | 50.0 |
| Case 6 | U2 | S5 | 0 | 187.0 | 187.0 |
| | U5 | S9 | 46.0 | 0 | 46.0 |
| | U21 | S47 | 0 | 680.0 | 1810 |
| | | S48 | 450 | 680.0 | |
| | U22 | S50 | 0 | 46.7 | 46.7 |
| Case 7 | U1 | S1 | 135.0 | 0 | 384.2 |
| | | S3 | 0 | 249.2 | |
| | U4 | S8 | 75.5 | 0 | 75.5 |
| | U7 | S17 | 93.8 | 0 | 93.8 |
| | U9 | S20 | 0 | 175.9 | 175.9 |
| | U10 | S21 | 36.9 | 66.1 | 103.0 |
| | | S22 | 37.4 | 100 | 137.4 |
| | U16 | S34 | 0 | 200.0 | 200.0 |
| | U17 | S35 | 0 | 540.0 | 905.0 |
| | | S36 | 365.0 | 0 | |
| | U19 | S42 | 0 | 27.4 | 27.4 |
| | U21 | S46 | 450.0 | 0 | 918.8 |
| | | S47 | 0 | 468.8 | |

(continued)

**Table 2** (continued)

Unique process constraint

| | Product | Process | $x_1^s$ | $x_2^s$ | $X^s$ |
|---|---|---|---|---|---|
| Case 4 | U1 | S3 | 0 | 310.0 | 310.0 |
| | U3 | S6 | 80 | 0 | 80.0 |
| | U5 | S9 | 0 | 160.0 | 160.0 |
| | U6 | S12 | 180.0 | 0 | 180.0 |
| | U7 | S17 | 100.0 | 0 | 100.0 |
| | U9 | S20 | 0 | 300.0 | 300.0 |
| | U11 | S23 | 250.0 | 0 | 250.0 |
| | U12 | S26 | 180.0 | 0 | 180.0 |
| | U14 | S30 | 135.0 | 0 | 135.0 |
| | U15 | S31 | 0 | 400.0 | 400.0 |
| | U19 | S41 | 25.0 | 0 | 25.0 |
| | U21 | S48 | 450.0 | 680.0 | 1130.0 |
| | U22 | S51 | 25.0 | 0 | 25.0 |

Without unique process constraint

| | Product | Process | $x_1^s$ | $x_2^s$ | $X^s$ |
|---|---|---|---|---|---|
| Case 8 | U2 | S5 | 95.0 | 0 | 95.0 |
| | U5 | S9 | 0 | 160.0 | 160.0 |
| | U6 | S12 | 132.4 | 0 | 132.4 |
| | U9 | S20 | 59.8 | 0 | 59.8 |
| | U10 | S22 | 0 | 100.0 | 100.0 |
| | U14 | S30 | 0 | 163.1 | 163.1 |
| | U15 | S31 | 0 | 275.0 | 275.0 |
| | U17 | S35 | 322.4 | 0 | 862.4 |
| | | S36 | 0 | 540.0 | |
| | U21 | S46 | 0 | 680.0 | 1973.9 |
| | | S47 | 0 | 613.9 | |
| | | S48 | 0 | 680.0 | |

# 4 Conclusion

In this article, we have proposed an adaptive penalty approach to handle the domain hole constraints that are encountered in the production planning problem to improve the efficiency of the meta-heuristics algorithms used to solve such optimization problems. It was observed that the proposed technique aids the optimization algorithms to discover better solutions. The performance of two recent meta-heuristics algorithms was evaluated, and it was observed that SOS is able to consistently outperform GWO. Future work can involve developing hybrid algorithms and incorporate multiple objectives.

# References

1. Alfares, H. and A. Al-Amer, *An optimization model for guiding the petrochemical industry development in Saudi Arabia.* Engineering Optimization, 2002. **34**(6): p. 671–687.
2. Kadambur, R. and P. Kotecha, *Optimal production planning in a petrochemical industry using multiple levels.* Computers & Industrial Engineering, 2016. **100**: p. 133–143.
3. Chauhan, S.S. and P. Kotecha. *Single level production planning in petrochemical industries using Moth-flame optimization.* in *2016 IEEE Region 10 Conference (TENCON).* 2016.
4. Kadambur, R. and P. Kotecha, *Multi-level production planning in a petrochemical industry using elitist Teaching–Learning-Based-Optimization.* Expert Systems with Applications, 2015. **42**(1): p. 628–641.
5. Mirjalili, S., S.M. Mirjalili, and A. Lewis, *Grey Wolf Optimizer.* Advances in Engineering Software, 2014. **69**: p. 46–61.
6. Cheng, M.-Y. and D. Prayogo, *Symbiotic Organisms Search: A new metaheuristic optimization algorithm.* Computers & Structures, 2014. **139**: p. 98–112.
7. *Supplementary information* https://goo.gl/mg91bG, *Last accessed July 30 2017.*

# *Special-q* Techniques for Number Field Sieve to Solve Integer Factorization

Tanmay Sharma and R. Padmavathy

**Abstract** Number Field Sieve is one of the best exist method for integer factorization. In this method, relation collection stage is the most time-consuming part and requires a large amount of memory. This paper reports comprehensive analysis between Pollard's algorithm and FK algorithm of *special-q* lattice sieving in Number Field Sieve. The experiments are performed using two widely used factoring tools CADO-NFS and GGNFS on the data sets ranging between 100–120 digits; since CADO-NFS is based on FK algorithm and GGNFS is based on Pollard's algorithm. Experiments are carried out using various parameters. The results are derived and influencing factors are reported. Through the results, it is shown that even though FK algorithm for sieving which avoids the lattice reduction compared with Pollard, the selection of *special-q*, influences on the performance of the sieving algorithm.

**Keywords** Integer factorization · Number Field Sieve · Function field sieve
*special-q* lattice sieving

## 1 Introduction

In the contemporary world of universal electronic connectivity, security and privacy are the major concern. Public Key cryptography (PKC) allows us to exchange information securely over an unsecure channel. PKC is the backbone of today's era of Internet.

T. Sharma (✉) · R. Padmavathy
National Institute of Technology, Warangal, India
e-mail: tanmayvdsharma@gmail.com

R. Padmavathy
e-mail: rpadma@nitw.ac.in

© Springer Nature Singapore Pte Ltd. 2019
B. K. Panigrahi et al. (eds.), *Smart Innovations in Communication and Computational Sciences*, Advances in Intelligent Systems and Computing 669, https://doi.org/10.1007/978-981-10-8968-8_40

471

## 1.1  Discrete Logarithm Problem and Integer Factorization

Diffie Hellman and RSA, introduced in 1970s, are the two revolutionized algorithm that have changed the face of digital world. Although both algorithms are different in their architecture as one is based on discrete log problem (DLP) while other is based on integer factorization problem (IFP), surprisingly there are a huge number of non-trivial similarities between them:

- There is no classical polynomial-time algorithm for both, but there exist a quantum algorithm that works for both.
- Both are used as basis for public key cryptography.
- Many classical algorithms for both like (Pollard's rhos, baby-step giant-step, Number Field Sieve and function field sieve ) are closely related to eachother.

## 1.2  Advance Methods to Solve DLP and IFP

Apart from the traditional generic methods with complexity $O(\sqrt{(G)})$ such as Shanks' baby-step giant-step, Pollard's rho, Index Calculus Method (ICM) is the method which uses extra available information about group structure wisely to get a faster algorithm. Index calculus technique is common to both algorithms for IFP and for DLP. The fastest known algorithms for solving the above two problems are based on ICM and they are Number Field Sieve (NFS-DL, NFS-IF) and Function Field Sieve (FFS).

Number Field Sieve has sub-exponential complexity $L_{\#G}(1/3, c)$ to compute discrete logarithms, and integer factorization and function field sieve (in a finite filed of small characteristic) have complexity $L_{\#G}(1/4 + \epsilon, c)$ to compute discrete logarithms [1].

In 2014, a heuristic approach has been proposed to compute discrete logarithms using FFS which have quasi-polynomial complexity of the form $n^{O(logn)}$ [2].

## 1.3  About This Paper

**Relation Collection**: This paper deals with the relation collection step of NFS. Relation collection is one of the main and most time-consuming step of NFS and FFS. The *special-q* lattice sieving method is the most efficient and widely used method for relation collection.

*special-q* **lattice sieving**: Previous methods of sieving use one-dimensional search to sieve relations, so they were inefficient for large numbers. But *special-q* lattice sieving method is not restricted to sieve relations in one dimension but allows freedom to search them in multiple dimensions; hence, it is very efficient. As per Pollard

[3] in ideal condition, one can get 83% of work output for just 8.6% of the amount of work with respect to the previous approach of sieving.

**Pollard's *special-q* Approach**: In 1993, J.M. Pollard [3] introduced this vibrant method. This method was adapted in various implementations of NFS and FFS, in 2002 and in 2005 by A. Joux to compute discrete logarithms in $GF(2^{521})$ [4] and $GF(2^{607})$ *and* $GF(2^{613})$ [5][1], respectively. In 2010 by T. Hayashi in solving a 676-bit discrete logarithm problem in $GF(3^{6n})$ [6], in 2013 by J.Detrey to compute discrete logarithms in $2^{809}$ [1, 7]. This shows wide application of this efficient method in computing various problems.

**Franke-Kleinjung's *special-q* Approach**: Later in 2005, Franke and Kleinjung [8] proposed improved version of Pollard's approach. FK point out that Pollard uses lattice basis reduction methods quite a lot, and these reduction methods are bit slower and consume a lot of time of algorithm. FK approach is based on their certain propositions and do not use any lattice basis reduction techniques instead it is based on truncated Euclidean algorithm.

In 2010, NFS factorization of 768-RSA uses this approach [9]. In 2016 -P. Gaudry proposed *three*-dimensional sieving approach for NFS based on this algorithm [10]. FK algorithm is started getting attention since its invention and it is being adapted widely in NFS class of algorithms.

**Our Contribution**: This paper reports comprehensive analysis between Pollard's algorithm and FK algorithm of *special-q* lattice sieving in NFS. The experiments are carried out using various parameters. The results are derived and the influencing parameters are reported through various experimental results and analysis. It is shown that even though the FK algorithm eliminates the reduction of basis compared with Pollard the selection of *special-q* influences a lot on the performance of the algorithm. This leads to portray that influencing parameters plays crucial role in the overall performance of relation collection step of ICM class of algorithms.

## *1.4  Application of Our Analysis*

Public key cryptography (PKC) is the backbone of today's secure communication over Internet. The security base of all PKC-based security protocols, such as SSL, IPsec, and certificate generation protocols, PKI, PGP, etc., is rely on the intractability of hard problems namely discrete logarithm problem and integer factorization problem. The cryptanalysis on these hard problems decides the key sizes to be used in PKC-based security protocols. The safe key size in the current scenario for these protocols is 2048 bits.

The Number Field Sieve and Function Field Sieve are best-known methods to solve these hard problems. The NFS is analyzed in this present work. The integer factorization on higher digits is possible by fine tuning the parameters used in the NFS algorithm. The analysis in the present work may aid to achieve computationally

---

[1]*GF* is Galois field.

fast NFS to solve integer factorization problem. This in turn may lead to decide the key sizes of PKC-based security protocols. For example, the popular RSA is based on the intractability of integer factorization. The analysis reported in the current study indirectly effects the security base of RSA. This shows the effective analysis on popular and efficient algorithms such as NFS/FFS may become a root cause for the improvement of the above algorithms and in turn rule the security base of all PKC methods used in the internet protocols.

**Outline of this paper**: In the coming section, we will discuss basics of Number Field Sieve algorithm (Sect. 2), and then we will lead toward *special-q* lattice sieving (Sect. 3). In the subsections of Sect. 3, we are focusing on Pollard and FK approach. In Sect. 4, we are highlighting the cruciality of parameters in the relation collection step through various test cases, and then in Sect. 5, we are proceeding to concluding remarks.

## 2 Number Field Sieve Algorithm

This is just a simple introduction of NFS and avoiding various conceptual complexity here to make our description easy. One can refer [11] for in depth details, it covers almost all information related to development of Number Field Sieve and the contributions of actual creator of this algorithm.

The NFS is the one of the most sophisticated factoring algorithm based on the concept that to factor a composite number $N$, find integers pairs $n_1, n_2$ such that $n_1^2 \equiv n_2^2 (mod N)$.

The NFS algorithm consists of five major steps:

 (i) Polynomial Selection
 (ii) Factorbase Construction
 (iii) Relation Collection
 (iv) Linear Algebra
 (v) Square Root.

### 2.1 Polynomial Selection

NFS works in an algebraic number field. To construct an algebraic number field in NFS, we need the following two free parameters:

- a polynomial $f : \mathbb{R} \to \mathbb{R}$ with integer coefficients.
- a natural number $m \in \mathbb{N}$ that satisfies $f(m) \equiv 0 (mod N)$.

To satisfy the above conditions easily, we first choose $m$, and the following method is use to find the polynomial $f(x)$, known as "*base-m*" method. In this method, we write $N$ in terms of $m$, where $m \in \mathbb{N}$.

$$N = c_d m^d + c_{d1} m^{d1} + \ldots + c_0 \tag{2.1.1}$$

Now, we are defining the function $f(x)$ as:

$$f(x) = c_d x^d + c_{d1} x^{d1} + \ldots + c_0 \tag{2.1.2}$$

As we can see that $f(m) = N$ and $f(m) \equiv 0 (mod N)$, so $f(x)$ and $m$ both are satisfying the above conditions. Throughout this paper, we will use this $f(x)$ and $m$ unless stated explicitly.

## 2.2 Mapping

NFS works over $\mathbb{Z} \times \mathbb{Z}[\alpha]$, the integers $\mathbb{Z}$ and over the ring $\mathbb{Z}[\alpha]$, where $\alpha \in \mathbb{C}$ is the root of polynomial $f(x)$ chosen above.

The ring $\mathbb{Z}[\alpha]$ is defined as follows:

$$\mathbb{Z}[\alpha] \cong \mathbb{Z}[x]/(f(x)) \cong \mathbb{Z}.1 \oplus \mathbb{Z}.\alpha \oplus \ldots \oplus \mathbb{Z}.\alpha^{d-1} \tag{2.2.1}$$

For the polynomial $f(x)$ and $m$, there exists a unique ring homomorphism mapping:

$$\phi : \mathbb{Z} \times \mathbb{Z}[\alpha] \longrightarrow (\mathbb{Z}/N\mathbb{Z} \times \mathbb{Z}/N\mathbb{Z}) \quad \text{satsfying :} \tag{2.2.2}$$

(i) $\phi(pq) = \phi(p)\phi(q) \quad \forall p, q \in \mathbb{Z}[\alpha]$
(ii) $\phi(p + q) = \phi(p) + \phi(q) \quad \forall p, q \in \mathbb{Z}[\alpha]$
(iii) $\phi(1) \equiv 1 (mod N)$
(iv) $\phi(\alpha) \equiv m (mod N)$
(v) $\phi(zp) = z\phi(p) \quad \forall z \in \mathbb{Z}, p \in \mathbb{Z}[\alpha]$

Now assume there exist a finite set $U$ which consist of integer pairs $(a, b)$ such that:

$$\prod_{(a,b) \in U} (a + b\alpha) = \beta^2 \quad \text{and} \quad \prod_{(a,b) \in U} (a + bm) = y^2 \tag{2.2.3}$$

where $y \in \mathbb{Z}$, $\beta \in \mathbb{Z}[\alpha]$.

Using above rules of mapping, we will get:

$$\phi : a + bm \times a + b\alpha \longrightarrow (a + bm (mod N) \times a + bm (mod N)) \tag{2.2.4}$$

## 2.3 Factor Base Construction

Factor base is a set which consists of prime numbers less than a certain bound, and this bound is called factor base bound or smoothness bound. Algebraic elements in NFS are of the form $a + bx \in \mathbb{Z}[x]$ $(a + ba \in \mathbb{Z}[\alpha]$ & $a + bm \in \mathbb{Z})$. Pair (a,b) should be coprime, otherwise multiples of the same (a,b) will be sieved again and again.

### 2.3.1 Algebraic and Rational Factor Bases

As we have discussed NFS works over $\mathbb{Z} x \mathbb{Z}[\alpha]$, so it will have two factor base one for the ring $\mathbb{Z}[\alpha]$ known as *algebraic factor base* and other for $\mathbb{Z}$ known as *rational factor base*.

- Algebraic Factor base: It consists of finite prime ideals (means ideals cannot be factor in non-trivial way) of the form $a + ba$ where $\alpha \in \mathbb{Z}[\alpha]$, and $a, b \in \mathbb{Z}$ less than a particular bound($B_\alpha$).
- Rational Factor base: It consists of finite consecutive prime numbers (for ex: 2, 3, 5...) less than an appropriate bound($B_m$).

Smoothness bound for a factor base usually chosen by considering various observations and experiments.

### 2.3.2 $\mathbb{Z}[\alpha]$ is inconvenient

$\mathbb{Z}[\alpha]$ is a complex algebraic structure to represent on a computer, so it is bit hard to represent element of algebraic factor base $a + ba \in \mathbb{Z}[\alpha]$ in memory. But $a + ba$ can be represent as $(p, r)$ in factor base as they are in bijective correspondence [12].

| prime p' | root r |
|---|---|
| $p'_1$ | $r_1^1, r_1^2, r_1^3, r_1^4$ |
| $p'_2$ | $r_2^1, r_2^2, r_2^3$ |
| $p'_3$ | $r_3^1, r_3^3 \ldots r_3^d$ |
| $p'_4$ | $r_4^1, r_4^2$ |
| $p'_5$ | $r_5^1, r_5^2, r_5^3$ |
| . | . |
| . | . |
| . | . |
| $p'_n < B_\alpha$ | $r_n^1, r_n^2, r_n^3$ |

| prime p : | $p_1$ | $p_2$ | $p_3$ | $p_4$ | $p_5$ | .......... | $p_a < B_m$ |
|---|---|---|---|---|---|---|---|

Rational Factor Base

Algebraic Factor Base

## 2.4 Relation Collection

Now, we will discuss commutative diagram of NFS which we used to refer as 'magic diagram'

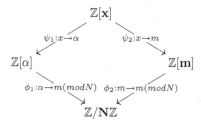

**Algebraic Side and Rational Side**: The main aim of the relation collection step is to find doubly smooth elements (smooth on both algebraic and rational side) of the form in $a + bx$ in the ring $\mathbb{Z}[x]$. Left-hand side of the diagram is known as algebraic side and element is of type $a + b\alpha$. Right-hand side is known as rational side and element is of type $a + bm$.

Thus, same element $a + bx$ is representing in two different ways $a + b\alpha$ and $a + bm$ in two different mathematical structure, and finally, they will map to ring $\mathbb{Z}/N\mathbb{Z}$ yielding a relation between two representation (if satisfying smoothness criteria).We call these $(a, b)$ pair as *relation*, and to find such relation, we use various sieving techniques.

## 3 Sieving Approaches

Relation collection step is to sieve a, b pair satisfying the following conditions:

(i) $gcd(a, b) = 1$,
(ii) rational number $a + bm$ is $B_m$-smooth on rational side,
(iii) Norm of algebraic element $\mathcal{N}(a + b\alpha)$ or $\mathcal{N}(a, b) = b^d f(a/b)$ is $B_\alpha$-smooth on algebraic side.

To solve $GF(2^{1039})$, we require one billion such $(a, b)$ pairs and we may get one smooth coprime pairs after sieving several millions such pairs, so total sieving space will be beyond $10^{15}(10^6 \times 10^9)$. Such a large sieving space cannot be searched by hit and trial search. There are several sieving algorithms, the popular and efficient algorithms are *special-q* lattice sieving algorithms.
*special-q* Lattice Sieving:

1. sieving by Pollard's Approach
2. sieving by Franke-Klienjung Approach.

## 3.1 Pollard's Approach of special-q lattice sieving

The sieving can be achieved by many traditional sieving algorithms like basic sieving, line sieving and they are suitable for small primes. The drawback of these algorithms is allocation of large amount of memory or inefficient for large primes. *special-q* lattice sieving approach overcomes the above problems, for details refer [3, 13, 14].

**A new approach**: In the previous sieving approaches, we fix one from $(a, b)$ (+ve) and vary other in both direction( +ve and −ve). Instead of sieivng in this way, Pollard [3] proposed a new algorithm *special-q* lattice sieving in two-dimensional space for NFS. In this approach, all the pairs (a,b) which are being examined will be divisible by a common prime number $q$, and this common prime is called as *special-q*.

**Setting ingredients for new algorithm**: Factor base $B_x$: $(x \in m, \alpha)$ will be divided into different parts small $S$, medium $M$ and large $L$ primes, we then choose the sieving limits for $a$ and $b$, and then we will choose *special-q* primes $\in M$ and sieve for pairs $(a, b)$ that satisfy

$$\mathcal{N}(a, b) \equiv 0(mod q) \text{ or } a + bs \equiv 0(mod q) \qquad (3.1.1)$$

Above 'or' depends on whether to sieve rational side first or algebraic, here we are sieving rational side first.

**Lattice $\mathcal{L}_q$**: Pairs (a, b) will form a two-dimensional lattice $\mathcal{L}_q$ in the $(a, b)plane$, and the basis of the lattice is $(q, 0), (s, 1)$ ($s$ is root w.r.t. prime $q$, i.e., $s = ab^{-1} mod q$).

$\mathcal{L}_q$ **basis reduction**: With that basis of lattice, we cannot sieve region efficiently; generally, we use reduced basis which we can get by using any of the lattice basis reduction algorithm like LLL or Gaussian lattice reduction. Suppose by applying any of the above algorithm, we get two reduced basis vector $V_1$ and $V_2$.

$$V_1 = (a_1, b_1), V_2 = (a_2, b_2)$$

So, a point on lattice $\mathcal{L}_q$ can be represented as linear combination of $V_1$ and $V_2$:

$$\begin{aligned} (a, b) &= c.V_1 + e.V_2 \\ &= c.(a_1, b_1) + e.(a_2, b_2) \\ &= (c.a_1 + e.a_2, c.b_1 + e.b_2) \end{aligned}$$

We will call pair $(c, e) \in \mathbb{Z}$ as a point in $(c, e)$ plane.

**Sieivng**: $(c, e)$: $(c, e)$ should be coprime in order to have $(a, b)$ to be coprime. Up till now, we have lattice $\mathcal{L}_q$ in $(a, b)$ plane, and each pair in that plane satisfies $a \equiv bs(mod q)$, reduced basis of lattice $V_1, V_2$. Since $V_1$ and $V_2$ are constant, so sieve of $(a, b)$ now depends on sieve of pair $(c, e)$. So, we will seek for smooth coprime pairs $(c, e)$ for primes $p \in B_x$. We sieve only those pairs by prime $p$ which satisfy:

$$c.u_1 + e.u_2 \equiv 0(modp)$$

where $u_1 = a_1 - b_1x$, $u_2 = a_2 - b_2x$ and $gcd(u_1, u_2) = q$.

$$c + e.(u_2/u_1) \equiv 0(modp)$$
$$c + e.(root) \equiv 0(modp)$$
$$c \equiv -e.(root)(modp)$$

where $root = (u2/u1)$ and $c = e.(\frac{a_2 - b_2\alpha}{\alpha.b_1 - a_1})(modp)$ in terms of $a$ and $b$.
If we observe careful, it will form a lattice $\mathcal{L}_p$ in $(c, e)plane$.

**Lattice $\mathcal{L}_p$ and $\mathcal{L}_{pq}$:** $\mathcal{L}_p$ is a lattice which consist of element of the form ($c + root.e \equiv 0(modp)$) and divisible by prime $p$ of factor base.
  We can sieve such $(c, e)$-pairs using two approaches:

(i) Sieving by rows: This method is nothing but line sieving, ticking every $p$th element. This method is good for small primes; so, we use it here only for small primes.

(ii) Sieving by vectors: This method is suitable for large primes. Now, we want those element of $\mathcal{L}_q$ elements which are divisible by $p$ as well. It means if we sieve for element in the sub-lattice $\mathcal{L}_{pq} = \mathcal{L}_p \cap \mathcal{L}_q$, then we will get the desirable elements, i.e., elements which are divisible by both $q$ and $p$.

  Basis of the sub-lattice $\mathcal{L}_{pq}$ will be $(p, 0)$ ,$(root, 1)$. Again, we will apply lattice reduction techniques and get two reduced basis $U_1$ and $U_2$:
  So, pair $(c, e)$ can be written as

$$(c, e) = k_1.U_1 + k_2.U_2$$

where $U_1 = (c_1, e_1)$ and $U_2 = (c_2, e_2)$ and $k_1, k_2 \in \mathbb{Z}$

**Exceptional case:** Since $c \equiv root.e(modp)$, so there might have a possibility that $root \equiv 0(modp)$ or $e \equiv 0(modp)$. We call such cases as exceptional and have to be dealt differently. We use sieving by rows approach to sieve pairs $(c, e)$ in these cases.

### 3.1.1 How Sieving by Vectors Works?

Pollard's and FK's approach basically defers here, in both approach for small prime line sieving is used.
**Segmentation of Sieving Region:** We will sieve pair $(c, e)$ in $(c, e)plane$, so firstly we set limits of $(c, e)$ to sieve. Let $|c| \leq C$ and $0 < e \leq E$ and the choice of $C$ and $E$ depends on $N$. We will take an array $A[c, e]$ which represent elements of $(c, e) plane$.

As limits $C$ and $E$ will be large, so $(2C + 1)xE$ not able to fit in memory at a time, thus we will split this area into smaller segments as $S_1, S_2...S_t$ as $S_i = (c, e) : c \leq |C|, E_{i-1} < e \leq E_i$ for $i = 1, 2...t$.

**Sieving on Rational Side:** We start with two nested for loops one varying $k_1$ and other varying $k_2$ and getting different linear combinations of $(c, e)$ pairs, considering all the equalities and inequalities between $C, E_{i-1}, E_i, U_1, U_2$. As we are sieving in the lattice $\mathcal{L}_{pq}$ and all the elements of this lattice are divisible by both $p$ and $q$, so we can add $logp$ in $A[c, e]$ to all the linear combinations of $(c, e)$ obtained in the above nested loops.

The above procedure repeat for multiple *special-q*, and for all the primes $p$ considered in the factor base. After sieving for sufficient *special-q* and with all $p$ considered in factor base $B_m$ in $\mathcal{L}_{pq}$, we check number of entries in $A[c, e]$ surpass the given threshold and mark those entries.

**Sieivng on Algebraic Side** Now those entries which were managed to cross a given threshold previously have to be checked on algebraic side. Same procedure as above is repeated on this side also but certainly with the well-known modifications. Now, we get candidate pairs $(c, e)$ which are likely to be smooth on both algebraic and rational side.

**Rigorous Test for smoothness** Again these candidates pairs are to be tested rigorously to find that whether the pairs are really smooth or not. Checking smoothness alone can take upto $1/4$th − $1/3$rd time of whole relation collection step. We can use Shank's squfof or ECM methods to factor elements in order to find smoothness.

If for some pairs, their one or two factors are greator than their corresponding factorbase bound then also we can allow such pairs. Such relief helps in producing large amount of relations without any significant loss in efficiency.

**Duplicate Relation** As we choose a number of different *special-q* while sieving, there may be a possibility that two *special-q* form same relation which result in duplication of relation. Such duplicate relation has to be removed in the sub-step between relation collection and linear algebra known as filtering step.

## 3.2   FK Approach of special-q Lattice Sieving

As relation collection is one of the most time-consuming step and real bottleneck of the NFS method. If we are able to improve any of its dominant part slightly, it can contribute to reduce overall time taken by the algorithm. Pollard's *special-q* sieving approach was a significant milestone as compared to previous sieving approach. In 2005, Franke and Kleinjung [8] proposed a method of *special-q* sieving which is better than Pollard's approach by a constant factor.

**New Approach:** Franke and Kleinjung point out that Pollard's uses lattice basis reduction methods quite a lot, and these reduction methods are bit slower, consumes a lot of time of algorithm. In the FK approach, they choose lattice basis which is based on their certain propositions and do not use any lattice basis reduction techniques instead it is based on truncated Euclidean algorithm, which is usually faster than those basis reduction algorithm. But FK approach works only for large primes. For small primes, we still stick to our line sieving algorithm.

In Pollard's approach, we have pair $(q, s)$, lattice $\mathcal{L}_q$, reduced basis $(a_1, b_1)$, $(a_2, b_2)$, and we try to find coprime smooth $(a, b)$ pairs as:

$$(a, b) = i.(a_1, b_1) + j.(a_2, b_2) \tag{3.2.1}$$

where $-I \leq i \leq I$ and $0 \leq j \leq J$ and $I$ and $J \in A$(notation as per [8]).

FK approach is based on the following propositions:

**Proposition 1** *For the lattice $\mathcal{L}_p$ form by a large prime $p \geq I$, there exist a unique basis $(\alpha, \beta), (\gamma, \delta)$ with the following properties:*

(i) $\beta > 0$ *and* $\delta > 0$
(ii) $-I < \alpha \leq 0 \leq \gamma < I$
(iii) $\gamma - \alpha \geq I$

*and for each $(i, j) \in \mathcal{L}_p$ with $-I \leq i \leq I$ and $0 \leq j \leq J$, then each such pair can be represented as linear combination of above uniqe basis, i.e., $(i, j) = k(\alpha, \beta) + l(\gamma, \delta)$ with $k \geq 0$ and $l \geq 0$.*

**Proof of Proposition 1** To prove above proposition, they are using argument that, for elements $i - j.r \equiv 0 (mod p)$ and $0 \leq r < p$, there exist a sequence of integer pairs which can be defined as $(i_0, j_0) = (-p, 0)$, $(i_1, j_1) = (r, 1)$ and $(i_{k+1}, j_{k+1}) = (i_{k-1}, j_{k-1}) + a_k(i_k, j_k)$ where $a_k = \lfloor \frac{-i_{k-1}}{i_k} \rfloor$ and this series stops when $i_k = \pm 1$.

**Decreasing Sequence** $((-1)^{k+1} i_k)$: In the above sequence, one can notice that $((-1)^{k+1} i_k)$ is a decreasing sequence; e.g., at $k = 0$, it is $-(-p) = p$ and $i$ is bounded by $I$ and $I < p$ so at $k = 1$ whatever it is, it will be less than $p$.

**Increasing Sequence** $j_k$: And $j_k$ is an increasing sequence, as initially at $k = 0, j = 0$; at $k = 1, j = 1$ and then each time some positive is added to it as $a_k$ is always positive.

- For odd $k$: we set $(\alpha, \beta) = (i_{k-1}, j_{k-1}) + a_k(i_k, j_k)$ and $(\gamma, \delta) = (i_k, j_k)$. It is because, for odd $k$, $i_{k-1}$ will always be negative and $i_{k-1} \geq a_k i_k$ always, so we will get $\alpha \leq 0$ always and $i_k$ will positive so that we can get $\delta \geq 0$.
- For even $k$: we set $(\alpha, \beta) = (i_k, j_k)$ and $(\gamma, \delta) = (i_{k-1}, j_{k-1}) + a_k(i_k, j_k)$. Reason for this is the converse of previous case.

So this proves that there exist a basis $(\alpha, \beta), (\gamma, \delta)$ which can generate all the $(i, j) \in \mathcal{L}_p$ by its linear combination.

**Proposition 2** *Let $(i, j) \in \mathcal{L}_p$ and $(i', j') \in \mathcal{L}_p$ and $-I \leq i \leq I$, $-I \leq i' \leq I$, $j' > j$, and $j'$ is as small as possible, then*

$$(i', j') = (i, j) + \begin{cases} (\alpha, \beta), & i \geq A - \alpha \\ (\gamma, \delta), & i < A + I - \gamma \\ (\alpha, \beta) + (\gamma, \delta) & A + I - \gamma \leq i < A - \alpha \end{cases}$$

*In this proposition, they are consider this equation:*

$$(i', j') = (i, j) + k(\alpha, \beta) + l(\gamma, \delta)$$

*We can easily observe that if k and l are nonzero and are of opposite sign then this may occur $\gamma - \alpha \geq I$ (for $k = 1$ and $l = -1$) but it contradicts with condition as $-I \leq i' \leq I$. If one of them is negative and other is non-positive, then the condition $j > 0$ is violated as both $\beta$ and $\delta$ are positive. So only possibility which is remaining is $k \geq 0$ and $l \geq 0$.*

Now for the first case, if we allow $(\gamma, \delta)$ as well then for $i = -\alpha$, we may get $\gamma - \alpha$ (i.e. $i' \geq I$) leads to contradiction. Alone $(\alpha, \beta)$ is appropriate. For second case, if $i < I - \gamma$ then $i' < I$ which is maintaining constraints. Similar arguments are for the third case as well. And in all the cases, we are getting $j' > j$.

So every constraint is adequate and not violating any conditions and every equation is fulfilling its criteria.

## 4 Experiments

We have considered two most widely used libraries for factoring CADO-NFS [15] and GGNFS [16]. CADO-NFS used FK algorithm for sieiving while GGNFS uses Pollard approach. We have performed several tests on system having configuration Intel Core i7-3770 CPU@ 3.40GHzx8 and 8 GB RAM.

$n(100$ *digits*) : 2881039827457895971881627053137530734638790825166127496066674320241571446494762386620442953820735453

$p_1(45$ *digits*) : 618162834186865969389336374155487198277265679

$p_2(55$ *digits*) : 4660648728983566373964395375209529291596595400646068307

| $n(100$ *digits*$) = p_1(45$ *digits*$) \times p_2(55$ *digits*$)$ | | | | |
|---|---|---|---|---|
| Library | Factor base bound | Lattice sieivng (min) | *special-q* Range | #Relation collected |
| GGNFS | $A : 1.8 \times 10^6, R : 1.8 \times 10^6$ | 50 | 900000–1300000 | 4548846 |
| CADO-NFS | $A : 2 \times 10^6, R : 10^6$ | 68 | 2000000–3344000 | 3653229 |

$n(103$ *digits*) : 11112345615175224390706325657636600577404170228204751628562724150101396235470745110620799925252477780247

$p_1(46$ *digits*) : 9001000021200032001100483237793193978070200007

$p_2(57$ *digits*) : 123456789123456789123456789101987654321987654321987654321

| $n(103$ *digits*$) = p_1(46$ *digits*$) \times p_2(57$ *digits*$)$ | | | | |
|---|---|---|---|---|
| Library | Factor base bound | Lattice sieivng | *special-q* Range | #Relation collected |
| GGNFS | $A : 2.3 \times 10^6, R : 2.3 \times 10^6$ | 71 min | 1150000–1750000 | 4570621 |
| CADO-NFS | $A : 4 \times 10^6, R : 2 \times 10^6$ | 2 h 4 min | 4000000–4692000 | 1193716 |

$n(106\ digits)$ : 2915355177292833688983861903977969897438031079667239430
7737872749766256791122059307153293916667601132580 01
$p_1(51\ digits)$ : 374831379939791939113997931991393133317939371999713
$p_2(52\ digits)$ : 7777777777733333337735555377352537735537735377337737777

$n(106\ digits) = p_1(51\ digits) \times p_2(55\ digits)$

| Library | Factor base bound | Lattice sieivng (min) | *special-q* Range | #Relation collected |
|---|---|---|---|---|
| GGNFS | $A : 2.5 \times 10^6, R : 2.5 \times 10^6$ | 74 | 1400000–2000000 | 4802162 |
| CADO-NFS | $A : 4 \times 10^6, R : 2 \times 10^6$ | 3 h 14 | 4000000–5228000 | 1198378 |

$n(110\ digits)$ : 11111122222233334455556666667788888900001122 22334579001257
5890233469122356666802455556900135789024677913581227
$p_1(52\ digits)$ : 1000000000000000000000000000000000000000000000000000121
$p_2(59\ digits)$ : 11111122222233334455556666667788888900001122223344445566787

$n(110\ digits) = p_1(52\ digits) \times p_2(59\ digits)$

| Library | Factor base bound | Lattice sieivng | *special-q* Range | #Relation collected |
|---|---|---|---|---|
| GGNFS | $A : 3.2 \times 10^6, R : 3.2 \times 10^6$ | 2 h 37 min | 1600000–2300000 | 8410476 |
| CADO-NFS | $A : 4 \times 10^6, R : 4 \times 10^6$ | 4 h | 4000000–5610000 | 6563326 |

$n(112\ digits)$ : 2000000000000002199999999999998000000000000002199999999990000
1999999998900000000000000099999999999998899999999999999
$p_1(54\ digits)$ : 199999999999999999999999999999999999999999999999999999
$p_2(59\ digits)$ : 10000000000000109999999999999900000000000000110000000000001

$n(112\ digits) = p_1(54\ digits) \times p_2(59\ digits)$

| Library | Factor base bound | Lattice sieivng | *special-q* Range | #Relation collected |
|---|---|---|---|---|
| GGNFS | $A : 3.5 \times 10^6, R : 3.5 \times 10^6$ | 3 h 5 min | 1750000–2650000 | 8026043 |
| CADO-NFS | $A : 4 \times 10^6, R : 2 \times 10^6$ | 4 h 50 min | 4000000–6510000 | 6785308 |

$n(118\ digits)$ : 9051123511605627633956614749254431036815180884513942958 0549
2158046992359307739672932592267609396239064233828538713 6391
$p_1(53\ digits)$ : 91757793193977937533333733533733333573977939139775719
$p_2(65\ digits)$ : 98641468986414689864146898641468986414689864146898641468986
414689

$n(118\ digits) = p_1(53\ digits) \times p_2(65\ digits)$

| Library | Factor base bound | Lattice Sieivng | *special-q* Range | **#Relation collected** |
|---|---|---|---|---|
| GGNFS | $A : 4.5 \times 10^6, R : 4.5 \times 10^6$ | 5 h 20 min | 1900000–3300000 | 10121558 |
| CADO-NFS | $A : 8 \times 10^6, R : 4 \times 10^6$ | 41 h 20 min | 8000000–12976000 | 13829754 |

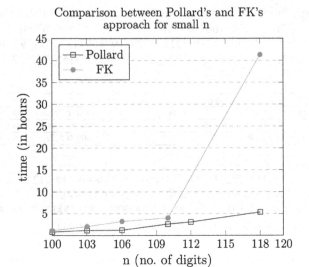

From the results it is shown that, even though CADO-NFS is following FK algorithm which avoids lattice basis reduction relatively takes more time compare to GGNFS. The intuition is the overall time of sieving increasing drastically due to large values of *special-q*. Also it is known, $q$ is one of the factor of $(a, b)$-pairs. On the other way, GGNFS is following Pollard's approach for sieving and the *special-q* values are lesser compared to CADO-NFS and producing more relations and completing sieving fastly. This may be because small factors repeat more frequently than larger one, so CADO-NFS larger factors repeating less frequently producing less relations and taking more time than GGNFS.

From the above result and discussion, it can be seen clearly that the selection of *special-q* plays a crucial role in sieving algorithms. But this discussion is upto numbers of size 140 digits number only because GGNFS does not support number larger than 140 digits, but CADO-NFS has 'state-of-the art' technique which make it works for higher digits.

## 5  Conclusion

In this paper, we discussed two well-known and most efficient sieving algorithms namely Pollard and FK algorithm. The performance of above two algorithms is analyzed based on the experiments performed using popular NFS tools GGNFS and CADO-NFS. Even though CADO-NFS following FK algorithm, which is improved version of Pollard it takes relatively more time for sieving compared to GGNFS which follows Pollard's approach for the data set ranging from 100 to 120 digits. The experimental results show that the selection of *special-q* in CADO-NFS is larger in size compared to GGNFS, and this may lead to increase sieving time of

CADO-NFS. This shows selection of *special-q* highly influencing the performance of sieving algorithms. To derive more concrete conclusions on various influencing parameters used by the sieving algorithms needs rigorous experiments on data set of large range and size of N.

# References

1. R. Barbulescu, C. Bouvier, J. Detrey, P. Gaudry, H. Jeljeli, E. Thome, M. Videau, and P. Zimmermann: Discrete logarithm in $GF(2^{809})$ with FFS.
2. R. Barbulescu, P. Gaudry, A. Joux, E. Thome: A quasi-polynomial algorithm for discrete logarithm in finite fields of small characteristic (2013), preprint, 8 pages, available at http://hal.inria.fr/hal-00835446
3. J.M. Pollard, The lattice sieve, 43–49 in [4].
4. A. Joux, R. Lercier,: Discrete logarithms in $GF(2^n)$ (521 bits) (Sep 2001), email to the NMBRTHRY mailing list. Available at http://listserv.nodak.edu/archives/nmbrthry.html
5. A. Joux, R. Lercier,: Discrete logarithms in $GF(2^{607})$ and $GF(2^{613})$. E-mail to the NMBRTHRY mailing list; http://listserv.nodak.edu/archives/nmbrthry.
6. T Hayashi, N Shinohara, L Wang, S Matsuo, M Shirase, T Takagi, Solving a 676-bit discrete logarithm problem in $GF(3^{6n})$ International Workshop on Public Key Cryptography, 351–367
7. J. Detrey, P. Gaudry, M. Videau,: Relation collection for the Function Field Sieve. In: Nannarelli, A., Seidel, P.M., Tang, P.T.P. (eds.) Proceedings of ARITH-21. pp. 201–210. IEEE (2013)
8. J. Franke, T. Kleinjung, Continued fractions and lattice sieving; Proceedings SHARCS 2005, available at http://www.ruhrunibochum.de/itsc/tanja/SHARCS/talks/FrankeKleinjung.pdf.
9. T. Kleinjung, K. Aoki, J. Franke, A.K. Lenstra, E. Thome, J. Bos, P. Gaudry, A. Kruppa, P.L. Montgomery, D.A. Osvik, H. te Riele, A. Timofeev, P. Zimmermann: Factorization of a 768-bit RSA modulus. In: Rabin, T. (ed.) Advances inCryptology - CRYPTO 2010. Lecture Notes in Comput. Sci., vol. 6223, p. 333–350. Springer-Verlag (2010)
10. P. Gaudry, L. Gremy, M. Videau (2016). Collecting relations for the number field sieve in $GF(p^6)$. LMS Journal of Computation and Mathematics, 19(A), 332–350. https://doi.org/10.1112/S1461157016000164
11. A.K. Lenstra, H.W. Lenstra, Jr. (editors), The development of the number field sieve, Springer-Verlag, LNM 1554, August 1993.
12. M. Case. A beginners guide to the general number field sieve(2003), available at: http://islab.oregonstate.edu/koc/ece575/03Project/Case/paper.pdf
13. A. Joux, Algorithmic Cryptanalysis Chapman and Hall/CRC 2009 Print ISBN: 978-1-4200-7002-6 eBook ISBN: 978-1-4200-7003-3 https://doi.org/10.1201/9781420070033
14. R. A. Golliver, A. K. Lenstra and K. S. McCurley, Lattice sieving and trial division, in: Algorithmic Number Theory (ed. by L. M. Adleman, M.-D. Huang), LNCS 877, Springer, 1994, 18–27.
15. CADO-NFS Library, available at http://cado-nfs.gforge.inria.fr/
16. GGNFS Library, available at http://gilchrist.ca/jeff/factoring/nfs-beginners-guide.html

# Author Index

Printed in the United States
By Bookmasters